Parallel Robots: Theory and Applications

Series editors

J. P. Merlet, Sophia Antipolis, France
Sébastien Briot, Institut de Recherche en Communications et Cybernétique de Nantes, CNRS, Nantes, France
P. Martinet, Nantes, France

Since a few decades, parallel robots have attracted the attention of industrials and researchers. For industrials, they are interesting due to their inherent advantages which are their large payload-to-mass ratio, intrinsic stiffness properties and great acceleration capacities. For researchers, their inherent complexity leading to issues in terms of design, modelling and control allowed the definition of many interesting scientific problems to be solved. We reached now a point where the research on parallel robots has led to the definition of a quantity of scientific results which, for the moment, miss to be compiled to be easily usable by researchers, teachers, professionals and students for their formation and future works. This collection aims at filling this gap, by proposing a series of book devoted to all the scientific and technological fields required for having a better understanding of the behavior of parallel robots and/or for helping the future parallel robot designer in his work. Therefore, we intend to cover all topics on parallel robots, from their (geometric, kinematic, dynamic, elastic, etc.) modelling, to their advanced control, by also considering the singularity analysis, the calibration and model parameter identification problems, the definition of advanced algebraic tools necessary for their study, their optimal design, and many other scientific and technological aspects. Springer and the series editors welcome book proposals. Potential authors are invited to contact Nathalie Jacobs, Senior Publishing Editor Engineering, at nathalie.jacobs@springer.com

More information about this series at http://www.springer.com/series/13855

Stefan Staicu

Dynamics of Parallel Robots

Springer

Stefan Staicu
Department of Mechanics
University Politehnica of Bucharest
Bucharest, Romania

ISSN 2524-6232 ISSN 2524-6240 (electronic)
Parallel Robots: Theory and Applications
ISBN 978-3-030-07604-7 ISBN 978-3-319-99522-9 (eBook)
https://doi.org/10.1007/978-3-319-99522-9

Softcover re-print of the Hardcover 1st edition 2019

This Springer imprint is published by the registered company Springer Nature Switzerland AG
The registered company address is: Gewerbestrasse 11, 6330 Cham, Switzerland

Preface

The first part of this book is a briefing on the *matrix mechanics* of the rigid body, from which are developed the following five chapters: Matrix kinematics of the rigid body, Matrix kinematics of composed motion, Kinetics of the rigid body, Dynamics of the rigid body, and Analytical Mechanics. This can constitute the theoretical basis necessary to establish recursive relations about the kinematics and the dynamics of *constrained robotic systems*.

By using a set of successive mobile frames, we will first analyze the geometrical properties and the kinematics of the vector system of velocities and accelerations for each element of the robot. The dynamics problem is solved in two energetic ways: by using an approach based on the principle of virtual work and by applying the formalism of Lagrange's equations of the second kind. When establishing the expressions for the active forces and the active torques in an inverse dynamics problem, these equations are in fact useful for real-time control of the robot's evolution. In the following, it will be shown how the recursive matrix method is applied to the kinematics and dynamics analysis of five distinct case studies: planar parallel manipulators, spatial parallel robots, planetary gear trains, mobile wheeled robots, and finally, two-module hybrid parallel robots.

Chapter 7 provides the fundamental knowledge needed for the kinematics analysis and the dynamics approach based on the principle of virtual work. Expressed for every independent loop of the mechanism, useful *conditions of connectivity* regarding the relative velocities and the relative accelerations are determined for either the direct or the inverse kinematics problem. Afterward, matrix relations written in a recursive compact form express precisely the *explicit dynamics equations* of the constrained robotic systems.

Chapter 8 is entirely devoted to planar robotic systems. Orthogonal matrices are used to represent coordinate transformations in the kinematics and dynamics analyses of some planar parallel manipulators. With a closed-loop structure, the 3-RRR planar parallel robot, for example, is a special symmetrical mechanism composed of three planar kinematical chains with identical topology, all connecting the base to the moving platform.

Chapter 9 addresses the dynamics of spatial parallel robots, applying both the principle of virtual work and the method of Lagrange's equations. The prototype of the Delta parallel robot with three revolute actuators, the Agile Wrist spherical parallel robot, the 6-6 Stewart parallel manipulator, and the 3-RPS parallel robot are spatial mechanisms which consists of a system of parallel closed kinematic chains connected to the moving platform.

Chapter 10 deals with the kinematics and inverse dynamics of some robotic wrist mechanisms. The mechanism of the orienting gear train, for example, is a parallel system with five moving links and three bevel gear pairs controlled by two electric motors. Using the theory of fundamental circuits, a systematic methodology for a complete mechanics of epicyclical gear trains and robotic wrist mechanisms is developed. Finally, it is shown that explicit equations of motion, based on the principle of virtual work, can be derived by applying Lagrange's formalism with multipliers.

Chapter 11 introduces the dynamics of special robotic mechanical systems, which include robots with conventional wheels. Although rolling robots are usually are non-holonomic mechanical systems, their mathematical modeling can formally be identical to those of parallel manipulators. The mobile robot considered consists of a platform, two cylindrical wheels, and a crank that controls the motion of a free rolling caster wheel.

Chapter 12, finally, concerns the kinematics and dynamics of a spatial two-module hybrid parallel robot. The dynamics analysis of this structure is complicated by the existence of multiple closed loops. Starting from the principle of virtual work, explicit matrix relations giving time-history evolution for the input forces and powers of six prismatic actuators are obtained over the course of the known evolution of two moving platforms.

This book is written for researches, engineers, teachers, and students at undergraduate and postgraduate levels.

Bucharest, Romania Stefan Staicu
January 2018

Contents

List of Figures

Chapter 1
Introduction

Robotics has grown strongly over the past years, fueled by rapid advances in sensor technology as well as in control and computer vision. A *robot* is basically an actuated mechanism programmable in two or more axes with a degree of autonomy, moving within its environment to perform intended tasks. The main task accomplished is to move an object, or move itself, according to the decisions of a high-level motion planner. Robotic mechanical systems are composed of several subsystems: mechanisms of connected rigid bodies, sensors, actuators, microprocessor-based controllers and information-processing parts.

1.1 Robotics Systems

Robot manipulators first appeared as mechanical systems constituted by a structure consisting of very robust links coupled by either revolute or prismatic joints. A manufacturing robot is typically an arm-type manipulator on a fixed base, and typically performs repetitive tasks within a local work cell. In recent decades, manipulators have been used primarily for various operations in hazardous environments including radioactive material handling, construction robots, flight and vehicle simulators, mining machines, pointing devices, intelligent machining centres, and the like. Because of the serial nature of the coupling of links in this type of manipulator, even though they are supplied with structurally robust links, their load-carrying capacity and stiffness are too low when compared with the same properties in other multi-axis machines. Recently, parallel robot manipulators have introduced by manufacturing industries for certain production operations such as machine loading and unloading, electronic assembly, spot welding, spray painting, packaging and palletizing.

Since robot manipulators are controlled by computers or microprocessors, they can easily be reprogrammed for different tasks. The output of robotic mechanical system is the actual task, which is monitored by the sensors, while they transmit

S. Staicu, *Dynamics of Parallel Robots*, Parallel Robots: Theory and Applications,
https://doi.org/10.1007/978-3-319-99522-9_1

task information in the form of feedback signal to be compared with the prescribed task. The errors between the prescribed and the actual task are then fed back into controller, which synthesizes the necessary corrective signals. Based on the information displayed, the human operator makes decisions about the corrections in order to accomplish the prescribed task.

Manipulators can be classified according to the type of kinematical chain used for their implementation. In general, most currently-existing industrial robots designed are built up either of serial or parallel and hybrid kinematical chains. In its most general form, a *serial robot*, called *robot arm*, is a set of several rigid bodies or *links* interconnected in series through various types of actuated joints, which are typically either *revolute* or *prismatic*. The first extremity of this serial kinematical chain is a fixed link attached to a supporting ground called *base* and the last extremity is the free *end-effector* as the output link, to which is attached a universal gripper device or a mechanical hand which, equipped with a tool to grasp, can lift and manipulate work-pieces in space or perform assembly tasks.

As a mechanical interface between the manipulator and its environment, equipped with force-sensor, the end-effector, which requires certain dexterity and versatility, is a specialized device that is used for handling objects of similar shape, size or weight in a repetitive operation. According to its topology, a manipulator is said to be an open-loop mechanism if its kinematical structure consists of a series of active joints connecting the base to the end-effector and takes the form of an *open-loop chain*. Based on open kinematical chains, serial manipulators generally have a great degree of dexterity and cover a large workspace with respect to the occupied floor space, but have low stiffness and highly nonlinear dynamic behaviour.

Due to their cantilever-type structure, serial robots have found tremendous development in industries with applications ranging from pick-and-place operations to welding and assembling. Although it is relatively simple to model their kinematics and dynamics, serial manipulators do have some major disadvantages: low stiffness causing some inaccuracies in end-effectors positions, relatively large positioning errors accumulated and amplified from link to link, low dynamic capabilities and inadequacy for very high-speed and high-acceleration applications.

Parallel robots are closed-loop mechanical structures whose mobile platforms are linked to the base by some independent kinematical chains, presenting very good potential in terms of accuracy, rigidity and ability to manipulate large loads with minimal positioning errors. In general, parallel manipulators consist of two main bodies coupled via numerous serial legs acting in parallel, providing closed kinematical loops in the system for better load-handling capacity and stiffness. One body is designated as fixed and is called the *base*, while the other is regarded as relatively movable body and hence is called the *moving platform* of the manipulator. Several legs made up as serial kinematical chains connect the moving platform to the fixed base. The links of the robot could be connected to each other by different kinds of known pairs: spherical joints, universal joints, cylindrical joints, revolute joints or prismatic joints.

Containing one or more loops, parallel robots can be regarded as a special class of mechanisms and are purposely designated to exploit the specific advantages offered by their closed chain structures. Multi-axis precision-positioning devices of rigid hexapods and tripods with fixed-length legs, for example, are based on parallel robots. Composed of a set of multiple parallel legs with active and passive joints, these manipulators are robotic devices that differ from the more traditional serial robotic manipulators by virtue of their kinematical structure. Most of their joints are not actuated and many of these passive joints have several degrees of freedom.

Some of the fastest velocities and accelerations recorded by industrial robots have been achieved by parallel robots, primarily by placing the actuators on the fixed platform and thereby minimizing the mass of all moving parts. It is true that the motors in many parallel robots are positioned on or near the fixed base, which leaves the mobile part relatively light. In contrast, serial manipulators are usually actuated at each joint along the serial linkage. For applications in which high precision is required or a relatively high load capacity in proportion to robot weight is important, parallel architectures are the prime choice. The particular nature of parallel manipulators has generated great interest in robotic applications where high loads and precision are required such as in machining processes and automated assembly operations. Generally, parallel architectures can be classified according to the criteria of number of degrees of freedom of the moving platform. The kinematical loops are formed by two or more chains that connect a moving platform to a base, where a single joint in the serial chain is actuated and the other joints are passive. Because the links of a parallel manipulator generally need not carry the load of the actuators, the links are made lighter than the links of a comparable serial manipulator. On the other hand, as each independent leg forms a closed loop and connects the base to the moving platform, parallel robots are able to achieve high stiffness and high force-to-weight ratio. However, parallel robots are known for their smaller and more restricted workspace and lower dexterity than serial manipulators.

Recently, the development of parallel robots has gained greater attention within the research community as their advantages become better known. These mechanisms can be found in technical applications in which a high-speed displacement of a rigid body within the robot's workspace is desired. Accuracy and precision in the execution of the tasks are essential; any error in the positioning of the tool could lead to costly damage.

Typically, a parallel mechanism is considered to be *symmetrical* if the following conditions are satisfied: (i) the number of legs is equal to the number of degrees of freedom (DOF) of the moving platform; (ii) one actuator controls each leg and the number of links in all kinematical chains is the same [1]. For three decades, parallel robots have attracted the attention of many studies that consider them as a valuable alternative design for robotic structures [2, 3]. Other types of mechanisms have therefore recently been studied and are being more regularly used within the industrial world, such as machine tools and industrial robots [4].

Parallel robots are being used in a large number of applications ranging from astronomy to flight simulators, and are becoming increasingly popular in the machine-tool industry. Typically equipped with hydraulic or pneumatic actuators, parallel robots have a robust built and can move bodies of considerable masses and dimensions at high speeds. Hydraulic actuators are used in flight simulators for which the nominal load is very light. Pneumatic actuators possess several advantages: they are light and their dynamics are very good, but the control is rather delicate.

A translational device being classified as a function of the type of motions of its platform is suitable for pick-and-place applications; a spherical mechanism is used for pointing applications such as orienting a camera, while a complex spatial structure includes, in general, a combination of both rotational and translational degrees of freedom. This is why the mechanisms which produce translation, rotation or spherical motion to a platform are based on the concept of a parallel robot.

Compared with serial manipulators, parallel robots have some special characteristics: very high accuracy, greater structural rigidity, stable functioning, better inertial distribution and larger dynamic load carrying capacity. Nowadays, parallel kinematic machines are applied widely since they possess many inherent advantages over their serial counterparts in terms of high speeds, lower moving masses, higher stiffness, simpler modular mechanical construction and the possibility to mount all actuators on the fixed base. However, a significant drawback of most parallel manipulators is that they have smaller and more complicated workspace volumes than serial manipulators of similar size, with singularities and highly non-isotropic input-output relations [5].

Considering the tradeoff between high operational precision and large working volume, parallel actuated mechanisms are becoming ideal devices for applications that require positional accuracy within a limited workspace.

1.2 Historical Development

While there are relatively fewer parallel robots than serial robots in use, the variety of parallel robots is greater. E. Gough built an octahedral telescoping-leg hexapod in 1950, the model that becomes the most common 6-DOF parallel robot. Presented by Merlet [3], this robot is called a universal tyre testing machine and was invented to address the problems associated with aircraft landing loads to determine the properties of tyres under combined loads.

The most widely-used parallel manipulator is the Stewart platform which consists of two rigid bodies connected by six linear parallel actuators. The motion of the platform with respect to the base is controlled by the displacements of the six actuators. The historic reference comes from 1965, when Stewart [6] proposed a parallel mechanism with six telescoping legs, differently from the octahedral hexapod, and some years later he built the six-degrees-of-freedom platform to be used as a flight simulator. Since then, and over the last three decades, parallel kinematic machines in general, and the Stewart platform in particular (Fig. 1.1), have been among the best-studied mechanisms. As a consequence of this plethora of research, several implementations have come to fruition in fields such us telescope precise positioning, flight simulation, medicine, and even six-axis machine tools.

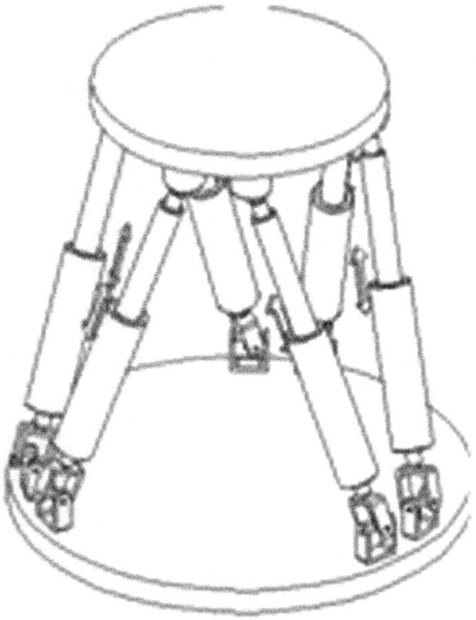

Fig. 1.1 Stewart parallel robot

Built by Chablat and Wenger [7], the prototype of the Orthoglide parallel robot (Fig. 1.2) has good kinematical performance and some technological advantages such as symmetrical design, regular workspace shape properties with a bounded velocity amplification factor and low inertia effects.

In the 1980s, Clavel [8] invented a parallel manipulator at the Federal Polytechnic Institute of Lausanne known as the Delta robot with three translational degrees of freedom dedicated to high-speed applications. The manipulator consists of multiple kinematical chains connecting the fixed base with the moving platform (Fig. 1.3).

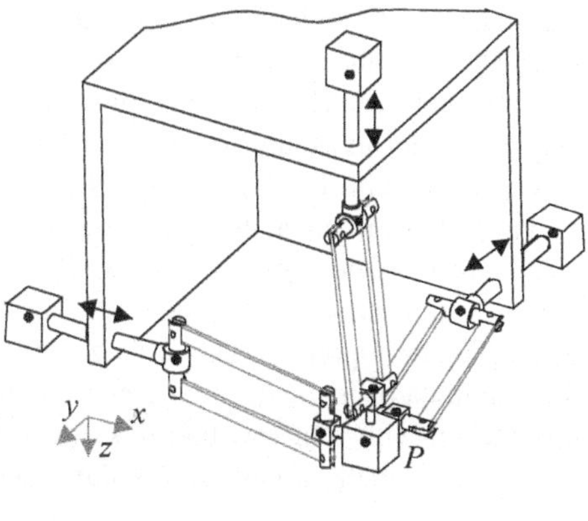

Fig. 1.2 Orthoglide parallel robot

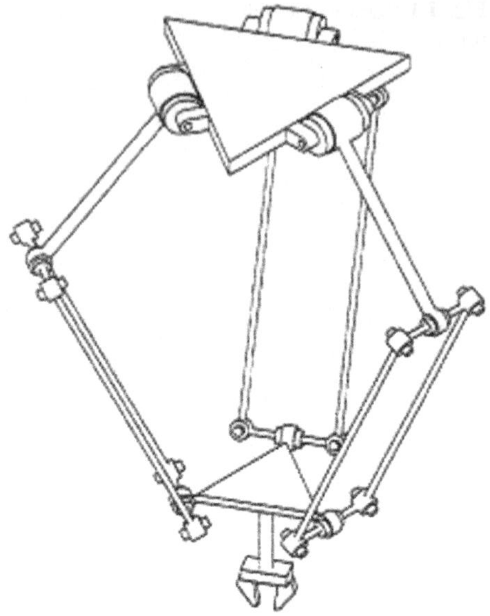

Fig. 1.3 Delta parallel robot

Due to its superior stiffness, domains that take advantage of the high speed of Delta robots are the packaging industry, surgery, medical and pharmaceutical industries. Other applications include high-precision assembly operations in a clean room for electronic components. Nowadays, as one of the best-known and most widely spread parallel robots, the Delta parallel mechanism has been successfully used in many fields, such as that of rapid picking, thanks to its outstanding features including light structure, high speed, high accuracy and simple control.

Over the last three decades, studies on the renowned Delta parallel mechanism have been conducted extensively and have covered almost all general aspects of kinematics, dynamics, control, singular configurations, workspace, calibration and mechanical design. We may mention here the prototype developed by Tsai and Stamper [9] at the University of Maryland. However, as a translational parallel manipulator, its application is limited because it lacks capability to change the orientation of the manipulated object. Besides various different designs of the original architecture, current industrial versions employ additional serial mechanisms in order to obtain rotational motions.

Equipped with three actuators, the Star parallel manipulator [10] can move its moving platform in a three-degrees-of-freedom translation (Fig. 1.4). The input mechanism of the robot is structured into three active revolute joints, while the output body is connected to the fixed base through a set of three variable length legs with identical topology. The architecture of one of the three kinematical chains of the manipulator is made of an electrical motor, a helical joint and an intermediary mechanism with four revolute links that connect four bars, which are parallel two by two, and finally a passive revolute joint of the moving platform [11].

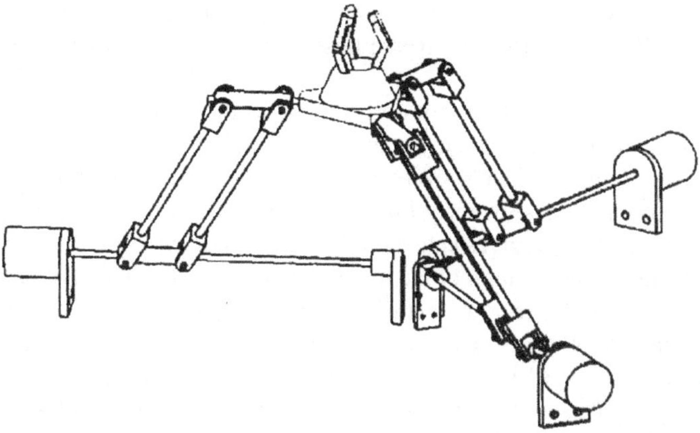

Fig. 1.4 Star parallel manipulator

Gosselin and Angeles [12], Wang and Gosselin [13] analysed the kinematics, the dynamics and the singularity loci of the Agile Wrist spherical parallel robot, which has three concurrent rotations. Major companies such as Giddings and Lewis, Ingersoll, Hexel and others have developed them as high-precision machine tools. In particular, the parallel manipulators have great potential applications in medical fields thank to their special characteristics.

1.3 Mechanics of Robots

A thorough understanding of the mechanics of robot manipulators is a necessary prerequisite to achieving better design, path planning and control of such electromechanical systems.

Statics deals with the relations of forces and moments that can produce equilibrium among all components of the robotic constrained system. The actions are represented by the forces of applied loads, gravitational forces, internal frictional forces in joints, and the like. Obviously, the forces of inertia are excluded from this analysis and the forces of equilibrium are not time-dependent, but rather depend strongly on the configuration of a robot manipulator.

Kinematics of parallel robots is concerned with relating the active joint variables to the platform's Cartesian pose variables. The variables that describe the actuated joints will be referred to as the *input variables* or alternatively as the active *joint variables*. On the other hand, the variables that describe fully the pose of the mobile platform will be referred to as *output variables*. The task of computing the input variables from the output variables will be referred to as the *inverse kinematics problem*, while the task of finding the valid set of output variables corresponding to a set of input variables will be referred to as the *direct kinematics problem*.

Unfortunately, there are also many passive intermediate joint variables, which are unknowns in both forward and inverse pose, adding more difficulties to the interrelation of kinematics problems.

The mobility of a parallel robot corresponds to the total degrees of freedom of the joints that can be independently actuated. In some cases, the number of actuated joint degrees of freedom may exceed the kinematics degree of freedom of the mechanism, in which case we say that the robot is redundantly actuated.

Inverse kinematics involves calculating the required active-joint variables, given the position and the orientation of the moving platform, when the end-effector of the robot can follow prescribed trajectories. Since the kinematical relationships are decoupled, this kinematics analysis of a parallel robot is simple to determine compared to the forward kinematics analysis, while the opposite situation obtains with serial manipulators. Direct kinematics of a parallel robot involves calculating the possible poses for the end-effector, given the active-joint variables. For most types of parallel robots, this task is very challenging and a numerical iterative method is most often used to solve the direct kinematics. Determining the pose of the end-effectors from the joint coordinates is equivalent to solving the system of equations that we have obtained for the inverse kinematics. Generally, the inverse kinematics problem is straightforward, while the forward kinematics problem is difficult, yielding a multitude of solutions.

The kinematics of parallel robots has been studied extensively over the past three decades. In the inverse kinematics problem, each joint variable may be computed independently for a given desired pose of the robot. Thus, the inverse geometric model can be written for each leg as $q = f(X)$, when q represents the vector of the joint active variables and X is the vector giving the position and the orientation of the moving platform. Based on the geometric model $X = f^{-1}(q)$, the forward pose kinematics allows the determination of the moving platform pose for a given set of control inputs, involving the solutions of a system of nonlinear coupled algebraic equations. Numerical approaches are provided to solve this problem either in a general or in a special architecture. The objective of the velocity and acceleration analysis problem is to find the instantaneous velocity and acceleration of the moving platform for a given set of joint variables.

Dynamics deals with the forces and torques required to cause the motion of a system of connected bodies. Typically, the end-effector is to be guided through a given path with certain prescribed motion characteristics. A set of forces and torque functions must be applied at actuated joints in order to produce that motion. These functions depend not only on the spatial and temporal attributes of the given path but also on the mass properties of the links, the payload, the externally applied forces, and the like.

The dynamics of robot manipulators can also be approached from two different points of view: dynamics analysis and dynamics synthesis.

Dynamics analysis of parallel manipulators presents an inherent complexity due to the closed-loop structure and the kinematic constraints. Nevertheless, when good dynamic performance and a precise positioning under high load are required, the dynamics modelling accepted is rather important for the real-time control of robot's motion, particularly because parallel robots are preferred in applications where precise positioning and suitable dynamic performance under high loads are the prime requirements.

Dynamics synthesis is the reverse of dynamics analysis. Much more complicated that the dynamics analysis problem, particularly in a dynamics synthesis problem, dynamics synthesis is the operation in which a new manipulator is to be designed with certain desired dynamical motion characteristics.

There are numerous well-established methods for modelling the dynamics of serial manipulators, but they are becoming practically unusable for parallel robots. This is mainly because of the presence of closed-loop constrained equations in parallel robots which makes the calculations involved in modelling of the dynamics very complex. The solutions are then combined with the help of the constrain equations to obtain the final actuator torques. In robotics it is essential that the computation time for dynamics model be very short, so that the actuator torques can be evaluated in real time; yet because of the complexity of calculation involved, very fast computation is difficult. Hence, the well established theory for parallel manipulators has to be suitably modified to fit the needs of parallel robots.

Dynamics analysis deals with derivation of the equation of motion for a given manipulator. There are two types of dynamics analysis problems: *direct dynamics* and *inverse dynamics*. The task of direct dynamics can be defined as follows: given a set of actuated joint force or torque functions, calculate the resulting motion of the end-effector as a function of time. The task of inverse dynamics can be defined as: given a trajectory of the end-effector as a function of time, find a set of actuated joint force or torque functions which will produce that motion. Because the inverse problem is purely algebraic, it is conceptually and computationally simpler to grasp than the forward problem. The computational efficiency of direct dynamics is not a critical factor, since it is an operation used primarily for computer simulations of a robot manipulator. On the other hand, efficient inverse dynamics modelling becomes extremely important for real-time model-based control of a robot manipulator.

In the inverse problem, a time-history of either the Cartesian or the joint coordinates is given, the architecture and inertial parameters of the system are known and the torque or force requirements at the different actuated joints are determined as time-histories as well. In the forward problem—the time-histories of externally-applied torques and forces being known along with the architecture and the mechanical parameters of the system—the values of the joint coordinates and their time-derivatives are computed at a later sampling instant by integration of the underlying system of nonlinear ordinary differential equations.

Dynamics plays an important role in the control of parallel robots for some applications, such as flight simulators or pick-and-place operations involving fast manipulators. It is therefore necessary to establish the dynamics relations for closed-loop mechanisms before we can establish a control scheme for fast manipulators. The dynamics analysis of parallel robots is usually implemented through analytical methods in classical mechanics [14] in which the projection and resolution of equations on the reference axes are written in a considerable number of cumbersome, scalar relations and in which the solutions are rendered by large-scale computation, together with time-consuming computer codes. Various methods of analysis, such as the *Newton-Euler* approach, the formalism of *Lagrange* equations and the *principle of virtual work* can be applied in the dynamics analysis of the constrained robotic systems, with these methods concentrating on recent advances in kinematics and dynamics of serial, parallel and hybrid manipulators.

Based on the Newton-Euler equations of rigid bodies, a recursive formulation has been modelled in the calculation of the joint torques of a six-revolute serial manipulator. Kane and Levinson [15] obtained some recursive relations concerning the equilibrium of the generalized forces that are applied to a serial robot arm. Kane's dynamic equations were described with the example of Stanford robot, and did not propose a general algorithm for the constrained robotic systems.

Recently, many efforts have been devoted to kinematics studies and dynamics analyses in the area of parallel robots and hybrid serial-parallel architectures. Among these, the class of manipulators known as the Stewart-Gough platform has enjoyed great attention and have attracted considerable research interests in the application field of manufacture [16]. It is these qualities that make it applicable to a wide range of engineering applications, ranging from flight simulators to high-speed milling machines and more recently to the design of parallel kinematic machines (PKMs). However, the main problem of this 6-UPS architecture is the power consumption of the manipulator, owing to the bulky prismatic actuators lying along the links.

Dasgupta and Mruthyunjaya [14] used the Newton-Euler approach to develop a set of closed-form dynamical equations of the Stewart platform, considering all dynamic and gravity effects as well as the viscous friction at joints. Sorli et al. [17] conducted the dynamics modelling for the Turin parallel robot, which has three identical legs but six degrees of freedom. Tsai [1] presented also an algorithm to solve the inverse dynamics for a Stewart platform, using the Newton-Euler procedure. It is commonly known that this approach results in a group of equations which are very difficult to use to derive control laws.

Under some simplifying assumptions on the geometry and the inertia distribution of the robot, Tsai and Stamper [9] and Geng et al. [18] developed Lagrange formalism to derive the equations of motion. The analytical calculi involved in the Lagrange method are too long to be used for every scheme of robots, and their runs the risk of making errors. Also, it has been proved that the time required for numerical calculus grows with the increasing number of bodies existing in robotic systems. Computationally, it is impossible to use this elegant approach for real-time control unless the numerous final equations are simplified.

The method based on the *principle of virtual work* implies that the total virtual work or power developed by all internal forces, external forces and also inertia forces acting on the mechanical system, corresponding to any set of virtual displacements which are consistent with the kinematics constraints imposed by the joints, will be zero [19–21].

The procedure for developing the inverse dynamics of a spatial parallel manipulator is to open up the closed-loops into individual open-loop structures and then to calculate the dynamics separately for each serial chain. The original structure of a parallel robot can artificially be transformed into a set of open kinematical chains subject to the applicable constraints. This is possible by virtually cutting each joint for the moving platform and taking its effect into account by introducing the corresponding constraint conditions. The first and most interesting open tree system can represent the moving platform only. The dynamics equations are derived from the original system by opening the loops at the level of the ball-and-socket joints or at certain passive frictionless joints.

Imagining several *virtual displacements*, which should be consistent with the virtual motions imposed by all kinematic constraints and by the joints at a given instant in time, some *connectivity relations* to the relative virtual velocities will be needed to be factored into dynamics equations expressed using the *principle of virtual work*. Yielding a set of *explicit recursive dynamics equations* for parallel robots, used to obtain compact results concerning the time-history evolution for input forces or torques of the actuators but also for external and internal reaction forces in joints, this energetic dynamics modelling is generally more efficient than the Newton-Euler method or Lagrange formalism.

References

1. Tsai, L.-W.: Robot Analysis: The Mechanics of Serial and Parallel Manipulators. Wiley, New York (1999)
2. Angeles, J.: Fundamentals of Robotic Mechanical Systems: Theory, Methods and Algorithms. Springer, New York (2002)
3. Merlet, J.-P.: Parallel Robots. Kluwer Academic, Dordrecht (2000)
4. Valenti, M.: Machine tools get smarter. ASME Mech. Eng. **17**, 70–75 (1995)
5. Bonev, I., Zlatanov, D., Gosselin, C.: Singularity analysis of 3-DOF planar parallel mechanisms via screw theory. J. Mech. Des. **125**(3), 573–581 (2003)
6. Stewart, D.: A platform with six degrees of freedom. Proc. Inst. Mech. Eng. Part. I **180**(15), 371–386 (1965)
7. Chablat, D., Wenger, P.: Architecture optimization of a 3-DOF parallel mechanism for machining applications: the Orthoglide. IEEE Trans. Robot. Autom. **19**(3), 403–410 (2003)
8. Clavel, R.: Delta: a fast robot with parallel geometry. In: Proceedings of 18th International Symposium on Industrial Robots, Sydney, pp. 91–100 (1988)
9. Tsai, L.-W., Stamper, R.: A parallel manipulator with only translational degrees of freedom. In: Proceedings of ASME Design Engineering Technical Conferences, Irvine, CA (1996)
10. Hervé, J.-M., Sparacino, F.: Star. A new concept in robotics. In: Proceedings of the Third International Workshop on Advances in Robot Kinematics, Ferrara, pp. 176–183 (1992)
11. Staicu, S.: Dynamics analysis of the Star parallel manipulator. Robot. Auton. Syst. **57**(11), 1057–1064 (2009)
12. Gosselin, C., Angeles, J.: The optimum kinematic design of a spherical three-degree-of-freedom parallel manipulator. ASME J. Mech. Transm. Automat. Des. **111** (2), 202–207 (1989)
13. Wang, J., Gosselin, C.: A new approach for the dynamic analysis of parallel manipulators. Multibody Syst. Dyn. **2**(3), 317–334 (1998)

14. Dasgupta, B., Mruthyunjaya, T.S.: A Newton-Euler formulation for the inverse dynamics of the Stewart platform manipulator. Mech. Mach. Theory **33**(8), 1135–1152 (1998)
15. Kane, R., Levinson, D.A.: Dynamics: Theory and Applications. McGraw-Hill, New York (1985)
16. Di Gregorio, R., Parenti Castelli, V.: Dynamics of a class of parallel wrists. J. Mech. Des. **126** (3), 436–441 (2004)
17. Sorli, M., Ferarresi, C., Kolarski, M., Borovac, B., Vucobratovic, M.: Mechanics of Turin parallel robot. Mech. Mach. Theory **32**(1), 51–77 (1997)
18. Geng, Z., Haynes, L.S., Lee, J.D., Carroll, R.L.: On the dynamic model and kinematic analysis of a class of Stewart platforms. Robot. Auton. Syst. **9**(4), 237–254 (1992)
19. Staicu, S., Zhang, D.: A novel dynamic modelling approach for parallel mechanisms analysis. Robot. Comput.-Integr. Manuf. **24**(1), 167–172 (2008)
20. Tsai, L.-W.: Solving the inverse dynamics of Stewart-Gough manipulator by the principle of virtual work. J. Mech. Des. **122**(1), 3–9 (2000)
21. Zhang, C.-D., Song, S.-M.: An efficient method for inverse dynamics of manipulators based on virtual work principle. J. Robot. Syst. **10**(5), 605–627 (1993)

Chapter 2
Matrix Kinematics of the Rigid Body

A *rigid body* is a mass point continuous system in which the mutual distances between the different constitutive particles are constant for any time, position or displacement of the body.

The motion of a rigid body with respect to a chosen base is completely known if it is possible to determine the position, velocity and acceleration of every point on the body at a given instant. To identify the location of a body in space, a reference coordinate system is established as a fixed frame. In the following we employ a Cartesian coordinate system to describe the position of the body.

In the kinematics analysis, we are primarily concerned with the geometrical aspects of the motion of the rigid body, regardless of the forces which produce that motion. The kinematics relationships are then developed and expressed in terms of the system degrees of freedom and their derivatives.

2.1 Position and Orientation of the Rigid Body

Having six degrees of freedom, the configuration of an unconstrained rigid body is identified using six coordinates. These independent scalar parameters would be sufficient to describe the location of the body in three-dimensional space.

As shown in Fig. 2.1, the general *displacement* of a rigid body can be conveniently described in an *inertial* $\Omega\xi\eta\varsigma$ Cartesian coordinate system, by introducing a local body $Oxyz$ Cartesian coordinate system whose axes and origin are rigidly connected to this moving body. On the other hand, in order to define the *position* of an arbitrary point P with respect to the rigid body, it is also necessary to use same frame $Oxyz$ and to refer to it as the *moving frame* which translates and rotates with the body itself.

The location of the rigid body can be considered as composed of the *position* of the origin O and the *orientation* of the moving frame with respect to the fixed base. Without loss of generality, it is possible sometimes to assume the origin of the local

S. Staicu, *Dynamics of Parallel Robots*, Parallel Robots: Theory and Applications,
https://doi.org/10.1007/978-3-319-99522-9_2

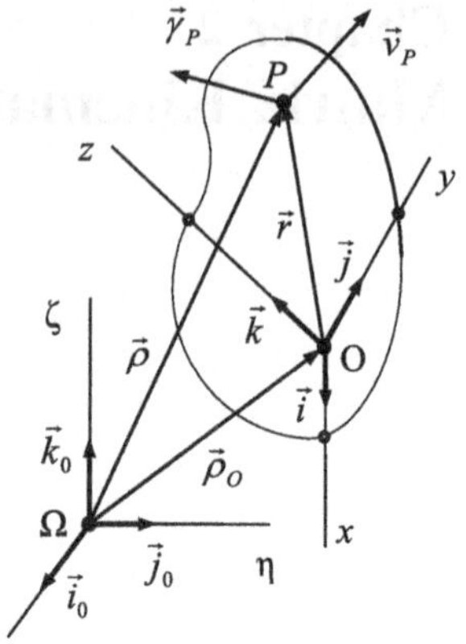

Fig. 2.1 Position and orientation of a rigid body

reference system O as coinciding with the centre of mass G of the body. Consequently, assuming that the moving frame initially coincides with the $\Omega\xi\eta\varsigma$ fixed frame, the absolute location of the moving frame and the spatial displacement of the rigid body from the initial coincident position are both equivalent.

An instantaneous position of the rigid body can then be given in terms of the displacement of the reference point O and also in terms of a set of coordinates that define the orientation of the moving frame with respect to the fixed frame of reference. Because the body-fixed frame $Oxyz$ is connected to the rigid body, the motion of the body can be also interpreted as the motion of this coordinate system.

The position of any point P with respect to the frame $Oxyz$ can be described by a *position vector*

$$\vec{r} = x\vec{i} + y\vec{j} + z\vec{k}, \tag{2.1}$$

where three constant coordinates x, y, z can determine a constant column matrix, which gives the distribution of the constitutive particles in the body of the rigid structure

$$\vec{r} = [x \quad y \quad z]^T. \tag{2.2}$$

In order to express the projections of three rotating unit vectors $\vec{i}, \vec{j}$ and $\vec{k}$ in terms of fixed unit vectors $\vec{i}_0, \vec{j}_0$ and $\vec{k}_0$, the following *nine direction cosines* of Ox, Oy and Oz axes can be expanded

$$\begin{aligned}
\alpha_{11} &= \vec{i}_0{,}^T\,\vec{i}, \quad \alpha_{12} = \vec{j}_0^T\,\vec{i}, \quad \alpha_{13} = \vec{k}_0^T\,\vec{i} \\
\alpha_{21} &= \vec{i}_0^T\,\vec{j}, \quad \alpha_{22} = \vec{j}_0^T\,\vec{j}, \quad \alpha_{23} = \vec{k}_0^T\,\vec{j} \\
\alpha_{31} &= \vec{i}_0^T\,\vec{k}, \quad \alpha_{32} = \vec{j}_0^T\,\vec{k}, \quad \alpha_{33} = \vec{k}_0^T\,\vec{k}.
\end{aligned} \tag{2.3}$$

These three orthogonal unit vectors, which define the directions and orientations of the moving axes, can permanently satisfy six orthogonality conditions

$$\begin{aligned} \vec{i}^T\vec{i} = 1, \quad \vec{j}^T\vec{j} = 1, \quad \vec{k}^T\vec{k} = 1 \\ \vec{i}^T\vec{j} = 0, \quad \vec{j}^T\vec{k} = 0, \quad \vec{k}^T\vec{i} = 0, \end{aligned} \tag{2.4}$$

which can be expressed as follows

$$\sum_{l=1}^{3} \alpha_{ml}\alpha_{nl} = \delta_{mn} \quad (m, n = 1, 2, 3), \tag{2.5}$$

using the Kronecker symbol δ_{mn}

$$\begin{aligned} \delta_{mn} &= 1 \;\; for \; m = n \\ \delta_{mn} &= 0 \;\; for \; m \neq n. \end{aligned} \tag{2.6}$$

The nine direction cosines define the orientation of the Ox, Oy, Oz axes relative to $\Omega\xi\eta\varsigma$ axes, but, considering the conditions (2.4), only three of the parameters (2.3) are independent. So, in the spatial kinematics, the unconstrained motion of a rigid body is described using six independent coordinates or *degrees of freedom*. Three of these degrees of freedom represent successive translations of the body along three fixed perpendicular axes, the other three represent successive rotations around another three concurrent axes passing through O.

The position vector $\vec{r}^{\circ}$ of the point P of the rigid body can also be expressed with respect to fixed frame $\Omega\xi\eta\varsigma$, in a matrix form as

$$\vec{r}^{\circ} = a^T \vec{r} \tag{2.7}$$

where a^T is the transpose of a 3 × 3 *transformation matrix* a of the moving frame with respect to the fixed base:

$$a = \begin{bmatrix} \alpha_{11} & \alpha_{12} & \alpha_{13} \\ \alpha_{21} & \alpha_{22} & \alpha_{23} \\ \alpha_{31} & \alpha_{32} & \alpha_{33} \end{bmatrix}, \quad \det a = 1. \tag{2.8}$$

Since the rows of a represent the projections of three moving unit vectors $\vec{i}, \vec{j}, \vec{k}$ expressed in the fixed base, it follows that the columns of a represent the projections expressed in the moving frame of another three unit vectors $\vec{i}_0, \vec{j}_0, \vec{k}_0$ defined along the coordinate axes of fixed base. The square orthogonal matrix a transforms the position vector $\vec{r}^{\circ}$ of any point P from the fixed frame to the moving frame, using an inversion of Eq. (2.7)

$$\vec{r} = a\vec{r}^{\circ} = aa^T\vec{r}. \tag{2.9}$$

Due to the above conditions (2.4), the inverse transformation of a rotation matrix is equal to the transposed matrix, as follows

$$aa^T = aa^{-1} = I; \quad a^T a = a^{-1} a = I, \tag{2.10}$$

where I is a 3×3 identity matrix.

When the origins of two coordinate systems do not coincide, as is the general case (Fig. 2.1), the location and the motion of the rigid body attached to the reference frame $Oxyz$ can now be described by reference to the position of the origin O and the rotation matrix a, which give the orientation of the moving frame with respect to the fixed base. In the course of time, the point in motion P changes its location, following the absolute path marked by the position vector $\overrightarrow{\Omega P} = \vec{\rho}$. The current absolute position of P is then defined for the location vector as a sum of two position vectors, expressed in the fixed global frame

$$\vec{\rho} = \vec{\rho}_O + a^T\vec{r}, \tag{2.11}$$

where $\vec{\rho}_O = \overrightarrow{\Omega O}$ is the vector from the fixed origin Ω to the origin of moving $Oxyz$ system. Pursuing *Chasles*' theorem, the first term on the right-hand side of (2.11) represents the contribution due to a translation along the direction of $\vec{\rho}_O$, while the second term represents the contribution due to three successive rotations of the rigid body about some concurrent axis.

2.2 Velocity Field

The velocity of a point P with respect to fixed frame is obtained by taking the derivative of the vector relation (2.11) with respect to time:

$$\dot{\vec{\rho}} = \dot{\vec{\rho}}_O + \dot{a}^T\vec{r}, \tag{2.12}$$

where $\dot{\vec{\rho}}_O$ is the velocity of the origin O, expressed in the fixed frame.

To obtain the components of these vectors along the axes Ox, Oy and Oz, we transform this relation, pre-multiplying by the operator a; it results that

$$\vec{v}_P = \vec{v}_O + a\dot{a}^T\vec{r}, \tag{2.13}$$

where

$$\vec{v}_P = a\dot{\vec{\rho}},\ \vec{v}_O = a\dot{\vec{\rho}}_O. \tag{2.14}$$

The components of the square matrix $a\dot{a}^T$, given by the derivatives of the orthogonality conditions (2.4), are expressed through only three independent elements of an axial vector $\vec{\omega}$ defined as *angular velocity* of the rigid body:

$$\vec{\omega} = \begin{bmatrix} \omega_x \\ \omega_y \\ \vec{\omega}_z \end{bmatrix} = \sum_l \begin{bmatrix} \alpha_{3l}\dot{\alpha}_{2l} \\ \alpha_{1l}\dot{\alpha}_{3l} \\ \alpha_{2l}\dot{\alpha}_{1l} \end{bmatrix}, \tag{2.15}$$

so that

$$a\dot{a}^I = -\dot{a}a^T = \tilde{\omega} = -\tilde{\omega}^T = \begin{bmatrix} 0 & -\omega_z & \omega_y \\ \omega_z & 0 & -\omega_x \\ -\omega_y & \omega_x & 0 \end{bmatrix}, \quad \det\tilde{\omega} = 0. \tag{2.16}$$

This significant linear operator $\tilde{\omega}$ represents the *skew-symmetric* matrix *associated* with the angular velocity $\vec{\omega}$ of the rigid body, expressed in the moving frame [1].

The velocity of P must therefore be equal to the vector sum

$$\vec{v}_P = \vec{v}_O + \tilde{\omega}\vec{r}. \tag{2.17}$$

This is *Euler*'s fundamental formula for the *velocity field* in a general spatial motion of the rigid body. The velocity of P is determined by considering the entire body to translate with a velocity $\vec{v}_O$ and rotating simultaneously about an instantaneous axis passing through the reference point O with an angular velocity $\vec{\omega}$.

The velocity equation (2.17) may be used to practical effect to study the general motion of a rigid body which is either in contact with or connected to other moving bodies. When applying this formula, the points P and O should generally be selected as points on the body which are connected to other bodies, or as points in contact with adjacent bodies which have known motions.

Remarks

1. The time derivative of any rotating vector $\vec{r}^{\circ}$ of constant magnitude as observed from the fixed base of reference, but projected on moving axes Ox, Oy and Oz, is equal to the change in direction of $\vec{r}$ caused by the rotation of moving frame

$$a\dot{\vec{r}}^{\circ} = \tilde{\omega}\vec{r}. \tag{2.18}$$

Consequently, the derivatives of rotating unit vectors $\vec{i}$, $\vec{j}$ and $\vec{k}$, lead to the commonly known formulae of *Poisson*

$$a\dot{\vec{i}}^{\circ} = \tilde{\omega}\vec{i}, \quad a\dot{\vec{j}}^{\circ} = \tilde{\omega}\vec{j}, \quad a\dot{\vec{k}}^{\circ} = \tilde{\omega}\vec{k}. \tag{2.19}$$

2. Working from (2.16), the time derivatives of the matrices a and a^T yield:

$$\dot{a} = \tilde{\omega}^T a, \quad \dot{a}^T = a^T \tilde{\omega}. \tag{2.20}$$

3. The relative velocity of two points of a rigid body is perpendicular to the line joining them.
4. Projection of the velocity of any point of a rigid body onto the direction of the vector $\vec{\omega}$ is a constant v_{ω}.
5. Projections of the velocities of two points of a rigid body along the line joining them are identical.

2.3 Acceleration Field

The acceleration of point P is a measure from the fixed frame of the change in time of its velocity and is determined by taking the time derivative of Eq. (2.12), which yields

$$\ddot{\vec{\rho}} = \ddot{\vec{\rho}}_O + \ddot{a}^T \vec{r}, \quad \ddot{a}^T = \dot{a}^T \tilde{\omega} + a^T \dot{\tilde{\omega}}. \tag{2.21}$$

To obtain the components of these vectors along the moving axes attached to the rigid body, we pre-multiply again by the operator a; this results in a matrix form of *Rivals*'s fundamental formula

$$\vec{\gamma}_P = \vec{\gamma}_O + \{\tilde{\omega}^2 + \tilde{\varepsilon}\}\vec{r}, \tag{2.22}$$

where we find the notations for accelerations of two points, symmetric square matrix $\tilde{\omega}^2$ and skew-symmetric matrix $\tilde{\varepsilon}$ associated with the angular acceleration $\vec{\varepsilon} = \dot{\vec{\omega}}$, but expressed in the moving frame as follows

$$\vec{\gamma}_P = a\ddot{\vec{\rho}}, \quad \vec{\gamma}_O = a\ddot{\vec{\rho}}_O, \quad \tilde{\omega}^2 = \tilde{\omega}\tilde{\omega}, \quad \tilde{\varepsilon} = \dot{\tilde{\omega}}. \tag{2.23}$$

For three-dimensional motion, the second term on the right-hand side of (2.22) measures the effect of change, in both magnitude and direction, of the angular velocity $\vec{\omega}$ on the acceleration $\vec{\gamma}_P$.

2.4 Twist of Velocity Field of a Rigid Body

A conventional *twist* $\vec{V}_O$ of the velocity field is defined in O as a six-dimensional array comprising three components of the velocity $\vec{v}_O$ of any point O of the rigid body and three components of the angular velocity $\vec{\omega}$. Assembling, these two vectors can constitute a column matrix of a vector called *dual velocity*:

$$\vec{V}_O = \begin{bmatrix} \vec{v}_O \\ \vec{\omega} \end{bmatrix}. \tag{2.24}$$

Introducing some new six-dimensional *polar vectors*

$$\vec{R}^{\circ} = \begin{bmatrix} \vec{\rho} \\ \vec{0} \end{bmatrix}, \quad \vec{R}^{\circ}_O = \begin{bmatrix} \vec{\rho}_O \\ \vec{0} \end{bmatrix}, \quad \vec{R} = \begin{bmatrix} \vec{r} \\ \vec{0} \end{bmatrix}, \tag{2.25}$$

a transformation of the vector $\vec{R}$ can be performed using a new operator A, as follows:

$$\vec{R}^{\circ} = \vec{R}^{\circ}_O + A^T\vec{R}, \quad A = \begin{bmatrix} a & 0 \\ 0 & a \end{bmatrix}. \tag{2.26}$$

Adding a column matrix $A^T\overrightarrow{\Omega}$, then the time derivation of (2.26) can be written as

$$\dot{\vec{R}}^{\circ} + A^T\overrightarrow{\Omega} = \dot{\vec{R}}^{\circ}_O + A^T\overrightarrow{\Omega} + \dot{A}^T\vec{R}, \tag{2.27}$$

with a new *axial vector* of the angular velocity

$$\overrightarrow{\Omega} = \begin{bmatrix} \vec{0} \\ \vec{\omega} \end{bmatrix}. \tag{2.28}$$

Pre-multiplying (2.27) by the matrix A, the vectors $\vec{v}_P = a\dot{\vec{\rho}}$ and $\vec{\omega}$ can also be merged into a 6×1 twist as $\vec{V}_P = \begin{bmatrix} \vec{v}_P^T & \vec{\omega}^T \end{bmatrix}^T$. Finally, we obtain a new formula for the *dual velocity* of the body in its point P

$$\vec{V}_P = \vec{V}_O + \widetilde{\Omega}\vec{R}, \tag{2.29}$$

with a new significant 6×6 skew-symmetric matrix of the angular velocity

$$\widetilde{\Omega} = A\dot{A}^T = \begin{bmatrix} \widetilde{\omega} & 0 \\ 0 & \widetilde{\omega} \end{bmatrix}. \tag{2.30}$$

2.5 Types of Rigid Body Motions

2.5.1 Translation

A motion is to be described as translation if any straight line inside the rigid body retains the same direction during the motion. Consequently, all straight lines in the body remain parallel to their initial orientation, $a = I$ being the unit diagonal matrix. It can also be observed that in a translation, all the particles forming the body move along congruent paths and they can be superimposed through a geometrical translation (Fig. 2.2). If these paths are straight lines, the motion is said to be a *rectilinear translation*; if the paths are curved lines, the motion is a *curvilinear translation*.

Differentiating the relative relation $\vec{\rho} = \vec{\rho}_O + \vec{r}$ with respect to time, we note that the vector $\vec{r}$ must maintain a constant direction and also a constant magnitude, since O and P belong to same rigid body. The derivative of $\vec{r}$ is zero and we have

$$\vec{v}_P = \vec{v}_O, \quad \vec{\gamma}_P = \vec{\gamma}_O. \tag{2.31}$$

Thus, when a rigid body is in translation, all the points of the body have the same velocity and same acceleration at any instant. In the case of curvilinear translation, the velocity and the acceleration change in direction, as well as in magnitude, at every instant. In the case of rectilinear translation, all particles of the body move along parallel straight lines, and the velocity and acceleration keep the same direction during the entire motion.

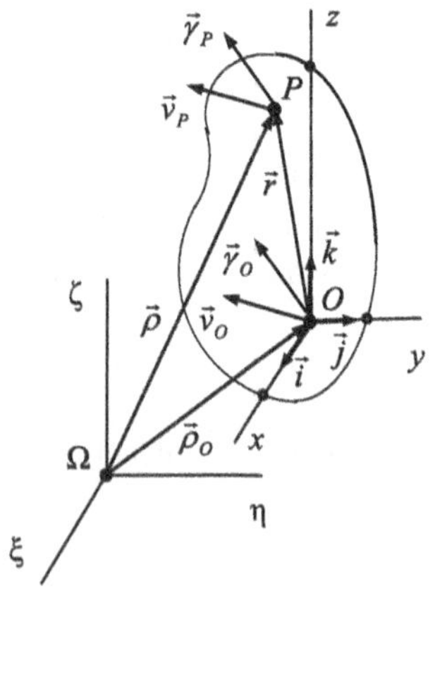

Fig. 2.2 Translation of the rigid body

2.5.2 *Rotation About a Fixed Axis*

A rotation around an axis takes place if there are two fixed points of the rigid body, during its motion. In this evolution, the particles forming the rigid body move in parallel planes along circles centred on the same fixed axis, called the *axis of rotation*. The points located on this axis have zero velocity and zero acceleration [2].

Consider a rigid body which rotates about a fixed $O\zeta$ axis. Let P be a point of the body, having $\overrightarrow{OP} = \vec{\rho}$ as position vector with respect to a fixed base $O\xi\eta\zeta$. For convenience, let us assume that the moving frame $Oxyz$ is linked with the rigid body and centred at two points O and O' on fixed axis. The angle $\varphi(t)$ between Ox and its initial position $O\xi$ defines the rotation of the rigid body around a fixed axis (Fig. 2.3).

The absolute position vector $\vec{\rho} = a_\varphi^T \vec{r}$ of the arbitrary point P is completely defined by this angle, known as the *angular coordinate* of the rigid body, through the rotation matrix a_φ:

$$a_\varphi = rot(z, \varphi) = \begin{bmatrix} \cos\varphi & \sin\varphi & 0 \\ -\sin\varphi & \cos\varphi & 0 \\ 0 & 0 & 1 \end{bmatrix}, \quad \vec{r} = \begin{bmatrix} x \\ y \\ z \end{bmatrix}. \tag{2.32}$$

The velocity of P is a vector perpendicular to the plane containing Oz and $\vec{r}$

$$\vec{v}_P = \tilde{\omega}\vec{r} = \tilde{r}^T \vec{\omega}, \tag{2.33}$$

where the *angular velocity* of the body, $\vec{\omega} = \dot{\varphi}\vec{k}$, is a vector directed along the axis of rotation and is associated with the skew-symmetric matrix $a_\varphi \dot{a}_\varphi^T = \tilde{\omega} = \dot{\varphi}\tilde{k}$.

Fig. 2.3 Rotation of the rigid body

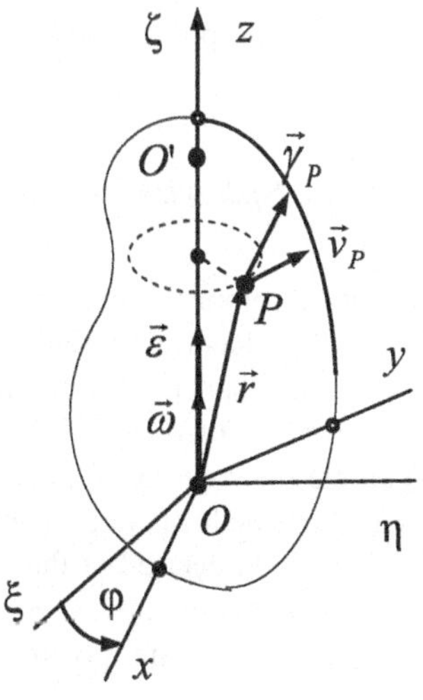

The acceleration of P can be obtained by time differentiating of (2.33), which yields

$$\vec{\gamma}_P = \{\tilde{\omega}^2 + \tilde{\varepsilon}\}\vec{r}. \tag{2.34}$$

The *angular acceleration* of a body rotating about a fixed axis, $\vec{\varepsilon} = \dot{\vec{\omega}} = \ddot{\varphi}\vec{k}$, is a vector likewise directed along the axis of rotation, associated with the skew-symmetric matrix $\tilde{\varepsilon} = \dot{\tilde{\omega}} = \ddot{\varphi}\tilde{k}$ and equal in magnitude to the rate of change $\dot{\omega} = \ddot{\varphi}$ of the angular velocity. We add that the acceleration $\vec{\gamma}_P$ is the sum of two vectors: the first vector is directed toward the centre of the circle described by the point P and therefore represents the *normal component* of the acceleration, while the second vector is tangent to the circle and represents the *tangential component*.

2.5.3 Cylindrical Motion of a Rigid Body

A rigid body is in a cylindrical motion or helical motion if there are two points of the body remaining permanently on a given straight line. This type of motion consists of a translation along a given line and which is superimposed on a rotation, taking place simultaneously around the same axis. Every point P describes helixes around the axis of translation-rotation.

Two *independent* parameters describing the position of the body are the *angular coordinate* φ, which defines the rotation of the body, and the *linear coordinate* λ, which represents the displacement of the body along same axis. Combining these motions of the rigid body, the velocity field and the acceleration field are calculated using two equations:

$$\vec{v}_P = \vec{v}_O + \tilde{\omega}\vec{r}, \quad \vec{\gamma}_P = \vec{\gamma}_O + \{\tilde{\omega}^2 + \tilde{\varepsilon}\}\vec{r}. \tag{2.35}$$

We remark that, at any given instant, the velocities and the accelerations are the same for the points chosen along a line parallel to the axis of translation-rotation.

2.5.4 Spherical Motion of a Rigid Body

If three arbitrary points of a rigid body move permanently on the surfaces of three spheres concentric to another fixed sphere, called the *sphere of reference*, then that body has a spherical motion. The shared centre O of these spheres is considered as the *fixed point* of the rigid body. The distance r from the centre O to a chosen particle P located on the body is the same for any time and any position of the body. Thus, the path of motion for the particle lies on the surface of a sphere having a radius r and centred at the fixed point O.

A general rotation is a displacement in which at least one point of the body remains in its initial position. *Euler's theorem* states that the general displacement of a rigid body with one point fixed is a simple rotation about some axis called the *screw axis*. We seek a description of the coordinate transformation in terms of the parameters of this rotation, namely, the angle of rotation and the direction cosines of the orientated axis of rotation [3].

Let the moving frame be rotated by angle φ about an axis passing through the origin O (Fig. 2.4). If the initial position of the vector $\overrightarrow{OP}$ is $\vec{r}_0 = \overrightarrow{OP_0}$, the final position is denoted $\vec{r}$. The unit vector along the (Δ) orientation axis of finite rotation is denoted by $\vec{u}$. A plane containing both points P_0 and P is perpendicular to the axis of rotation. The point of intersection between this plane and the axis of rotation is taken as M. Let N be the orthogonal projection of P onto vector $\overrightarrow{MP_0}$.

To derive a relation between $\overrightarrow{OP}$ and $\overrightarrow{OP_0}$, we express the following vector relations:

$$\begin{aligned}\overrightarrow{MP_0} &= \overrightarrow{OP_0} - \overrightarrow{OM} = \vec{r}_0 - (\vec{u}^T\vec{r}_0)\vec{u} \\ \overrightarrow{MP} &= \overrightarrow{MN} + \overrightarrow{NP} = \overrightarrow{MP_0}\cos\varphi + \tilde{u}\vec{r}_0\sin\varphi.\end{aligned} \tag{2.36}$$

Substitution of (2.36) into expression of the vector $\vec{r} = \overrightarrow{OM} + \overrightarrow{MP}$, together with a slight rearrangement of terms, leads to the known *Rodrigues* formula for a finite spherical rotation of a rigid body:

$$\vec{r} = \vec{r}_0\cos\varphi + (\vec{u}^T\vec{r}_0)\vec{u}(1-\cos\varphi) + \tilde{u}\vec{r}_0\sin\varphi. \tag{2.37}$$

According to Euler's theorem, a rigid body can be guided into any orientation in space through a single rotation. The variable axis and the angle of this rotation can be used to parameterise the possible body orientations, based on the quaternion and Rodrigues parameters. However, it is often much more practical to generate the new

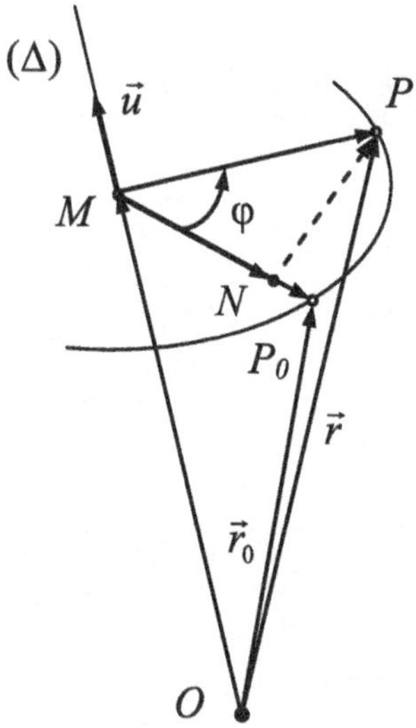

Fig. 2.4 Vector diagram for *Rodrigues* formula

orientation of the body, by using one or more rotations about the moving axes of some frames of reference.

The above cosine representation of the orientation of a rigid body generally contains nine parameters. Since a general rotation is a spherical motion with three degrees of freedom, a set of three independent parameters is sufficient to describe the orientation of this rigid body in space. In general, three rotations about some of the coordinate axes are needed to bring the body into an arbitrary orientation. The most commonly used set of three-parameter representation is given by the *Euler angles*.

The orientation of the body-fixed axes at any time can be obtained by an imaginary rotation of these axes from an orientation coincident with the global fixed axes. In this modelling, three successive rotations about the axes of some moving coordinate systems centred in O are used to describe the orientation of a rigid body.

Choosing the origin of the fixed Cartesian $O\xi\eta\varsigma$ frame in the fixed centre, we introduce three basic rotation matrices from which an *Euler angles representation* can be derived in terms of their product (Fig. 2.5). The coordinate transformation associated with a change of orientation of the rigid body is completely given by the proper orthogonal rotation matrix a. Starting from an initial position coincident with $O\xi\eta\zeta$, the absolute body posture is obtained by composing three successive relative rotations, namely: first rotation of the $O\xi y_1 z_1$ frame of angle ψ around the $O\xi$ axis, expressed by a rotation matrix a_ψ, followed by the second rotation of the $Ox_2 y_1 z$ frame of angle θ around the rotated Oy_1 axis, expressed by a rotation matrix a_θ, and finally followed by the third rotation of angle φ around the Oz axis of the frame $Oxyz$ attached to rigid body, expressed by a rotation matrix a_φ. The rotation about $O\xi$ fixed axis is called a *roll*, the rotation about the Oy_1 moving axis is called a *pitch* and the rotation about Oz axis, which is linked with the rigid body, is called a *yaw*.

In order to construct a unique rotation matrix a with respect to the fixed global coordinate system $O\xi\eta\varsigma$, the general transformation matrix can be obtained by pre-multiplying three well-known orthogonal rotation matrices as follows:

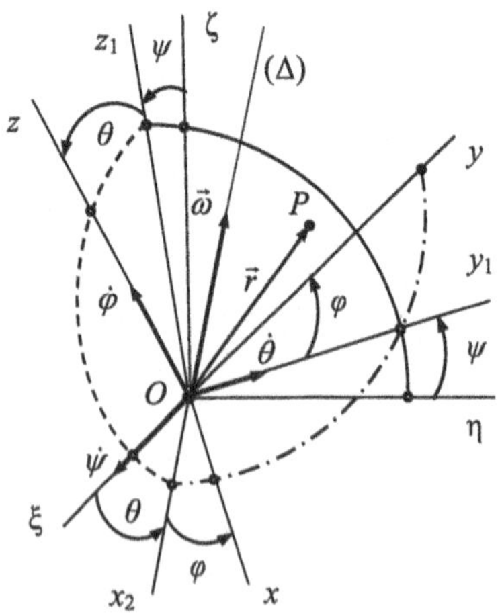

Fig. 2.5 Euler angles representation

$$a = a_\varphi a_\theta a_\psi, \tag{2.38}$$

where

$$\begin{aligned}
a_\psi = rot(x, \psi) &= \begin{bmatrix} 1 & 0 & 0 \\ 0 & \cos\psi & \sin\psi \\ 0 & -\sin\psi & \cos\psi \end{bmatrix} \\
a_\theta = rot(y, \theta) &= \begin{bmatrix} \cos\theta & 0 & -\sin\theta \\ 0 & 1 & 0 \\ \sin\theta & 0 & \cos\theta \end{bmatrix} \\
a_\varphi = rot(z, \varphi) &= \begin{bmatrix} \cos\varphi & \sin\varphi & 0 \\ -\sin\varphi & \cos\varphi & 0 \\ 0 & 0 & 1 \end{bmatrix}.
\end{aligned} \tag{2.39}$$

Developing this, nine components (2.3) of the general transformation matrix $a = a_\varphi a_\theta a_\psi$ are the following

$$\begin{aligned}
\alpha_{11} &= \cos\theta\cos\varphi \\
\alpha_{12} &= \sin\psi\sin\theta\cos\varphi + \cos\psi\sin\varphi \\
\alpha_{13} &= -\cos\psi\sin\theta\cos\varphi + \sin\psi\sin\varphi \\
\alpha_{21} &= -\cos\theta\sin\varphi \\
\alpha_{22} &= -\sin\psi\sin\theta\sin\varphi + \cos\psi\cos\varphi \\
\alpha_{23} &= \cos\psi\sin\theta\sin\varphi + \sin\psi\cos\varphi \\
\alpha_{31} &= \sin\theta \\
\alpha_{32} &= -\sin\psi\cos\theta \\
\alpha_{33} &= \cos\psi\cos\theta.
\end{aligned} \tag{2.40}$$

It should be kept in mind that the product of the matrices (2.38) is not commutative and when this product is read from right to left, the three rotations are successively performed, in that order, about the axes of three *absolutely distinct frames*. Actually, the rotations of the rigid body are taken during the same interval of time.

Given the roll, pitch and yaw, we can compute the rotation matrix a from (2.38) and (2.39). On the other hand, given a rotation matrix a, we can compute the equivalent roll-pitch-yaw angles as follows:

$$\begin{aligned}
\theta &= \sin^{-1}(a_{31}) \\
\psi &= a\tan 2(-a_{32}/\cos\theta, a_{33}/\cos\theta) \\
\varphi &= a\tan 2(-a_{21}/\cos\theta, a_{11}/\cos\theta),
\end{aligned} \tag{2.41}$$

provided that $\cos\theta \neq 0$, where $a\tan 2(x, y)$ is a two-argument arctangent function that yields one *unique* solution for the angle. Hence, corresponding to a given rotation matrix, there are generally two possible solutions of the roll-pitch-yaw angles. If the singular rotation angle $\theta = \pm\pi/2$ exist, the solutions of Eq. (2.41) will, however, degenerate. This stems from the fact that two elementary rotation axes coincide and thus one-degree freedom in the rotation is lost. For this special case, only the sum or the difference of ψ and φ can be computed.

Now, if $\vec{\rho}$ is a vector whose coordinates are expressed in the fixed frame, the location of the particle P of the rigid body is completely expressed in the rotated body-frame $Oxyz$ by the position vector

$$\vec{r} = a\vec{\rho}. \tag{2.42}$$

In the fixed base $O\xi\eta\varsigma$ the position of the same point P is given by the relation

$$\vec{\rho} = a^T\vec{r}, \quad a^T = a_\psi^T a_\theta^T a_\varphi^T. \tag{2.43}$$

Since the motion of the body is continuous, its orientation will change and the transformation matrix must be a continuous function of time. The velocity of any point P on a body rotating about a fixed point can be determined using the same methods as for a body rotating about a fixed axis

$$\vec{v}_P = \tilde{\omega}\vec{r} = \tilde{r}^T\vec{\omega}, \tag{2.44}$$

where the skew-symmetric matrix

$$\tilde{\omega} = a\dot{a}^T = a_\varphi a_\theta \tilde{\omega}_\psi a_\theta^T a_\varphi^T + a_\varphi \tilde{\omega}_\theta a_\varphi^T + \tilde{\omega}_\varphi \tag{2.45}$$

is associated with the angular velocity of the rigid body

$$\vec{\omega} = a_\varphi a_\theta \vec{\omega}_\psi + a_\varphi \vec{\omega}_\theta + \vec{\omega}_\varphi. \tag{2.46}$$

The vectors $\vec{\omega}_\psi$, $\vec{\omega}_\theta$ and $\vec{\omega}_\phi$ represent precisely the angular velocities of three successive rotations along the axes $O\xi$, Oy_1 and Oz, respectively, about which the Euler angles are performed

$$\begin{aligned}
\vec{\omega}_\psi &= \dot{\psi}\vec{u}_1 = \dot{\psi}[1 \quad 0 \quad 0]^T \\
\vec{\omega}_\theta &= \dot{\theta}\vec{u}_2 = \dot{\theta}[0 \quad 1 \quad 0]^T \\
\vec{\omega}_\varphi &= \dot{\varphi}\vec{u}_3 = \dot{\varphi}[0 \quad 0 \quad 1]^T.
\end{aligned} \tag{2.47}$$

The angular velocity vector $\vec{\omega}$ defined in the body-fixed frame can be expressed in terms of Euler angles and their time derivatives as

$$\vec{\omega} = \begin{bmatrix} \cos\theta\cos\varphi & \sin\varphi & 0 \\ -\cos\theta\sin\varphi & \cos\varphi & 0 \\ \sin\theta & 0 & 1 \end{bmatrix} \begin{bmatrix} \dot{\psi} \\ \dot{\theta} \\ \dot{\varphi} \end{bmatrix}. \tag{2.48}$$

However, useful projections of the same vector about fixed axes of $O\xi\eta\varsigma$ frame are the followings

$$\vec{\omega}^{\circ} = a^T\vec{\omega} = \begin{bmatrix} 1 & 0 & \sin\theta \\ 0 & \cos\psi & -\cos\theta\sin\psi \\ 0 & \sin\psi & \cos\theta\cos\psi \end{bmatrix} \begin{bmatrix} \dot{\psi} \\ \dot{\theta} \\ \dot{\varphi} \end{bmatrix}. \tag{2.49}$$

The velocity field (2.44) is identical with that of a rotation as if the body were to rotate about an *instantaneous axis of rotation* (Δ). We note that the orientation of this rotating axis, the direction and the intensity of the angular velocity change all from one instant to the next.

Knowing the matrices $\tilde{\omega}_{\psi} = \dot{\psi}\tilde{u}_1$, $\tilde{\omega}_{\theta} = \dot{\theta}\tilde{u}_2$ and $\tilde{\omega}_{\varphi} = \dot{\varphi}\tilde{u}_3$, the derivatives with respect to time of (2.45) and (2.46) lead to the angular acceleration $\vec{\varepsilon}$ and its associated skew-symmetric matrix $\tilde{\varepsilon}$, as follows:

$$\begin{aligned} \vec{\varepsilon} = \dot{\vec{\omega}} &= \ddot{\psi}a_{\varphi}a_{\theta}\vec{u}_1 + \ddot{\theta}a_{\varphi}\vec{u}_2 + \ddot{\varphi}\vec{u}_3 \\ &\quad - \dot{\psi}\dot{\theta}a_{\varphi}\tilde{u}_2a_{\theta}\vec{u}_1 - \dot{\psi}\dot{\varphi}\tilde{u}_3a_{\varphi}a_{\theta}\vec{u}_1 - \dot{\theta}\dot{\varphi}\tilde{u}_3a_{\varphi}\vec{u}_2 \\ \tilde{\varepsilon} = \dot{\tilde{\omega}} &= \ddot{\psi}a_{\varphi}a_{\theta}\tilde{u}_1a_{\theta}^Ta_{\varphi}^T + \ddot{\theta}a_{\varphi}\tilde{u}_2a_{\varphi}^T + \ddot{\varphi}\tilde{u}_3 \\ &\quad + \dot{\psi}\dot{\theta}(a_{\varphi}a_{\theta}\tilde{u}_1a_{\theta}^T\tilde{u}_2a_{\varphi}^T - a_{\varphi}\tilde{u}_2a_{\theta}\tilde{u}_1a_{\theta}^Ta_{\varphi}^T) \\ &\quad + \dot{\psi}\dot{\varphi}(a_{\varphi}a_{\theta}\tilde{u}_1a_{\theta}^Ta_{\varphi}^T\tilde{u}_3 - \tilde{u}_3a_{\varphi}a_{\theta}\tilde{u}_1a_{\theta}^Ta_{\varphi}^T) \\ &\quad + \dot{\theta}\dot{\varphi}(a_{\varphi}\tilde{u}_2a_{\varphi}^T\tilde{u}_3 - \tilde{u}_3a_{\varphi}\tilde{u}_2a_{\varphi}^T), \end{aligned} \tag{2.50}$$

where $\tilde{u}_1$, $\tilde{u}_2$ and $\tilde{u}_3$ represent the skew-symmetric matrices associated with the vectors $\vec{u}_1$, $\vec{u}_2$ and $\vec{u}_3$, respectively.

If the angular velocity $\vec{\omega}$, the angular acceleration $\vec{\varepsilon}$ and their associated skew-symmetric matrices $\tilde{\omega}$ and $\tilde{\varepsilon}$ are known at any given instant, we can compute the significant square matrix $\tilde{\omega}^2 + \tilde{\varepsilon}$ and the acceleration of P by differentiating (2.44) with respect to time:

$$\begin{aligned} \vec{\gamma}_P &= \{\tilde{\omega}^2 + \tilde{\varepsilon}\}\vec{r} \\ \tilde{\omega}^2 + \tilde{\varepsilon} &= \ddot{\psi}a_{\varphi}a_{\theta}\tilde{u}_1a_{\theta}^Ta_{\varphi}^T + \ddot{\theta}a_{\varphi}\tilde{u}_2a_{\varphi}^T + \ddot{\varphi}\tilde{u}_3 \\ &\quad + \dot{\psi}^2a_{\varphi}a_{\theta}\tilde{u}_1\tilde{u}_1a_{\theta}^Ta_{\varphi}^T + \dot{\theta}^2a_{\varphi}\tilde{u}_2\tilde{u}_2a_{\varphi}^T + \dot{\varphi}^2\tilde{u}_3\tilde{u}_3 \\ &\quad + 2\dot{\psi}\dot{\theta}a_{\varphi}a_{\theta}\tilde{u}_1a_{\theta}^T\tilde{u}_2a_{\varphi}^T + 2\dot{\theta}\dot{\varphi}a_{\varphi}\tilde{u}_2a_{\varphi}^T\tilde{u}_3 + 2\dot{\varphi}\dot{\psi}a_{\varphi}a_{\theta}\tilde{u}_1a_{\theta}^Ta_{\varphi}^T\tilde{u}_3. \end{aligned} \tag{2.51}$$

This equation is *formally* the same as that (2.34), above developed for the point P located on a body subjected to rotation about a fixed axis.

2.5.5 Planar Motion of a Rigid Body

When three points of a rigid body, those points not being situated on the same straight line, move along paths which are contained in parallel planes to another fixed *reference* plane, the body is said to undergo *planar motion*. A section of the body demarcated by the reference plane is itself a two-dimensional body called a rigid *lamina* [4].

A general planar motion can always be considered as a combination of translation and rotation; the translation occurs within a reference plane, while the rotation occurs about an axis perpendicular to the reference plane.

In order to specify the configuration or state of a planar body, it is first necessary to define the coordinates that give the location of the body. A rigid body in a planar motion has three degrees of freedom: two linear coordinates of any one point on the rigid body and one angular coordinate to define the angle of rotation with respect to two fixed orthogonal axes. The two translational components and the rotation are three independent variables, since any one of them can be changed arbitrarily while keeping the other two coordinates fixed.

Let the coordinate system $\Omega\xi\eta$ be a global fixed *reference* plane (Fig. 2.6). Defining a body-fixed $Oxyz$ frame embedded in the body, two global coordinates ξ_O, η_O of the origin O and the angle φ of rotation of this system relative to the global coordinate system $\Omega\xi\eta\varsigma$ can define the absolute location of the rigid body.

The matrix relation between the local and global coordinates of point P is

$$\vec{\rho} = \vec{\rho}_O + a_\varphi^T \vec{r}, \tag{2.52}$$

where a_φ is the same 3×3 rotational transformation (2.32) and the position vectors are expressed as follows

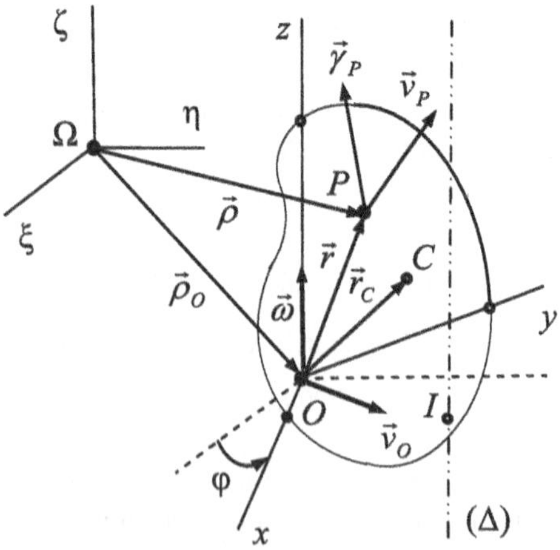

Fig. 2.6 Planar motion of the rigid body

$$\vec{\rho}_O = [\xi_O \quad \eta_O \quad \varsigma_O]^T, \quad \vec{r} = [x \quad y \quad z]^T. \tag{2.53}$$

The velocity field is generally given in the local frame by Euler's formula (2.17)

$$\vec{v}_P = \vec{v}_O + \tilde{\omega}\vec{r}, \tag{2.54}$$

where

$$\vec{v}_O = [v_{Ox} \quad v_{Oy} \quad 0]^T, \quad \vec{\omega} = \dot{\varphi}\vec{u}_3, \quad \tilde{\omega} = \dot{\varphi}\tilde{u}_3. \tag{2.55}$$

Written in terms of the angular velocity vector $\vec{\omega}$, the matrix relation (2.54) indicates that the velocity of any point P on the rigid body represents a vector sum of the velocity of a reference point O and the relative velocity between the two points. We add that the vector $\vec{\omega}$ is permanently perpendicular to the reference plane and that it is independent of the chosen reference point.

If $\vec{\omega} \neq \vec{0}$, there is a straight line (Δ) permanently parallel to $\vec{\omega}$ along which the velocities are zero $\vec{v}_O + \tilde{\omega}\vec{r}_I = \vec{0}$. This axis is called *instantaneous axis of rotation.* At any given instant, the velocities of various particles of the body are the same as they would be if the body were rotating about the instantaneous axis of rotation. This axis intersects the reference plane of the planar motion at a point I whose velocity is zero at the instant under consideration, called the *instantaneous centre of rotation.*

The centre I of the body in a planar motion can be located on or outside the lamina-body. If it is located on the body, the particle coinciding with the instantaneous centre at a given instant must have zero velocity at that instant. It should be noted that the point chosen as the instantaneous centre of zero velocity for the body is valid and can only be used for a given instant of time, since the body changes its position from one instant to the next.

To locate a centre I, we can use the fact that the velocity of a point on the body is always perpendicular to the relative position vector extending from I to the point considered. If the directions of the velocities $\vec{v}_A$ and $\vec{v}_B$ of two points A and B of a lamina body are known and if they differ, then the instantaneous centre I is obtained by drawing the perpendicular to $\vec{v}_A$ through A and the perpendicular to $\vec{v}_B$ through B and determining the point at which these two lines intersect. As a special case, note that if $\vec{v}_A$ and $\vec{v}_B$ were parallel, the instantaneous centre I would be located at an infinite distance and $\vec{\omega}$ would be zero; the body is translating and all its points would have the same velocity. If the velocities of two points are perpendicular to their common line AB and if the magnitudes of these velocities are known, the instantaneous centre can graphically be found by intersecting this line AB with the line joining the extremities of two vectors $\vec{v}_A$ and $\vec{v}_B$ (Fig. 2.7).

As the motion of the body proceeds, the instantaneous centre moves in space. The locus of points which define the location of I with respect to the fixed frame $\Omega\xi\eta$ is a planar curve in this frame called the *fixed centrode* (B) and so each point of the centrode acts as the instantaneous centre for the body only for an instant. Two coordinates of I are expressed in the fixed base by following relations:

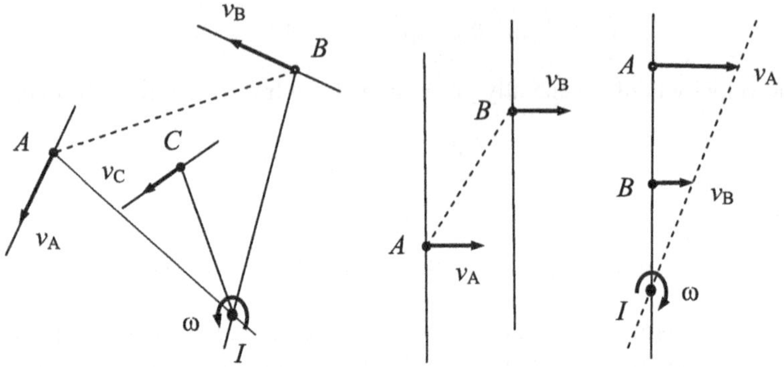

Fig. 2.7 Location of the instantaneous centre of rotation

$$\begin{aligned}\xi_I &= \xi_O + x_I \cos\varphi - y_I \sin\varphi \\ \eta_I &= \eta_O + x_I \sin\varphi + y_I \cos\varphi,\end{aligned} \tag{2.56}$$

where

$$x_I = -\frac{v_{Oy}}{\dot{\varphi}}, \quad y_I = \frac{v_{Ox}}{\dot{\varphi}}. \tag{2.57}$$

The locus of the same centre I with respect to the fixed-body frame $Oxyz$ is also a planar curve on the body called the *moving centrode* (R). It can be shown that at any time t these two centrodes are tangent at a common point, precisely at the instantaneous centre I, and that the moving centrode appears to roll without slipping on the fixed centrode, during the motion of the body (Fig. 2.8).

The acceleration of an arbitrary point P fixed on the rigid body is expressed in the local frame by differentiating the velocity equation (2.54) with respect to time:

$$\vec{\gamma}_P = \vec{\gamma}_O + \{\tilde{\omega}^2 + \tilde{\varepsilon}\}\vec{r}, \tag{2.58}$$

where $\vec{\gamma}_O = [\gamma_{Ox} \ \gamma_{Oy} \ 0]^T$ is the acceleration of the reference point O and $\tilde{\varepsilon} = \dot{\tilde{\omega}} = \ddot{\varphi}\,\tilde{u}_3$ is the skew-symmetric matrix associated with the angular acceleration $\vec{\varepsilon} = \dot{\vec{\omega}} = \ddot{\varphi}\vec{u}_3$. At any instant, if $\det\{\tilde{\omega}^2 + \tilde{\varepsilon}\} \neq 0$, there is a point J in the representative lamina body, for which the acceleration is zero. This point is called the *pole of accelerations*. The acceleration field in a planar motion of a rigid body is identical with the field of acceleration in a rotation, as if the body were rotating about an axis perpendicular to the fixed reference plane and passing through J. Note that the centre I and the pole J are generally two different points at every instant during the evolution of the rigid body.

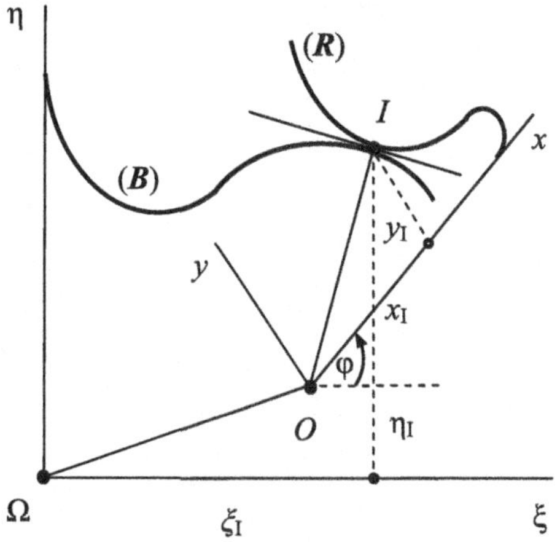

Fig. 2.8 Centrodes of planar motion of a rigid body

References

1. Staicu, S.: Mecanica Teoretica. Edit. Didactica & Pedagogica, Bucharest (1998)
2. Beer, F.P., Johnston Jr., E.R.: Vector Mechanics for Engineers, Dynamics. McGraw-Hill Ryerson (1988)
3. Hibbeler, R.C.: Engineering Mechanics, Dynamics. Prentice-Hall, Inc. (1997)
4. Duffy, J.: Statics and Kinematics with Applications in Robotics. Cambridge University Press, New York (1996)

Chapter 3
Matrix Kinematics of Composed Motion

In the computing of the velocities and the accelerations, in engineering problems, is useful to introduce one or more moving frames in addition to the fixed frame. The motion of the latter moving frame can either be predefined as a target motion or is obtained directly as a particular solution from the equations of motion.

3.1 Kinematics of Composed Motion of a Point

The position, the velocity and the acceleration of a point depend on the reference coordinate system, relative to which the motion of the point is analysed. The motion of a point will be described differently by two observers which are moving relative one to another.

We consider a fixed frame of reference $\Omega\,\xi\eta\varsigma$ and a second frame $Oxyz$ which move in a known, but arbitrary, fashion with respect to a fixed base (Fig. 3.1). Let P be a particle moving in both spaces of the two frames, as its position is defined at any instant by the vector $\vec{\rho}$ in the fixed frame and by the vector $\vec{r}$ in the moving frame. By denoting $\vec{\rho}_O$ the position vector of the origin O of the moving frame with respect to the fixed frame, we obviously have

$$\vec{\rho} = \vec{\rho}_O + a^T \vec{r}, \tag{3.1}$$

where a is the transformation matrix of the moving frame with respect to the fixed base, and

$$\vec{r} = [x \quad y \quad z]^T \tag{3.2}$$

is the *variable* column matrix of the relative position vector of P, evaluated in the space of the moving frame.

S. Staicu, *Dynamics of Parallel Robots*, Parallel Robots: Theory and Applications,
https://doi.org/10.1007/978-3-319-99522-9_3

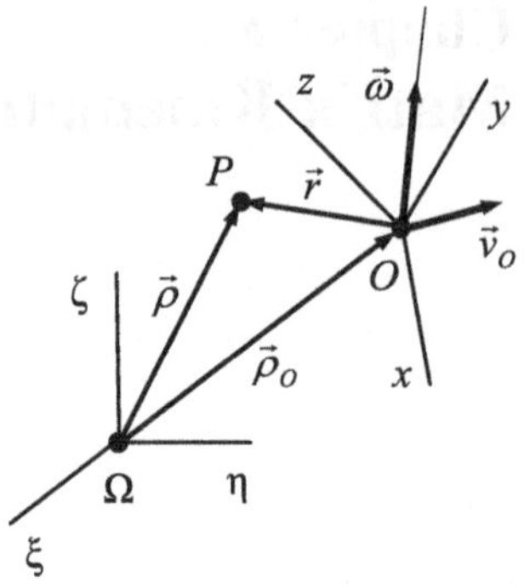

Fig. 3.1 Composed motion of a point

3.1.1 Absolute Velocity

The velocity of the point P, measured by first observer from $\Omega\,\xi\eta\varsigma$, is determined by taking the time derivative of (3.1), which yields

$$\dot{\vec{\rho}} = \dot{\vec{\rho}}_O + \dot{a}^T\vec{r} + a^T\dot{\vec{r}}. \tag{3.3}$$

Pre-multiplying by the general rotation matrix a, we obtain the components of these vectors in the mobile coordinate system $Oxyz$

$$\vec{v}_a^P = \vec{v}_O + \tilde{\omega}\vec{r} + \dot{\vec{r}}, \tag{3.4}$$

where the first term in the right-hand member $\vec{v}_O = a\dot{\vec{\rho}}_O$ represents the velocity of the origin O of the moving axes and $\tilde{\omega} = a\dot{a}^T$ is the skew-symmetric matrix associated with the angular velocity $\vec{\omega}$ of the moving frame. On the other hand, the vector

$$\vec{v}_r^P = \dot{\vec{r}} = [\dot{x} \quad \dot{y} \quad \dot{z}]^T \tag{3.5}$$

is the *relative velocity* of P with respect to axes $Oxyz$, and the remaining vector sum

$$\vec{v}_t^P = \vec{v}_O + \tilde{\omega}\vec{r} \tag{3.6}$$

represents precisely the velocity of a point P', belonging to the space of moving frame and coincident at a given instant with P. This second vector is called *guidance velocity* or *velocity of transport*.

We obtain the fundamental relation

$$\vec{v}_a^P = \vec{v}_t^P + \vec{v}_r^P. \tag{3.7}$$

The absolute velocity of a point is equal to the vector sum of the velocity of transport and of the relative velocity.

3.1.2 Absolute Acceleration

The absolute acceleration of P is defined as the rate of change of the absolute velocity with respect to the fixed frame $\Omega\xi\eta\varsigma$. By performing the time derivative of (3.3), we write

$$\ddot{\vec{\rho}} = \ddot{\vec{\rho}}_O + \ddot{a}^T\vec{r} + 2\dot{a}^T\dot{\vec{r}} + a^T\ddot{\vec{r}}. \tag{3.8}$$

Since $\dot{a}^T = a^T\tilde{\omega}$ and $\ddot{a}^T = a^T\{\tilde{\omega}\tilde{\omega} + \tilde{\varepsilon}\}$, the components of all vectors in the moving frame $Oxyz$ are obtained by pre-multiplying by the operator a:

$$\vec{\gamma}_a^P = \vec{\gamma}_O + \{\tilde{\omega}^2 + \tilde{\varepsilon}\}\vec{r} + 2\tilde{\omega}\,\vec{v}_r^P + \ddot{\vec{r}}, \tag{3.9}$$

where $\vec{\gamma}_O = a\ddot{\vec{\rho}}_O$ represents the acceleration of the origin O and $\tilde{\varepsilon} = \dot{\tilde{\omega}}$ is the skew-symmetric matrix associated with the angular acceleration $\vec{\varepsilon} = \dot{\vec{\omega}}$ of the moving frame.

We note that the vector sum

$$\vec{\gamma}_t^P = \vec{\gamma}_O + \{\tilde{\omega}^2 + \tilde{\varepsilon}\}\vec{r} \tag{3.10}$$

is the *guidance acceleration* or *acceleration of transport* and that the latter term in the right hand member of (3.9), namely $\vec{\gamma}_r^P = \dot{\vec{v}}_r^P = \ddot{\vec{r}}$, represents the *relative acceleration* of P with respect to $Oxyz$ coordinate system. On the other hand, the additional term from (3.9)

$$\vec{\gamma}_c^P = 2\tilde{\omega}\,\vec{v}_r^P \tag{3.11}$$

defines a complementary acceleration called the *Coriolis acceleration*. Further, we add that $\vec{\gamma}_c^P$ is perpendicular to the plane containing the two vectors $\vec{v}_r^P$, $\vec{\omega}$ and that the orientation of the vector of the Coriolis acceleration can be obtained by rotating the vector $\vec{\omega}$ around $\vec{v}_r^P$, in the positive sense rule.

Pursuing the fundamental relation

$$\vec{\gamma}_a{}^P = \vec{\gamma}_t^P + \vec{\gamma}_r{}^P + \vec{\gamma}_c^P, \tag{3.12}$$

the absolute acceleration of a point is a vector sum of the acceleration of transport and of relative acceleration, where is necessary to add a complementary term representing the Coriolis acceleration $\vec{\gamma}_c^P$.

3.2 Kinematics of Composed Motion of a Rigid Body

A mechanical system is defined as an assembly of rigid bodies connected by points, where some or all of the bodies can move relatively one to another. A free body has six degrees of freedom, but the number of positional parameters is reduced by the existence of some constraints which are imposed by coupling mechanisms that introduce restrictions on the relative motions. Of primary importance in a kinematic analysis is the *connectivity* of the bodies, expressed being by the type and the location of the joints into the system. The purpose of the joints is to transmit the motion in certain fashion.

The most general way to analyze the tree-dimensional motion of a rigid body $T_n(O_n x_n y_n z_n)$ requires the use of some intermediate frames that translate and rotate relative to fixed base. This analysis will also provide a means for determining of motions of the points located on separate members of a spatial or planar mechanism.

3.2.1 *Velocity Field*

We consider a fixed frame $T_0(O_0 x_0 y_0 z_0)$ and some rigid bodies linked with their moving frames $T_k(O_k x_k y_k z_k, k = 1, 2, \ldots, n-1)$ with respect to which move the last rigid body $T_n(O_n x_n y_n z_n)$ (Fig. 3.2). The location and the position of an arbitrary rigid body, which is attached to T_k frame with respect to T_{k-1} frame, are known by the relative coordinates $x_{k,k-1}, y_{k,k-1}, z_{k,k-1}$ of the origin O_k and the Euler angles $\psi_{k,k-1}, \theta_{k,k-1}, \varphi_{k,k-1}$ through two column matrices $\vec{r}_{k.k-1}$, $\vec{\lambda}_{k.k-1}$ and the transformation matrix $a_{k,k-1}$

$$\vec{r}_{k,k-1} = \begin{bmatrix} x_{k,k-1} \\ y_{k,k-1} \\ z_{k,k-1} \end{bmatrix}, \quad \vec{\lambda}_{k,k-1} = \begin{bmatrix} \psi_{k,k-1} \\ \theta_{k,k-1} \\ \varphi_{k,k-1} \end{bmatrix}, \quad a_{k,k-1} = a^{\varphi}_{k,k-1} a^{\theta}_{k,k-1} a^{\psi}_{k,k-1} \tag{3.13}$$
$$(k = 1, 2, \ldots, n-1)$$

where $a^{\psi}_{k,k-1} = rot(x, \psi_{k,k-1})$, $a^{\theta}_{k,k-1} = rot(y, \theta_{k,k-1})$, $a^{\varphi}_{k,k-1} = rot(z, \varphi_{k,k-1})$.

A position vector can be a function of time in one reference frame, but constant if is measured in another coordinate system. If the relative position of a point P_n in the fixed-body frame T_n is specified by the constant position vector $\vec{r}_n{}^{P_n} = \left[x_n{}^{P_n} \quad y_n{}^{P_n} \quad z_n{}^{P_n} \right]^T$, then a vector addition will express the absolute position vector of the same point with respect to the fixed reference frame T_0

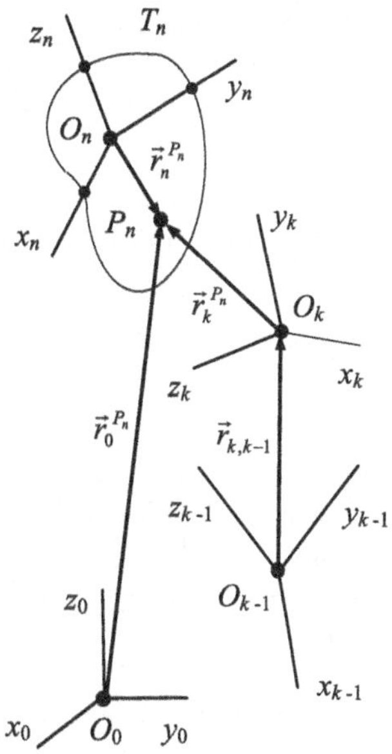

Fig. 3.2 Composed motion of a rigid body

$$\vec{r}_0^{\,P_n} = \vec{r}_{10} + \sum_{k=1}^{n-1} a_{k0}^T \vec{r}_{k+1,k} + a_{n0}^T \vec{r}_n^{\,P_n}, \tag{3.14}$$

with the notations

$$a_{k0}^T = \prod_{s=1}^{k} a_{s,s-1}^T, \quad a_{n0}^T = \prod_{s=1}^{n} a_{s,s-1}^T. \tag{3.15}$$

Especially, the position vector of the origin O_n is expressed as follows

$$\vec{r}_{n0} = \vec{r}_{10} + \sum_{k=1}^{n-1} a_{k0}^T \vec{r}_{k+1,k}. \tag{3.16}$$

By performing the derivative of (3.14) with respect to time, we obtain

$$\dot{\vec{r}}_0^{\,P_n} = \dot{\vec{r}}_{10} + \sum_{k=1}^{n-1} \dot{a}_{k0}^T \vec{r}_{k+1,k} + \sum_{k=1}^{n-1} a_{k0}^T \dot{\vec{r}}_{k+1,k} + \dot{a}_{n0}^T \vec{r}_n^{\,P_n}, \tag{3.17}$$

where $\dot{\vec{r}}_{k,k-1} = a_{k,k-1}^T \vec{v}_{k,k-1}$ is the relative velocity of O_k, measured in the frame T_{k-1}.

Pre-multiplying by the matrix

$$a_{n0} = \prod_{s=1}^{n} a_{n-s+1,n-s} = a_{n,n-1}a_{n-1,n-2}\ldots a_{21}a_{10}, \tag{3.18}$$

we have a general formula for the velocity field

$$\vec{v}_0^{P_n} = \sum_{k=1}^{n} a_{nk}\vec{v}_{k,k-1} + \sum_{k=1}^{n-1} a_{nk}\tilde{\omega}_{k,k-1}\sum_{q=k}^{n-1} a_{qk}^T\vec{r}_{q+1,q} + \sum_{k=1}^{n} a_{nk}\tilde{\omega}_{k,k-1}a_{nk}^T\vec{r}_n^{P_n}, \tag{3.19}$$

where $\tilde{\omega}_{k,k-1} = a_{k,k-1}\dot{a}_{k,k-1}^T$ represents the skew-symmetric matrix associated with the relative angular velocity $\vec{\omega}_{k,k-1}$, here expressed in the frame T_k. Using this formula, we can compute primarily the velocities of two distinct points A_n and B_n located on the body T_n and which, afterwards, we substitute in the known vector relation

$$\vec{v}_0^{B_n} - \vec{v}_0^{A_n} = \tilde{\omega}_{n0}\left(\vec{r}_0^{B_n} - \vec{r}_0^{A_n}\right). \tag{3.20}$$

By rearranging, we obtain the skew-symmetric matrix

$$\tilde{\omega}_{n0} = \sum_{k=1}^{n} a_{nk}\tilde{\omega}_{k,k-1}a_{nk}^T, \tag{3.21}$$

which is associated with the *absolute angular velocity* of the body T_n:

$$\vec{\omega}_{n0} = \sum_{k=1}^{n} a_{nk}\vec{\omega}_{k,k-1}. \tag{3.22}$$

Expressing the *absolute velocity* of the origin O_n:

$$\vec{v}_{n0} = \sum_{k=1}^{n} a_{nk}\vec{v}_{k,k-1} + \sum_{k=1}^{n-1} a_{nk}\tilde{\omega}_{k,k-1}\sum_{q=k}^{n-1} a_{qk}^T\vec{r}_{q+1,q}, \tag{3.23}$$

we can again compute the velocity $\vec{v}_0^{P_n}$ in the boy-frame T_n

$$\vec{v}_0^{P_n} = \vec{v}_{n0} + \tilde{\omega}_{n0}\vec{r}_n^{P_n}. \tag{3.24}$$

Remarks

1. Alternate approaches for formulating the kinematics equations of the constraining mechanical systems are the *recursive methods*. Used in the solving differential equations, a numerical procedure is simpler against the procedure based on above extended relations.

In order to illustrate the development of the recursive kinematics equations, we consider the two successive bodies T_{n-1} and T_n, which are connected along the kinematical chain. Starting from (3.21) to (3.23) and rearranging the terms, we obtain rapidly three significant *recursive relations*:

$$\begin{aligned}
\tilde{\omega}_{n0} &= a_{n,n-1}\tilde{\omega}_{n-1,0}a^T_{n,n-1} + \tilde{\omega}_{n,n-1} \\
\vec{\omega}_{n0} &= a_{n,n-1}\vec{\omega}_{n-1,0} + \vec{\omega}_{n,n-1} \\
\vec{v}_{n0} &= a_{n,n-1}\vec{v}_{n-1,0} + a_{n,n-1}\tilde{\omega}_{n-1,0}\vec{r}_{n,n-1} + \vec{v}_{n,n-1}.
\end{aligned} \tag{3.25}$$

2. Recursive relations (3.25) can be merged and written in a compact form, using the dual expression of the absolute velocity twist $\vec{V}_{n0} = \left[\vec{v}_{n0}{}^T \quad \vec{\omega}^T_{n0}\right]^T$, as follows:

$$\vec{V}_{n0} = A_{n,n-1}\vec{V}_{n-1,0} + A_{n,n-1}\tilde{\Omega}_{n-1,0}\vec{R}_{n,n-1} + \vec{V}_{n,n-1}, \tag{3.26}$$

where

$$\begin{aligned}
A_{n,n-1} &= \begin{bmatrix} a_{n,n-1} & 0 \\ 0 & a_{n,n-1} \end{bmatrix}, \quad \vec{R}_{n,n-1} = \begin{bmatrix} \vec{r}_{n,n-1} \\ 0 \end{bmatrix} \\
\tilde{\Omega}_{n-1,0} &= \begin{bmatrix} \tilde{\omega}_{n-1,0} & 0 \\ 0 & \tilde{\omega}_{n-1,0} \end{bmatrix}, \quad \vec{V}_{n,n-1} = \begin{bmatrix} \vec{v}_{n,n-1} \\ \vec{\omega}_{n,n-1} \end{bmatrix}.
\end{aligned} \tag{3.27}$$

3.2.2 Acceleration Field

By differentiating Eq. (3.19) with respect to time, one obtains the general formula for the acceleration field

$$\begin{aligned}
\vec{\gamma}_0^{P_n} = {} & \sum_{k=1}^{n} a_{nk}\vec{\gamma}_{k,k-1} + \sum_{k=1}^{n-1} a_{nk}\{\tilde{\omega}^2_{k,k-1} + \tilde{\varepsilon}_{k,k-1}\}\sum_{q=k}^{n-1} a^T_{qk}\vec{r}_{q+1,q} \\
& + 2\sum_{k=1}^{n-1} a_{nk}\tilde{\omega}_{k,k-1}\left\{\sum_{q=k+1}^{n} a^T_{qk}\vec{v}_{q,q-1} + \sum_{q=k+1}^{n-1} a^T_{qk}\tilde{\omega}_{q,q-1}\sum_{s=q}^{n-1} a^T_{sq}\vec{r}_{s+1,s}\right\} \\
& + \left\{\sum_{k=1}^{n} a_{nk}\{\tilde{\omega}^2_{k,k-1} + \tilde{\varepsilon}_{k,k-1}\}a^T_{nk} + 2\sum_{k=1}^{n-1} a_{nk}\tilde{\omega}_{k,k-1}\sum_{q=k+1}^{n} a^T_{qk}\tilde{\omega}_{q,q-1}a^T_{nq}\right\}\vec{r}_n{}^{P_n}.
\end{aligned} \tag{3.28}$$

By giving the relative velocity field and the relative acceleration field, the relations (3.19) and (3.28) represent two matrix fundamental formulae of the general composed motion of a rigid body [1].

The derivative according the time of Eq. (3.22), leads to the expression of the *absolute angular acceleration*

$$\vec{\varepsilon}_{n0} = \sum_{k=1}^{n-1} a_{nk}\tilde{\omega}_{k,k-1} \sum_{q=k+1}^{n} a_{qk}^{T}\vec{\omega}_{q,q-1} + \sum_{k=1}^{n} a_{nk}\vec{\varepsilon}_{k,k-1}, \tag{3.29}$$

which is associated with the following skew-symmetric matrix

$$\begin{aligned}\tilde{\varepsilon}_{n0} = &\sum_{k=1}^{n-1}\sum_{q=k+1}^{n} a_{nk}\left\{\tilde{\omega}_{k,k-1}a_{qk}^{T}\tilde{\omega}_{q,q-1}a_{qk} - a_{qk}^{T}\tilde{\omega}_{q.q-1}a_{qk}\tilde{\omega}_{k,k-1}\right\}a_{nk}^{T}\\ &+\sum_{k=1}^{n} a_{nk}\tilde{\varepsilon}_{k,k-1}a_{nk}^{T}.\end{aligned} \tag{3.30}$$

We compute the accelerations of the same two distinct points A_n and B_n located on the body T_n, using the general formula (3.28), and which we substitute in the known vector relation

$$\vec{\gamma}_0^{B_n} - \vec{\gamma}_0^{A_n} = \{\tilde{\omega}_{n0}^2 + \tilde{\varepsilon}_{n0}\}(\vec{r}_0^{B_n} - \vec{r}_0^{A_n}). \tag{3.31}$$

By rearranging, we obtain a significant *square matrix*:

$$\begin{aligned}\tilde{\omega}_{n0}^2 + \tilde{\varepsilon}_{n0} = &\sum_{k=1}^{n} a_{nk}\{\tilde{\omega}_{k,k-1}^2 + \tilde{\varepsilon}_{k,k-1}\}a_{nk}^{T}\\ &+2\sum_{k=1}^{n-1} a_{nk}\tilde{\omega}_{k,k-1}\sum_{q=k+1}^{n} a_{qk}^{T}\tilde{\omega}_{q,q-1}a_{nq}^{T}.\end{aligned} \tag{3.32}$$

By expressing the *absolute acceleration* of the origin O_n

$$\begin{aligned}\vec{\gamma}_{n0} = &\sum_{k=1}^{n} a_{nk}\vec{\gamma}_{k,k-1} + \sum_{k=1}^{n-1} a_{nk}\{\tilde{\omega}_{k,k-1}^2 + \tilde{\varepsilon}_{k,k-1}\}\sum_{q=k}^{n-1} a_{qk}^{T}\vec{r}_{q+1,q}\\ &+2\sum_{k=1}^{n-1} a_{nk}\tilde{\omega}_{k,k-1}\left\{\sum_{q=k+1}^{n} a_{qk}^{T}\vec{v}_{q,q-1} + \sum_{q=k+1}^{n-1} a_{qk}^{T}\tilde{\omega}_{q,q-1}\sum_{s=q}^{n-1} a_{sq}^{T}\vec{r}_{s+1,s}\right\},\end{aligned} \tag{3.33}$$

we can again compute the acceleration $\vec{\gamma}_0^{P_n}$ in the boy-frame T_n

$$\vec{\gamma}_0^{P_n} = \vec{\gamma}_{n0} + \{\tilde{\omega}_{n0}^2 + \tilde{\varepsilon}_{n0}\}\vec{r}_n^{P_n}. \tag{3.34}$$

Starting from (3.29), (3.30), (3.32) and (3.33) and rearranging the terms, we obtain once four usual *recursive relations*

$$\begin{aligned}
\vec{\varepsilon}_{n0} &= a_{n,n-1}\vec{\varepsilon}_{n-1,0} + \vec{\varepsilon}_{n,n-1} + a_{n,n-1}\tilde{\omega}_{n-1,0}a_{n,n-1}^T\vec{\omega}_{n,n-1}\\
\tilde{\varepsilon}_{n0} &= a_{n,n-1}\tilde{\varepsilon}_{n-1,0}a_{n,n-1}^T + \tilde{\varepsilon}_{n,n-1}\\
&\quad + a_{n,n-1}\tilde{\omega}_{n-1,0}a_{n,n-1}^T\tilde{\omega}_{n,n-1} - \tilde{\omega}_{n,n-1}a_{n,n-1}\tilde{\omega}_{n-1,0}a_{n,n-1}^T\\
\tilde{\omega}_{n0}^2 + \tilde{\varepsilon}_{n0} &= a_{n,n-1}\{\tilde{\omega}_{n-1,0}^2 + \tilde{\varepsilon}_{n-1,0}\}a_{n,n-1}^T + \tilde{\omega}_{n,n-1}^2 + \tilde{\varepsilon}_{n,n-1}\\
&\quad + 2a_{n,n-1}\tilde{\omega}_{n-1,0}a_{n,n-1}^T\tilde{\omega}_{n,n-1}\\
\vec{\gamma}_{n0} &= a_{n,n-1}\vec{\gamma}_{n-1,0} + a_{n,n-1}\{\tilde{\omega}_{n-1,0}^2 + \tilde{\varepsilon}_{n-1,0}\}\vec{r}_{n,n-1}\\
&\quad + 2a_{n,n-1}\tilde{\omega}_{n-1,0}a_{n,n-1}^T\vec{v}_{n,n-1} + \vec{\gamma}_{n,n-1}.
\end{aligned} \tag{3.35}$$

The set of the recursive formulae (3.25) and (3.35) will be widely applied in kinematics analysis of the constrained robotic systems.

3.3 Applications to Kinematics Analysis of Mechanisms

Having a topological structure which serves a certain purpose, a mechanical system is made up of several links connected by the joints. In kinematics we remark a distinction between the free systems with elements that can move without any restrictions and the constrained multi-body systems. In a planar or a spatial mechanism, the bodies are not free to have arbitrary displacements, because they are connected by joints or internal force elements such as springs and dampers, with the remark than an internal force element does not reduce the number of independent coordinates required to describe the configuration of the system.

If every body is connected to at least two other bodies, the chain forms one or more closed loops. A *kinematical chain* of the rigid bodies may be composed up as *open-loop* chain, *closed-loop* chain or *hybrid* chain, as a combination of open and closed loops. A closed chain of which one link is made stationary is called a *mechanism*. Moreover, a link that is held fixed is called frame of reference and the relative motion between any pair of components is independent of choice of this reference frame.

The *degree of freedom* (DOF) of a mechanical system, defined to specify completely its configuration with respect to a reference frame, is equal to the number of the independent parameters or the inputs which are needed to uniquely describe its configuration in space at any instant in time.

Except for some special cases, a general expression for the degree of freedom of a mechanism depends on the number of links and also on the number and the types of joints used for the construction of the mechanical system. Following equation, known as the *Grübler-Kutzbach criterion*, establish the degree of freedom or the mobility of a mechanism [2]

$$F = \lambda(n - j - 1) + \sum_i f_i, \tag{3.36}$$

with the notations

- F degrees of freedom
- λ degrees of freedom of the space in which the end-effectors are intended to function: $\lambda = 6$ for spatial mechanisms, $\lambda = 3$ for planar and spherical mechanisms
- n number of rigid bodies in the mechanism, including the fixed frame
- j number of joints in the mechanism
- f_i degrees of relative motion permitted by the joint i

For the closed-loop mechanisms, the number and the locations of the actuated joints should be chosen carefully such the end-effectors can be controlled as will. The number of the actuated joints should be equal to the number of degrees of freedom of the mechanism and the locations of the actuated joints should be chosen such that they form a set of independent coordinates. If the number of actuators is less than the number of degrees of freedom, the output body cannot be controlled at will. On the other hand, if the number of actuators is greater than the number of degrees of freedom, the motion of these actuators must be coordinated in accordance with their kinematic constraints.

A joint provides some physical constraints into the relative motion between two members. The mechanical joints as prismatic joints, revolute joints, cylindrical and spherical joints are used to allow motion only in certain directions, because these constraints prevent the motion in some directions and, consequently, reduce the number of the independent coordinates of the systems.

The kind of the relative motion permitted by a joint is governed by the form of the contact surface between the bodies. There are six basic types contact of the *lower pairs*, than two elements contact each another with a substantial surface area [2–4].

A *revolute* (R) *joint* permits two paired elements to rotate with respect to each another about an axis of an angle φ. Because allows a relative rotation only, a revolute joint impose five constraints between the two bodies and has one degree of freedom.

A *prismatic* (P) *translational joint* allows a relative translation only between two bodies, which slide with respect to each another along the axis that is defined by the geometry of the joint of a displacement λ. Hence, the prismatic joint imposes five constraints between the paired elements and is a 1-DOF joint.

A *helical* (H) *joint* allows two paired elements to rotate about and translate along an axis defined by the geometry of the joint. However, the translation displacement λ is related to the rotation of angle φ by the pitch of a screw. The helical joint or the *screw pair* imposes five constraints into the paired elements and is a 1-DOF joint.

A *cylindrical* (C) *joint* allows body to translate along and rotate about an axis which is fixed on another body of a displacement λ and an angle φ. This joint allows, also, relative translation and rotation between two bodies along the joint axis. Having two independent motions, this joint imposes four constraints on the paired and has two degrees of freedom, eliminating the freedom of four possible independent relative displacements.

The simple-degree-of-freedom prismatic joint can be obtained as a special case of the cylindrical joint by elimination the freedom of the relative rotation between the bodies about their common joint axis, while the revolute joint can be considered as a case of the cylindrical joint, eliminating the freedom of the relative translation between the adjacent bodies.

A *spherical* (S) *joint* eliminates the freedom of the relative translations between the two bodies and allows one element to rotate freely with respect to the other about the centre of the sphere in all possible orientations. Having the freedom to rotate simultaneously about three concurrent axes of angles ψ, θ and φ, this joint imposes three constraints on the paired elements and it allows three degrees of freedom of relative rotations only.

The *universal joint* is a two-degrees-of-freedom pair, where the coordinates of the point of intersection of the two orthogonal bars of the *joint cross* are constant in the frames of two adjacent bodies.

In a *serial* planar mechanism the end-effector or some form of gripping device is attached at the free end. Each revolute joint is actuated so that all relative displacements are controlled independently. This open device is capable of performing a multitude of tasks in contrast to the *closed mechanism*, which is capable of only cyclic movement. However, note that a single input drive is required to actuate a closed planar mechanism with one degree of freedom, whereas a drive is required to actuate each joint in a serial chain.

Especially, a component body of a *planar mechanism* is said to possess generally *three degrees of freedom* with respect to a fixed plane. But, the moving of a lamina body could not possess three DOF. It may be constrained, for example, to rotate about an axis normal to the reference plane by a revolute kinematical pair, hinge and turning joint, which permits only one freedom of the moving body relative to the fixed plane.

In a planar mechanical system of three component ($n = 3$), for example, the bodies may be interconnected by one or more revolute joints or sliding prismatic pairs. Revolute and prismatic joints are lower-pair kinematic joints, used in construction of the mechanism. Consequently, the relative motion of each component body is described by one parameter, either a translational coordinate or one rotation angle.

Considering the universal *Hooke* joint and three known planar mechanisms, *slider-crank mechanism*, *four-bar linkage* and *quick-return mechanism*, vector equations on the positions, the velocities and the accelerations of the connecting joints are given by the relations (3.16), (3.23) and (3.33).

3.3.1 Universal Hooke Joint

The *universal Hooke joint* is essentially a combination of two intersecting revolute joints [5]. Composed of a joint *cross* and two bars connected to cross by revolute joints, this special mechanism has finally *one degree of freedom* and allows successive relative rotations about two perpendicular cross's arms, uniquely described by two Cardan angles as coordinates.

Attached to frame $A_1x_1^Ay_1^Az_1^A$, a lower body **1a** of length $A_1A_2 = l_1$, mass m_1^A and tensor of inertia $\hat{J}_1^A$, under the active torque m_{10}^A, has a rotation about z_1^A axis with the angle φ_{10}^A, the angular velocity $\omega_{10}^A = \dot{\varphi}_{10}^A$ and the angular acceleration $\varepsilon_{10}^A = \ddot{\varphi}_{10}^A$. Linked at the local frame $A_2x_2^Ay_2^Az_2^A$, the cross **2a** consists of two identical orthogonal arms of total mass m_2^A and tensor of inertia $\hat{J}_2^A$, having a relative rotation with the angle φ_{21}^A, the angular velocity $\omega_{21}^B = \dot{\varphi}_{21}^B$ and the angular acceleration $\varepsilon_{21}^A = \ddot{\varphi}_{21}^A$. The upper rotor **1b** of length $B_1B_2 = l_2$ is inclined with respect to **1a** of a constant angle α and is attached to frame $B_1x_1^By_1^Bz_1^B$, rotating around the joint B_1 with the angle φ_{10}^B, the angular velocity $\omega_{10}^B = \dot{\varphi}_{21}^B$ and the angular acceleration $\varepsilon_{10}^B = \ddot{\varphi}_{10}^B$. It has the mass m_1^B and the tensor of inertia $\hat{J}_1^B$, with respect to his frame. Finally, the same cross **2b** of mass $m_2^B = m_2^A$ and a new tensor of inertia $\hat{J}_2^B$ is attached to the frame $B_2x_2^By_2^Bz_2^B$ and rotates relatively about z_2^B with the angle φ_{21}^B, the angular velocity $\omega_{21}^B = \dot{\varphi}_{21}^B$ and the angular acceleration $\varepsilon_{21}^B = \ddot{\varphi}_{21}^B$ (Fig. 3.3).

Pursuing along two independent ways OA_1A_2 and OB_1B_2 of this spatial mechanism, we define the following transformation matrices

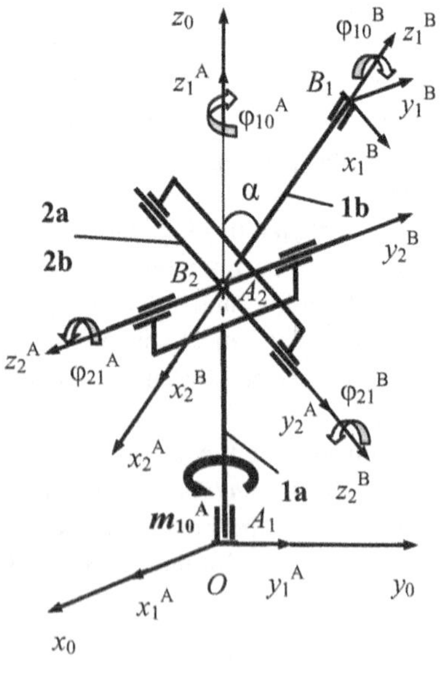

Fig. 3.3 Universal *Hooke* joint

$$a_{10} = a_{10}^{\varphi}, \quad a_{21} = a_{21}^{\varphi} a_{\alpha}^{T} \theta_1, \quad b_{10} = b_{10}^{\varphi} b_{\alpha} \theta_2, \quad b_{21} = b_{21}^{\varphi} \theta_1, \tag{3.37}$$

where we denote the matrices

$$\begin{aligned} a_{10}^{\varphi} &= rot(z, \varphi_{10}^{A}), \quad a_{21}^{\varphi} = rot(z, \varphi_{21}^{A}) \\ b_{10}^{\varphi} &= rot(z, \varphi_{10}^{B}), \quad b_{21}^{\varphi} = rot(z, \varphi_{21}^{B}) \\ a_{\alpha} &= rot(z, \alpha), \quad b_{\alpha} = rot(y, \alpha) \\ \theta_1 &= rot\left(y, \frac{\pi}{2}\right), \quad \theta_2 = rot\left(z, \frac{\pi}{2}\right). \end{aligned} \tag{3.38}$$

The angles of rotation φ_{21}^{A}, φ_{10}^{B} and φ_{21}^{B} can be computed as function of φ_{10}^{A}, starting from following kinematic relations of constraint

$$b_{21} b_{10} = a_{BA} a_{21} a_{10}, \quad a_{BA} = \begin{bmatrix} 1 & 0 & 0 \\ 0 & 0 & -1 \\ 0 & 1 & 0 \end{bmatrix}. \tag{3.39}$$

We choose the following three independent equations

$$\begin{aligned} &\tan(\varphi_{21}^{A} - \alpha) = -\tan\alpha \cos\varphi_{10}^{A} \\ &\tan\varphi_{10}^{B} = \cos\alpha \tan\varphi_{10}^{A}, \quad \sin\varphi_{21}^{B} = -\sin\alpha \sin\varphi_{10}^{A}. \end{aligned} \tag{3.40}$$

The *conditions of connectivity* of the velocities are given by the derivative with respect to time of the above matrix geometrical conditions (3.39)

$$\begin{aligned} &\omega_{21}^{A} \vec{u}_i^T a_{10}^T a_{21}^T \vec{u}_3 - \omega_{10}^{B} \vec{u}_i^T b_{10}^T \vec{u}_3 - \omega_{21}^{B} \vec{u}_i^T b_{10}^T b_{21}^T \vec{u}_3 \\ &= -\omega_{10}^{A} \vec{u}_i^T a_{10}^T \vec{u}_3 \quad (i = 1, 2, 3). \end{aligned} \tag{3.41}$$

where $\vec{u}_1$, $\vec{u}_2$ and $\vec{u}_3$ are three orthogonal unit vectors; we obtain the angular velocities

$$\omega_{21}^{A} = \frac{\sin\varphi \sin 2\varphi_{10}^{B}}{2\cos\varphi_{10}^{A}} \omega_{10}^{A}, \quad \omega_{10}^{B} = \frac{\sin 2\varphi_{10}^{B}}{\sin 2\varphi_{10}^{A}} \omega_{10}^{A}, \quad \omega_{21}^{B} = -\omega_{10}^{A} \sin\alpha \cos\varphi_{10}^{B}. \tag{3.42}$$

By differentiating the above Eq. (3.41), we obtain the matrix *conditions of connectivity* at level of accelerations, from which the angular accelerations ε_{21}^{A}, ε_{10}^{B} and ε_{21}^{B} are finally derived:

$$\begin{aligned} &\varepsilon_{21}^{A} \vec{u}_i^T a_{10}^T a_{21}^T \vec{u}_3 - \varepsilon_{10}^{B} \vec{u}_i^T b_{10}^T \vec{u}_3 - \varepsilon_{21}^{B} \vec{u}_i^T b_{10}^T b_{21}^T \vec{u}_3 \\ &= -\varepsilon_{10}^{A} \vec{u}_i^T a_{10}^T \vec{u}_3 - \omega_{10}^{A} \omega_{21}^{A} \vec{u}_i^T a_{10}^T \tilde{u}_3 a_{21}^T \vec{u}_3 \\ &\quad + \omega_{10}^{B} \omega_{21}^{B} \vec{u}_i^T b_{10}^T \tilde{u}_3 b_{21}^T \vec{u}_3 \quad (i = 1, 2, 3). \end{aligned} \tag{3.43}$$

We add other useful vector expressions of angular velocities and angular accelerations

$$\begin{aligned}
\vec{\omega}_{10}^{A} &= \omega_{10}^{A}\vec{u}_3, \quad \vec{\omega}_{20}^{A} = \omega_{10}^{A}a_{21}\vec{u}_3 + \omega_{21}^{A}\vec{u}_3 \\
\vec{\varepsilon}_{10}^{A} &= \varepsilon_{10}^{A}\vec{u}_3, \quad \vec{\varepsilon}_{20}^{A} = \varepsilon_{10}^{A}a_{21}\vec{u}_3 + \varepsilon_{21}^{A}\vec{u}_3 + \omega_{10}^{A}\omega_{21}^{A}a_{21}\tilde{u}_3 a_{21}^{T}\vec{u}_3 \\
\vec{\omega}_{10}^{B} &= \omega_{10}^{B}\vec{u}_3, \quad \vec{\omega}_{20}^{B} = \omega_{10}^{B}b_{21}\vec{u}_3 + \omega_{21}^{B}\vec{u}_3 \\
\vec{\varepsilon}_{10}^{B} &= \varepsilon_{10}^{B}\vec{u}_3, \quad \vec{\varepsilon}_{20}^{B} = \varepsilon_{10}^{B}b_{21}\vec{u}_3 + \varepsilon_{21}^{B}\vec{u}_3 + \omega_{10}^{B}\omega_{21}^{B}b_{21}\tilde{u}_3 b_{21}^{T}\vec{u}_3.
\end{aligned} \tag{3.44}$$

3.3.2 *Slider-Crank Mechanism*

Used in several engineering applications such as automobile engines and pumps, the architecture of the mechanism consists of four bodies: fixed cylinder frame, crankshaft **1a**, slider-block which represents the piston **1b** and intermediate connecting rod **2b** (Fig. 3.4). This system has one degree of freedom and the motion of all components can be controlled and described using only one independent variable, which is selected to be the angular orientation of the crankshaft [1]. A specified input rotary motion to the crankshaft, for example, produces a desired rectilinear motion of the piston. If the rectilinear motion of the piston is selected to be the independent variable, the force that acts on the piston can be chosen such that a desired output rotary motion of the crankshaft can be achieved. There are one active revolute joint A_1, one passive prismatic pair jointed at the point B_1 and two internal passive revolute joints B_2 and $A_2 \equiv B_3$.

For the purpose of analysis, we assign a fixed Cartesian coordinate system $Ox_0y_0z_0$ at the point O. Initially, the crankshaft and the connecting rod are rotated with respect to fixed base of the angles α_1 and α_2, respectively. We also consider that the slider block is initially starting from the position $OB_1 = d$.

Attached to rotating frame $A_1x_1^Ay_1^Az_1^A$, the crankshaft of length $A_1A_2 = l_1$, mass m_1^A and tensor of inertia $\hat{J}_1^A$, has a rotation about z_1^A axis with the angle φ_{10}^A, the angular velocity $\omega_{10}^A = \dot{\varphi}_{10}^A$ and the angular acceleration $\varepsilon_{10}^A = \ddot{\varphi}_{10}^A$. Acted permanently by a resistant horizontal force f_{10}^B, the slider-block is as a piston of mass m_1^B linked at the $B_1x_1^By_1^Bz_1^B$ translating frame, having a rectilinear motion of

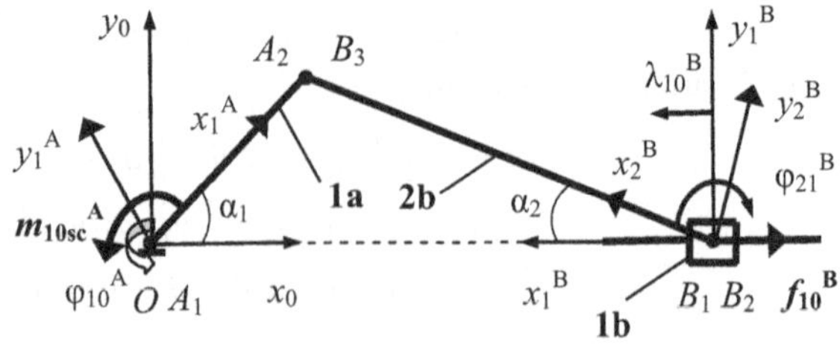

Fig. 3.4 Slider-crank mechanism

displacement λ_{10}^{B} and the velocity $v_{10}^{B} = \dot{\lambda}_{10}^{B}$. Finally, the intermediate rod of length $B_2B_3 = l_2$, mass m_2^B and tensor of inertia $\hat{J}_2^B$ with respect to a local frame $B_2x_2^By_2^Bz_2^B$, rotates around the joint B_2 with the angle φ_{21}^{B} and the angular velocity $\omega_{21}^{B} = \dot{\varphi}_{21}^{B}$.

Starting from O and pursuing along two independent ways OA_1A_2 and $OB_1B_2B_3$, we define the following transformation matrices

$$a_{10} = a_{10}^{\varphi}a_1^{\alpha}, \quad b_{10} = \theta, \quad b_{21} = b_{21}^{\varphi}a_2^{\alpha}, \quad b_{20} = b_{21}b_{10}, \tag{3.45}$$

where we denote the matrices

$$\begin{aligned} a_{10}^{\varphi} &= rot(z, \varphi_{10}^{A}), \quad b_{21}^{\varphi} = rot(z, \varphi_{21}^{B}) \\ \theta &= rot(y, \pi), \quad a_{\sigma}^{\alpha} = rot(z, \alpha_{\sigma}) \ (\sigma = 1, 2). \end{aligned} \tag{3.46}$$

The planar mechanism can virtually be transformed in two *open kinematical chains* OA_2 and $OB_1B_2B_3$ subject to the constraints, by cutting the joint A_2, for example. Knowing the rotation motion of the crankshaft through the function $\varphi_{10}^{A} = \varphi_{10}^{A}(t)$, other two remaining variables $\lambda_{10}^{B}, \varphi_{21}^{B}$ will be determined from the vector kinematic relationships of constraint

$$\vec{r}_0^{A_2} = \vec{r}_0^{B_3} = a_{10}^T\vec{r}_{21}^{A} = \vec{r}_{10}^{B} + b_{10}^T\vec{r}_{21}^{B} + b_{20}^T\vec{r}_{32}^{B}, \tag{3.47}$$

where

$$\begin{aligned} &\vec{r}_{21}^{A} = l_1\vec{u}_1, \quad \vec{r}_{10}^{B} = (d - \lambda_{10}^{B})\vec{u}_1, \quad \vec{r}_{21}{}^{B} = \vec{0}, \quad \vec{r}_{32}{}^{B} = l_2\vec{u}_1 \\ &l_1 \sin\alpha_1 = l_2 \sin\alpha_2, \quad l_1 \cos\alpha_1 + l_2 \cos\alpha_2 = d. \end{aligned} \tag{3.48}$$

Developing, we have the following analytical expressions

$$\begin{aligned} \sin\phi_2 &= \frac{l_1}{l_2}\sin\phi_1, \quad \lambda_{10}^{B} = d - l_1\cos\phi_1 - l_2\cos\phi_2 \\ \phi_1 &= \varphi_{10}^{A} + \alpha_1, \quad \phi_2 = \varphi_{21}^{B} + \alpha_2. \end{aligned} \tag{3.49}$$

By differentiating the geometrical constraints (3.47), we obtain the matrix *conditions of connectivity* for the velocities

$$v_{10}^{B}\vec{u}_i^T\vec{u}_1 - l_2\omega_{21}^{B}\vec{u}_i^Tb_{20}^T\vec{u}_2 = -l_1\omega_{10}^{A}\vec{u}_i^Ta_{10}^T\vec{u}_2 \quad (i = 1, 2), \tag{3.50}$$

where $\vec{u}_1$ and $\vec{u}_2$ are two planar orthogonal unit vectors; it results the intensities of the two velocities

$$\omega_{21}^{B} = \omega_{10}^{A}\frac{l_1\cos\phi_1}{l_2\cos\phi_2}, \quad v_{10}^{B} = l_1\omega_{10}^{A}\sin\phi_1 + l_2\omega_{21}^{B}\sin\phi_2. \tag{3.51}$$

The *conditions of connectivity* of the accelerations will be given by the derivative with respect to time of above Eq. (3.50):

$$\begin{aligned}\gamma_{10}^{B}\vec{u}_i^T\vec{u}_1 - l_2\varepsilon_{21}^{B}\vec{u}_i^T b_{20}^T\vec{u}_2 &= -l_1\varepsilon_{10}^{A}\vec{u}_i^T a_{10}^T\vec{u}_2 \\ &+ l_1\omega_{10}^{A}\omega_{10}^{A}\vec{u}_i^T a_{10}^T\vec{u}_1 - l_2\omega_{21}^{B}\omega_{21}^{B}\vec{u}_i^T b_{20}^T\vec{u}_1 \quad (i = 1,2),\end{aligned} \tag{3.52}$$

from which we obtain the angular acceleration ε_{21}^{B} of the connecting rod and the acceleration γ_{10}^{B} of the piston:

$$\begin{aligned}\varepsilon_{21}^{B} &= \frac{l_1\varepsilon_{10}^{A}\cos\phi_1 - l_1\omega_{10}^{A}\omega_{10}^{A}\sin\phi_1 + l_2\omega_{21}^{B}\omega_{21}^{B}\sin\phi_2}{l_2\cos\phi_2} \\ \gamma_{10}^{B} &= l_1\varepsilon_{10}^{A}\sin\phi_1 + l_2\varepsilon_{21}^{B}\sin\phi_2 + l_1\omega_{10}^{A}\omega_{10}^{A}\cos\phi_1 + l_2\omega_{21}^{B}\omega_{21}^{B}\cos\phi_2.\end{aligned} \tag{3.53}$$

The above Eqs. (3.47), (3.50) and (3.52) can easily be programmed in MATLAB, in order to obtain some graphs for the coordinates, the velocities and the accelerations of the components of this interesting planar mechanism, adding some needful expressions of kinematics

$$\begin{aligned}&\vec{v}_{10}^{A} = \vec{0},\quad \vec{\omega}_{10}^{A} = \dot{\varphi}_{10}^{A}\vec{u}_3,\quad \tilde{\omega}_{10}^{A} = \dot{\varphi}_{10}^{A}\tilde{u}_3,\quad \vec{\gamma}_{10}^{A} = \vec{0},\quad \vec{\varepsilon}_{10}^{A} = \ddot{\varphi}_{10}^{A}\vec{u}_3,\quad \tilde{\varepsilon}_{10}^{A} = \ddot{\varphi}_{10}^{A}\tilde{u}_3 \\ &\vec{v}_{10}^{B} = \dot{\lambda}_{10}^{B}\vec{u}_1,\quad \vec{\omega}_{10}^{B} = \vec{0},\quad \tilde{\omega}_{10}^{B} = 0,\quad \vec{\gamma}_{10}^{B} = \ddot{\lambda}_{10}^{B}\vec{u}_1,\quad \vec{\varepsilon}_{10}^{B} = \vec{0},\quad \tilde{\varepsilon}_{10}^{B} = 0 \\ &\vec{v}_{20}^{B} = b_{21}\vec{v}_{10}^{B},\quad \vec{\omega}_{20}^{B} = \vec{\omega}_{21}^{B} = \dot{\varphi}_{21}^{B}\vec{u}_3,\quad \tilde{\omega}_{20}^{B} = \tilde{\omega}_{21}^{B} = \dot{\varphi}_{21}^{B}\tilde{u}_3 \\ &\vec{\gamma}_{20}^{B} = b_{21}\vec{\gamma}_{10}^{B},\quad \vec{\varepsilon}_{20}^{B} = \vec{\varepsilon}_{21}^{B} = \ddot{\varphi}_{21}^{B}\vec{u}_3,\quad \tilde{\varepsilon}_{20}^{B} = \tilde{\varepsilon}_{21}^{B} = \ddot{\varphi}_{21}^{B}\tilde{u}_3.\end{aligned} \tag{3.54}$$

As an application, let us consider a planar mechanism which has following geometrical and mechanical characteristics:

$$l_1 = l_2 = 0.1m,\quad \varphi_{10}^{A*} = \frac{\pi}{12},\quad \alpha_1 = \alpha_2 = \frac{\pi}{12},\quad \Delta t = 6s. \tag{3.55}$$

The displacements, the velocities and the accelerations for the slider-block **1b** and the connecting rod **2b** are calculated by the program and plotted versus time as follows: Figs. 3.5, 3.6, 3.7, 3.8, 3.9 and 3.10, supposing that the angle of rotation of the crankshaft **1a** is expressed by the following analytical function

$$\varphi_{10}^{A} = \varphi_{10}^{A*}\left(1 - \cos\frac{\pi}{6}t\right). \tag{3.56}$$

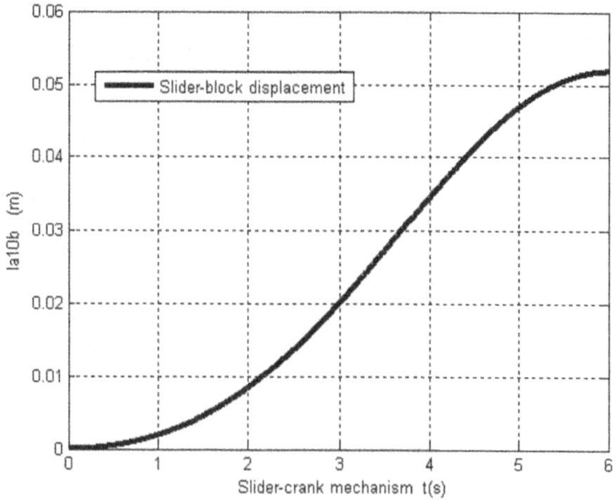

Fig. 3.5 Slider-block displacement λ_{10}^{B}

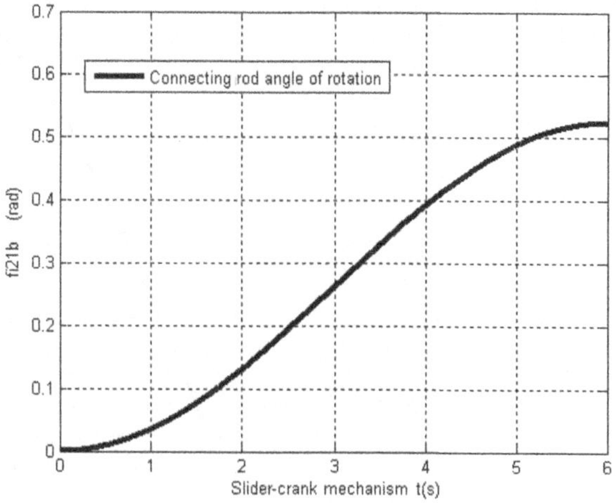

Fig. 3.6 Connecting rod angle of rotation φ_{21}^{B}

3.3.3 Four-Bar Mechanism

Having only one degree of freedom, this mechanism is used in many industrial and technological applications. The linkage is constituted of the following elements: fixed frame or ground, crankshaft **1a**, coupler **2a** and rocker **1b**. The motion of the components of the mechanism can be controlled by using one torque input, such as driving the crankshaft using a motor located at point A_1. A desired planar motion of the coupler link can be obtained, for example, when the crankshaft rotates uniformly one complete cycle. There are one active revolute joint A_1 and three passive revolute joints A_2, B_1 and $B_2 \equiv A_3$ (Fig. 3.11).

Relative to a fixed Cartesian frame $Ox_0y_0z_0$, centred at the point O, the crankshaft, the coupler and the rocker are initially rotated of three given angles

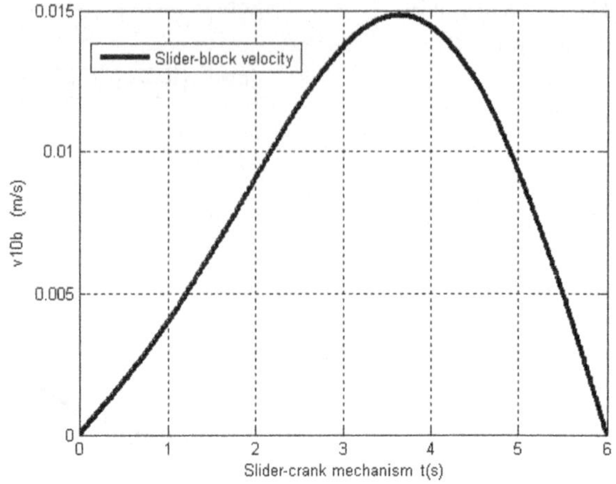

Fig. 3.7 Slider-block velocity v_{10}^{B}

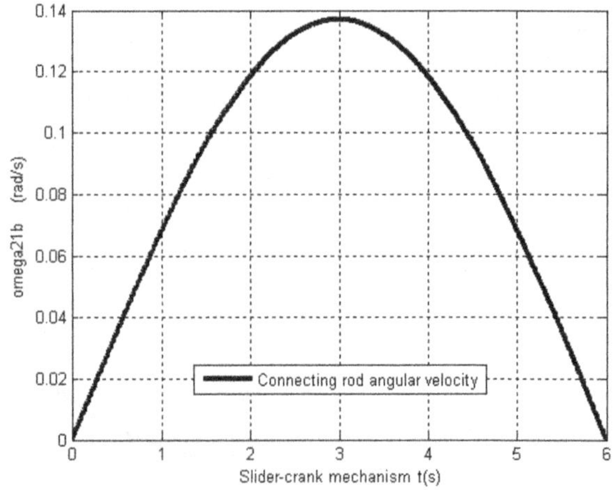

Fig. 3.8 Connecting rod angular velocity ω_{21}^{B}

α_σ $(\sigma = 1, 2, 3)$. We also consider as known in fixed base, the position $OB_1 = d$ of external revolute joint B_1.

Active crankshaft of length $A_1A_2 = l_1$, mass m_1^A and tensor of inertia $\hat{J}_1^A$ is attached to frame $A_1x_1^Ay_1^Az_1^A$ and has a rotation about z_1^A axis with the angle φ_{10}^A, the angular velocity $\omega_{10}^A = \dot{\varphi}_{10}^A$ and the angular acceleration $\varepsilon_{10}^A = \ddot{\varphi}_{10}^A$. Coupler **2** of length $A_2A_3 = l_2$, mass m_2^A and tensor of inertia $\hat{J}_2^A$ is linked at $A_2x_2^Ay_2^Az_2^A$ moving frame and rotates relative with an angle φ_{21}^A and the angular velocity $\omega_{21}^A = \dot{\varphi}_{21}^A$. Finally, the rocker of length $B_1B_2 = l_3$ and its local frame $B_1x_1^By_1^Bz_1^B$ can rotate around the joint B_1 with the angle φ_{10}^B and the angular velocity $\omega_{10}^B = \dot{\varphi}_{10}^B$. It has the mass m_1^B and the tensor of inertia $\hat{J}_1^B$ with respect to his frame. Knowing the angular orientation $\varphi_{10}^A = \varphi_{10}^A(t)$, the angular velocity and the angular acceleration of the

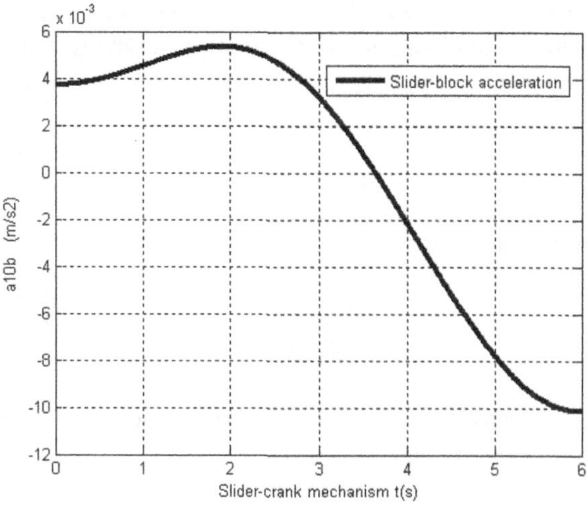

Fig. 3.9 Slider-block acceleration γ_{10}^{B}

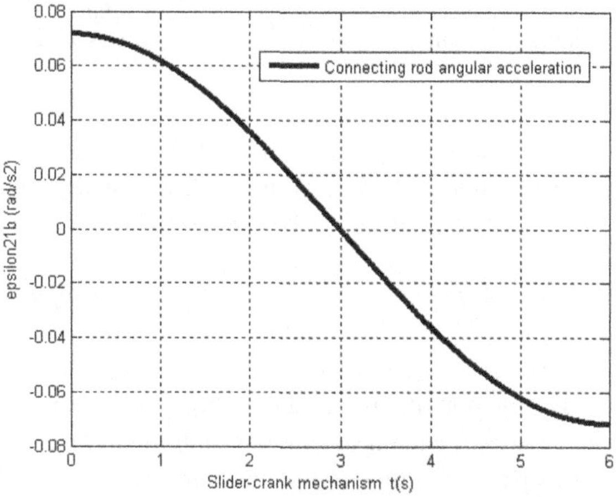

Fig. 3.10 Connecting rod angular acceleration ε_{21}^{B}

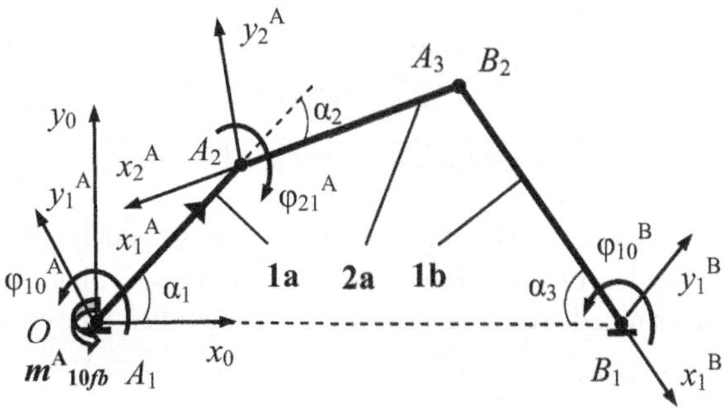

Fig. 3.11 Four-bar mechanism

crankshaft, we will determine the values of the angular coordinates, the velocities and the accelerations of the coupler and the rocker.

Pursuing along two independent circuits $OA_1A_2A_3$ and OB_1B_2, we define primarily the transformation matrices

$$a_{10} = a^{\varphi}_{10}a^{\alpha}_{1}, \quad a_{21} = a^{\varphi}_{21}a^{\alpha}_{2}\theta, \;\; b_{10} = b^{\varphi}_{10}a^{\alpha T}_{3}, \tag{3.57}$$

where we denote

$$\begin{aligned} a^{\varphi}_{10} &= rot(z, \varphi^{A}_{10}), \quad a^{\varphi}_{21} = rot(z, \varphi^{A}_{21}), \;\; b^{\varphi}_{10} = rot(z, \varphi^{B}_{10}) \\ \theta &= rot(y, \pi), \quad a^{\alpha}_{\sigma} = rot(z, \alpha_{\sigma}) \;\; (\sigma = 1, 2, 3). \end{aligned} \tag{3.58}$$

Starting from the following kinematic constraining relations, we determine the variables φ^{A}_{21} and φ^{B}_{10}

$$\begin{gathered} \vec{r}^{A_3}_{0} = \vec{r}^{B_2}_{0} = a^{T}_{10}\vec{r}^{A}_{21} + a^{T}_{10}a^{T}_{21}\vec{r}^{A}_{32} = \vec{r}^{B}_{10} + b^{T}_{10}\vec{r}^{B}_{21} \\ \vec{r}^{A}_{21} = l_1\vec{u}_1, \quad \vec{r}^{A}_{32} = -l_2\vec{u}_1, \;\; \vec{r}^{B}_{10} = d\vec{u}_1, \;\; \vec{r}^{B}_{21} = -l_3\vec{u}_1 \\ l_1 \sin\alpha_1 + l_2 \sin(\alpha_1 - \alpha_2) = l_3 \sin\alpha_3, \quad l_1 \cos\alpha_1 + l_2 \cos(\alpha_1 - \alpha_2) + l_3 \cos\alpha_3 = d. \end{gathered} \tag{3.59}$$

The *conditions of connectivity* of the velocities are given by the derivative with respect to time of (3.59), from which we obtain the angular velocities ω^{A}_{21} and ω^{B}_{10} for the coupler and the rocker, respectively:

$$\omega^{A}_{21}\vec{u}^{T}_{i}a^{T}_{10}a^{T}_{21}\tilde{u}_3\vec{r}^{A}_{32} - \omega^{B}_{10}\vec{u}^{T}_{i}b^{T}_{10}\tilde{u}_3\vec{r}^{B}_{21} = -\omega^{A}_{10}\vec{u}^{T}_{i}a^{T}_{10}\tilde{u}_3(\vec{r}^{A}_{21} + a^{T}_{21}\vec{r}^{A}_{32}) \;\; (i = 1, 2). \tag{3.60}$$

By differentiating the above Eq. (3.60), we obtain the matrix *conditions of connectivity* for the accelerations and, finally, the angular accelerations ε^{A}_{21} and ε^{B}_{10} of two passive bodies in the mechanism:

$$\begin{aligned} &\varepsilon^{A}_{21}\vec{u}^{T}_{i}a^{T}_{10}a^{T}_{21}\tilde{u}_3\vec{r}^{A}_{32} - \varepsilon^{B}_{10}\vec{u}^{T}_{i}b^{T}_{10}\tilde{u}_3\vec{r}^{B}_{21} \\ &\quad = -\varepsilon^{A}_{10}\vec{u}^{T}_{i}a^{T}_{10}\tilde{u}_3(\vec{r}^{A}_{21} + a^{T}_{21}\vec{r}^{A}_{32}) - 2\omega^{A}_{10}\omega^{A}_{21}\vec{u}^{T}_{i}a^{T}_{10}\tilde{u}_3a^{T}_{21}\tilde{u}_3\vec{r}^{A}_{32} \\ &\quad\quad - \omega^{A}_{10}\omega^{A}_{10}\vec{u}^{T}_{i}a^{T}_{10}\tilde{u}_3\tilde{u}_3(\vec{r}^{A}_{21} + a^{T}_{21}\vec{r}^{A}_{32}) - \omega^{A}_{21}\omega^{A}_{21}\vec{u}^{T}_{i}a^{T}_{10}a^{T}_{21}\tilde{u}_3\tilde{u}_3\vec{r}^{A}_{32} \\ &\quad\quad + \omega^{B}_{10}\omega^{B}_{10}\vec{u}^{T}_{i}b^{T}_{10}\tilde{u}_3\tilde{u}_3\vec{r}^{B}_{21} \quad (i = 1, 2). \end{aligned} \tag{3.61}$$

The matrix equation (3.61) can be solved for the two unknowns, since all the variables on the right-hand side of these equations are either given or can be determined from the position and velocities analyses.

We can add other useful expressions of velocities and accelerations

$$
\begin{aligned}
&\vec{v}^{A}_{10} = \vec{0},\quad \vec{\omega}^{A}_{10} = \dot{\varphi}^{A}_{10}\vec{u}_3,\quad \tilde{\omega}^{A}_{10} = \dot{\varphi}^{A}_{10}\tilde{u}_3,\quad \vec{\gamma}^{A}_{10} = \vec{0},\quad \vec{\varepsilon}^{A}_{10} = \ddot{\varphi}^{A}_{10}\vec{u}_3,\quad \tilde{\varepsilon}^{A}_{10} = \ddot{\varphi}^{A}_{10}\tilde{u}_3\\
&\vec{\omega}^{A}_{21} = \dot{\varphi}^{A}_{21}\vec{u}_3,\quad \tilde{\omega}^{A}_{21} = \dot{\varphi}^{A}_{21}\tilde{u}_3,\quad \vec{\varepsilon}^{A}_{21} = \ddot{\varphi}^{A}_{21}\vec{u}_3,\quad \tilde{\varepsilon}^{A}_{21} = \ddot{\varphi}^{A}_{21}\tilde{u}_3\\
&\vec{\omega}^{A}_{20} = (\dot{\varphi}^{A}_{21} - \dot{\varphi}^{A}_{10})\vec{u}_3,\quad \tilde{\omega}^{A}_{20} = (\dot{\varphi}^{A}_{21} - \dot{\varphi}^{A}_{10})\tilde{u}_3,\\
&\vec{\varepsilon}^{A}_{20} = (\ddot{\varphi}^{A}_{21} - \ddot{\varphi}^{A}_{10})\vec{u}_3,\quad \tilde{\varepsilon}^{A}_{20} = (\ddot{\varphi}^{A}_{21} - \ddot{\varphi}^{A}_{10})\tilde{u}_3\\
&\vec{v}^{A}_{20} = a_{21}\tilde{\omega}^{A}_{10}\vec{r}^{A}_{21},\quad \vec{\gamma}^{A}_{20} = a_{21}\{\tilde{\omega}^{A}_{10}\tilde{\omega}^{A}_{10} + \tilde{\varepsilon}^{A}_{10}\}\vec{r}^{A}_{21},\quad \vec{v}^{B}_{10} = \vec{0}\\
&\vec{\omega}^{B}_{10} = \dot{\varphi}^{B}_{10}\vec{u}_3,\quad \tilde{\omega}^{B}_{10} = \dot{\varphi}^{B}_{10}\tilde{u}_3,\quad \vec{\gamma}^{B}_{10} = \vec{0},\quad \vec{\varepsilon}^{B}_{10} = \ddot{\varphi}^{B}_{10}\vec{u}_3,\quad \tilde{\varepsilon}^{B}_{10} = \ddot{\varphi}^{B}_{10}\tilde{u}_3.
\end{aligned}
\tag{3.62}
$$

3.3.4 Quick-Return Mechanism

This third known planar mechanism is composed of the following elements: fixed frame or ground, crankshaft **1a**, oscillatory rocker **1b** and slider-block **2b** (Fig. 3.12).

As a mechanical system having also one degree of freedom, the motion of all bodies can be described, for example, using the angular orientation of the crankshaft only. There are one active revolute joint A_1, two passive revolute joints B_1, A_2 and one passive prismatic pair jointed at $A_2 \equiv B_2$.

With respect to the frame $Ox_0y_0z_0$, the crankshaft and the rocker are initially rotated of two given angles $\alpha_\sigma(\sigma = 1, 2)$. We also consider as known in fixed base, the position $OB_1 = d$ of external revolute joint B_1 and the initial relative distance $B_1B_0 = l_0$ of the slider-block. The crankshaft of length $A_1A_2 = l_1$, mass m^A_1 and tensor of inertia $\hat{J}^A_1$ is attached at the coordinate system $A_1x^A_1y^A_1z^A_1$ and has a rotation about z^A_1 axis with the angle φ^A_{10}, the angular velocity $\omega^A_{10} = \dot{\varphi}^A_{10}$ and the angular acceleration $\varepsilon^A_{10} = \ddot{\varphi}^A_{10}$. Linked at $B_1x^B_1y^B_1z^B_1$, the rocker **2** of length $B_1B_3 = l_2$, mass m^B_1 and tensor of inertia $\hat{J}^B_1$ rotates with an angle φ^B_{10} and the angular velocity $\omega^B_{10} = \dot{\varphi}^B_{10}$. Finally, starting from B_0, the slider-block **2b** of mass m^B_2 and its local frame $B_2x^B_2y^B_2z^B_2$ have a relative translation along the rocker of displacement λ^B_{21} and the velocity $v^B_{21} = \dot{\lambda}^B_{21}$.

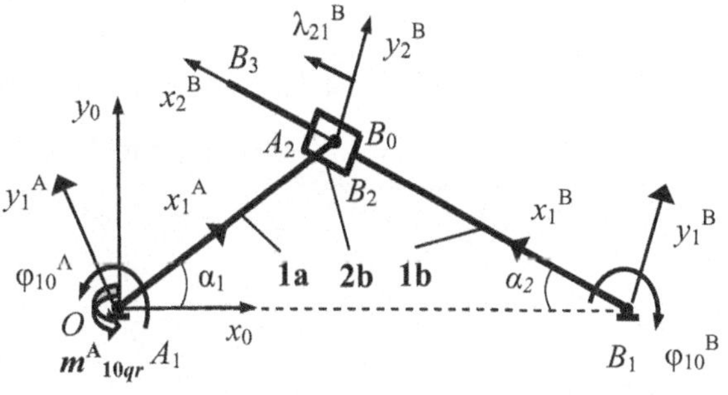

Fig. 3.12 Quick-return mechanism

Along two different circuits OA_1A_2 and OB_1B_2, we define following transformation matrices

$$a_{10} = a^{\varphi}_{10}a^{\alpha}_{1}, \quad b_{10} = b^{\varphi}_{10}a^{\alpha}_{2}\theta, \;\; b_{21} = I, \tag{3.63}$$

where we denote

$$a^{\varphi}_{10} = rot(z, \varphi^{A}_{10}), \quad b^{\varphi}_{10} = rot(z, \varphi^{B}_{10}), \quad \theta = rot(y, \pi), \quad a^{\alpha}_{\sigma} = rot(z, \alpha_{\sigma}). \\ (\sigma = 1, 2) \tag{3.64}$$

The rotational rod A_1A_2 is assumed to have a specified motion $\varphi^{A}_{10}(t)$ and, consequently, the system has one degree of freedom, which we select to be just the angular orientation φ^{A}_{10}. The angle of rotation φ^{B}_{10} and the relative displacement λ^{B}_{21} can be computed as function of φ^{A}_{10}, starting from kinematical relation of constraint

$$\vec{r}^{\,A_2}_{0} = \vec{r}^{\,B_2}_{0} = a^{T}_{10}\vec{r}^{\,A}_{21} = \vec{r}^{\,B}_{10} + b^{T}_{10}\vec{r}^{\,B}_{21}, \quad \vec{r}^{\,A}_{21} = l_1\vec{u}_1, \;\; \vec{r}^{\,B}_{10} = d\vec{u}_1, \;\; \vec{r}^{\,B}_{21} = (l_0 + \lambda^{B}_{21})\vec{u}_1. \tag{3.65}$$

Developing, we have the following analytical expressions

$$\begin{aligned} &(l_0 + \lambda^{B}_{21})\sin\phi_2 = l_1 \sin\phi_1, \quad (l_0 + \lambda^{B}_{21})\cos\phi_2 = d - l_1\cos\phi_1, \\ &\phi_1 = \varphi^{A}_{10} + \alpha_1, \quad \phi_2 = \varphi^{B}_{21} + \alpha_2 \\ &l_1 \sin\alpha_1 = l_0 \sin\alpha_2, \quad l_1\cos\alpha_1 + l_0\cos\alpha_2 = d. \end{aligned} \tag{3.66}$$

The derivative with respect to time of (3.65), leads to some *conditions of connectivity* for velocities, from which we obtain the angular velocity ω^{B}_{10} and the relative velocity v^{B}_{21}:

$$(l_0 + \lambda^{B}_{21})\omega^{B}_{10}\vec{u}^{T}_{i}b^{T}_{10}\vec{u}_2 + v^{B}_{21}\vec{u}^{T}_{i}b^{T}_{10}\vec{u}_1 = l_1\omega^{A}_{10}\vec{u}^{T}_{i}a^{T}_{10}\vec{u}_2 \;\; (i = 1, 2). \tag{3.67}$$

The *conditions of connectivity* for the accelerations will be given by differentiating (3.67), from which we obtain the angular acceleration ε^{B}_{10} of the rocker and the relative acceleration γ^{B}_{21} of the slider-block:

$$\begin{aligned} &(l_0 + \lambda^{B}_{21})\varepsilon^{B}_{10}\vec{u}^{T}_{i}b^{T}_{10}\vec{u}_2 + \gamma^{B}_{21}\vec{u}^{T}_{i}b^{T}_{10}\vec{u}_1 \\ &\quad = l_1\varepsilon^{A}_{10}\vec{u}^{T}_{i}a^{T}_{10}\vec{u}_2 - l_1\omega^{A}_{10}\omega^{A}_{10}\vec{u}^{T}_{i}a^{T}_{10}\vec{u}_1 \\ &\quad\quad + (l_0 + \lambda^{B}_{21})\omega^{B}_{10}\omega^{B}_{10}\vec{u}^{T}_{i}b^{T}_{10}\vec{u}_1 \\ &\quad\quad - 2\omega^{B}_{10}v^{B}_{21}\vec{u}^{T}_{i}b^{T}_{10}\vec{u}_2 \quad (i = 1, 2). \end{aligned} \tag{3.68}$$

Finally, we add new relations for the velocities and accelerations

$$\begin{aligned}
&\vec{v}^{A}_{10} = \vec{0}, \quad \vec{\omega}^{A}_{10} = \dot{\varphi}^{A}_{10}\vec{u}_3, \quad \tilde{\omega}^{A}_{10} = \dot{\varphi}^{A}_{10}\tilde{u}_3, \quad \vec{\gamma}^{A}_{10} = \vec{0}, \quad \vec{\varepsilon}^{A}_{10} = \ddot{\varphi}^{A}_{10}\vec{u}_3, \quad \tilde{\varepsilon}^{A}_{10} = \ddot{\varphi}^{A}_{10}\tilde{u}_3 \\
&\vec{v}^{B}_{10} = \vec{0}, \quad \vec{\omega}^{B}_{10} = \vec{\omega}^{B}_{20} = \dot{\varphi}^{B}_{10}\vec{u}_3, \quad \tilde{\omega}^{B}_{10} = \tilde{\omega}^{B}_{20} = \dot{\varphi}^{B}_{10}\tilde{u}_3 \\
&\vec{\gamma}^{B}_{10} = \vec{0}, \quad \vec{\varepsilon}^{B}_{10} = \vec{\varepsilon}^{B}_{20} = \ddot{\varphi}^{B}_{10}\vec{u}_3, \quad \tilde{\varepsilon}^{B}_{10} = \tilde{\varepsilon}^{B}_{20} = \ddot{\varphi}^{B}_{10}\tilde{u}_3 \\
&\vec{v}_{21}{}^{B} = \dot{\lambda}^{B}_{21}\vec{u}_1, \quad \vec{v}^{B}_{20} = b_{21}\tilde{\omega}^{B}_{10}\vec{r}_{21}{}^{B} + \vec{v}^{B}_{21} \\
&\vec{\gamma}_{21}{}^{B} = \ddot{\lambda}^{B}_{21}\vec{u}_1, \quad \vec{\gamma}^{B}_{20} = b_{21}\{\tilde{\omega}^{B}_{10}\tilde{\omega}^{B}_{10} + \tilde{\varepsilon}^{B}_{10}\}\vec{r}_{21}{}^{B} + 2b_{21}\tilde{\omega}^{B}_{10}b^{T}_{21}\vec{v}^{B}_{21} + \vec{\gamma}^{B}_{21}.
\end{aligned} \tag{3.69}$$

The set of these formulae will afterwards be used in the dynamics analysis of the planar mechanisms.

References

1. Staicu, S.: Mecanica Teoretica. Edit. Didactica & Pedagogica, Bucharest (1998)
2. Tsai, L.-W.: Robot Analysis: The Mechanics of Serial and Parallel Manipulators. Wiley, New York (1999)
3. Merlet, J.-P.: Parallel Robots. Kluwer Academic, Dordrecht (2000)
4. Nikravesh, P.: Computer-Aided Analysis of Mechanical Systems. Prentice-Hall International, Inc. (1988)
5. Shabana, A.: Computational Dynamics. Wiley, New York (1994)

Finally, we find new relations for the velocities and accelerations

Chapter 4
Kinetics of the Rigid Body

Study of the distribution of mass around the centre of gravity of a mass point system is a key subject in dynamics. Here a development of kinetics for a system of particles and a rigid body with respect to a spatially fixed frame is provided. We will define and analyze various entities: static moment, inertia moment, kinetic impulse, kinetic moment, kinetic energy, power and work.

4.1 Centre of Mass and Tensor of Static Moments of a Rigid Body

Centre of gravity C is a point which locates the resultant weight of a system of particles of total mass m. Provided that the acceleration $\vec{g}$ due to gravity for every particle is constant, then the location of the centre of gravity coincides with that of the *centre of mass* [1]. This centre is an intrinsic point whose position is independent of gravity and choice of reference frame. For example, it is important to define the centre of mass of a group of bodies representing the planets of our solar system.

Considering the system of a set of particles A_i of masses m_i and coordinates x_i, y_i, z_i fixed within a region of space, the location of the centre of mass C with respect to a Cartesian frame $Oxyz$ (Fig. 4.1) is given by the position vector $\vec{r}_C$ or three coordinates x_C, y_C, z_C, as follows:

$$\vec{r}_C = \frac{\sum_i m_i \vec{r}_i}{m}, \quad x_C = \frac{\sum_i m_i x_i}{m}, \quad y_C = \frac{\sum_i m_i y_i}{m}, \quad z_C = \frac{\sum_i m_i z_i}{m}. \tag{4.1}$$

The above three terms $S_{yOz} = \sum_i m_i x_i$, $S_{zOx} = \sum_i m_i y_i$, $S_{xOy} = \sum_i m_i z_i$ are the *mass first moments* called *static moments* of the mass point system with respect to the planes Oyz, Ozx and Oxy, respectively. These scalar quantities can define the *tensor of static moments* thus

S. Staicu, *Dynamics of Parallel Robots*, Parallel Robots: Theory and Applications,
https://doi.org/10.1007/978-3-319-99522-9_4

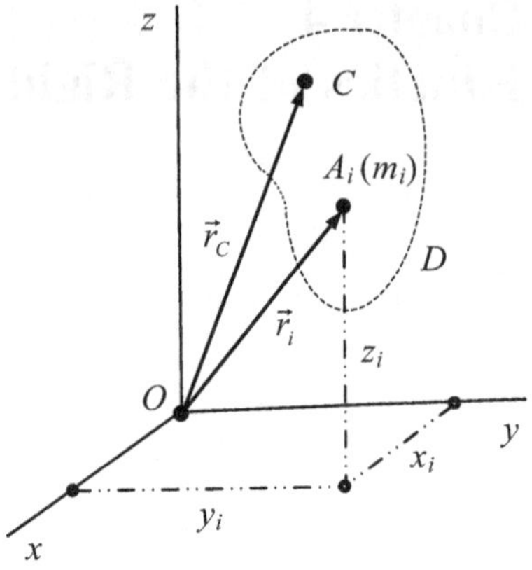

Fig. 4.1 Position of a system of particles

$$\tilde{S}_O = m\tilde{r}_C = \sum_i m_i\tilde{r}_i = \begin{bmatrix} 0 & -S_{xOy} & S_{xOz} \\ S_{yOx} & 0 & -S_{yOz} \\ -S_{zOx} & S_{zOy} & 0 \end{bmatrix}, \tag{4.2}$$

whose *skew-symmetric* matrix is associated with the vector $\vec{S}_O = \sum_i m_i\vec{r}_i = m\vec{r}_C$.

For a rigid body of mass m, considered as a continuum system composed of an infinite number of particles, it becomes necessary to use integration rather than a discrete summation of the terms

$$\vec{r}_C = \frac{\int_D \rho\vec{r}dV}{m}, \tag{4.3}$$

where $\vec{r}$ represents the position vector of an arbitrary point A of elementary mass $dm = \rho dV$, ρ is the mass density or mass per unit volume and D denotes the region of the three-dimensional space occupied by the body (Fig. 4.2).

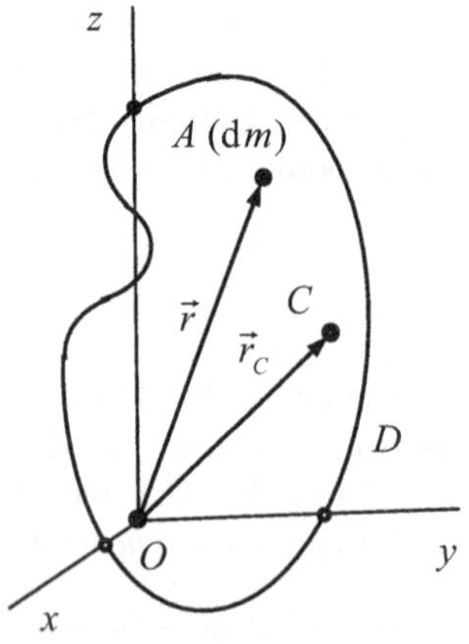

Fig. 4.2 Position of a rigid body

In particular, if the material comprising a body of volume V is uniform or homogenous, the density ρ will be constant throughout the body and therefore this term will factor out of the integrals and will cancel out from all expressions and, finally, we obtain $V\vec{r}_C = \int_D \vec{r}dV$. *Centroid* C is a point which defines the *geometric centre* of a *homogenous* object. Its location is determined from above formulae, similar to those used to determine the centre of mass [2].

Remarks

1. The centre of mass of a system is a point C with respect to which the tensor of static moments $\tilde{S}_C$ is equal to zero.
2. If the mass points constitute a planar rigid body, the terms $S_x = \int_D ydm$ and $S_y = \int_D xdm$ are called static moments with respect to axes Ox and Oy.
3. The areas of surfaces of revolution and the volumes of bodies of revolution can be computed using the theorems of *Guldinus-Pappus* (Fig. 4.3), as follows:

   - The area of a surface generated by a planar curve Γ, by its rotating about a nonintersecting fixed axis Ox in the plane of the curve, is equal to the product of the length L of the generating curve and the distance travelled by the centre C of the curve while the surface is being generated: $A = 2\pi y_C L$.
   - The volume of a body generated by a planar surface Σ, by its revolving about a nonintersecting fixed axis Ox in the plane of the surface, is equal to the product of the area A of the generating surface and the distance travelled by the centre C of the surface while the volume is being generated: $V = 2\pi y_C A$.

4. The elements of the tensor of static moments $\tilde{S}_O = \int_D \tilde{r}dm$ of a *rigid body* are defined relative to O in a coordinate system $Oxyz$ whose origin and axes are attached to an arbitrary point on the body. In this coordinate system, let $\vec{r}$ be the local position vector of an arbitrary point A on the body. With respect to another reference frame $\Omega\xi\eta\varsigma$, the position vector of same point A can be written as shown in Fig. 4.4:

$$\vec{\rho} = \vec{\rho}_O + a^T\vec{r}, \quad \tilde{\rho} = \tilde{\rho}_O + a^T\tilde{r}a, \tag{4.4}$$

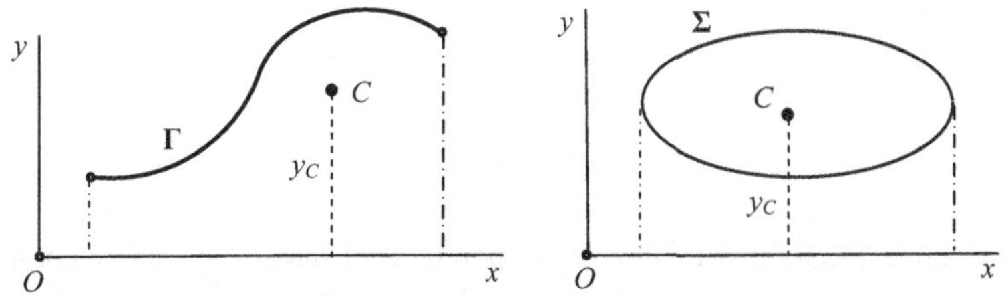

Fig. 4.3 Theorems of Pappus-Guldinus

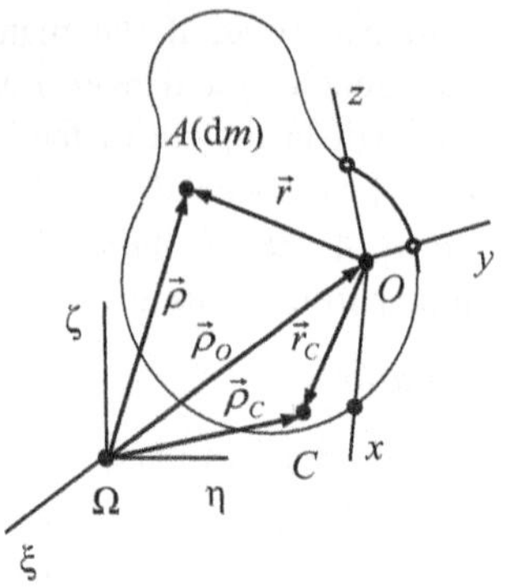

Fig. 4.4 Variation of tensors $\tilde{S}_\Omega$ and $\hat{J}_\Omega$ of a rigid body

where $\tilde{\rho}_O$ is the skew-symmetric matrix associated with the vector $\vec{\rho}_O$ and a represents the general transformation matrix between the two frames. Starting from the equation $\tilde{S}_\Omega = \int_D \tilde{\rho} dm$, we obtain a new skew-symmetric tensor, relative to Ω [3]:

$$\tilde{S}_\Omega = a^T \tilde{S}_O a + m\tilde{\rho}_O. \tag{4.5}$$

4.2 Moments of Inertia of a Rigid Body

Moments of inertia are geometrical terms characterising the mass distribution of a rigid or deformable system of particles.

4.2.1 Tensor of Inertia

Consider a Cartesian coordinate frame $Oxyz$ and a mass point system A_i $(i = 1, 2, \ldots, n)$ (Fig. 4.1). Let $\vec{r}_i$ be the position vector of A_i and the *associated* skew-symmetric matrix $\tilde{r}_i$, as follows

$$\vec{r}_i = \begin{bmatrix} x_i \\ y_i \\ z_i \end{bmatrix}, \quad \tilde{r}_i = \begin{bmatrix} 0 & -z_i & y_i \\ z_i & 0 & -x_i \\ -y_i & x_i & 0 \end{bmatrix}. \tag{4.6}$$

In dynamics, mass distribution of the points A_i of masses m_i can be completely defined by the *mass second moments* called *moments of inertia*. These scalar quantities can be expressed by the *mass tensor of inertia* $\hat{J}_O$ relative to the pole O, using following components:

$$J_{jk} = \sum_{l} m_l (x_l^2 \delta_{jk} - x_j x_k) \quad (j, k = 1, 2, 3), \tag{4.7}$$

where δ_{jk} is the known symbol of Kronecker.

The *tensor of inertia* $\hat{J}_O$ relative to O of a system of particles is expressed using a 3×3 positive-definite symmetric matrix [3]:

$$\hat{J}_O = \sum_{i} m_i \tilde{r}_i \tilde{r}_i^T, \tag{4.8}$$

which can easily be written in expanded form as

$$\hat{J}_O = \begin{bmatrix} J_x & -J_{xy} & -J_{xz} \\ -J_{yx} & J_y & -J_{yz} \\ -J_{zx} & -J_{zy} & J_z \end{bmatrix}. \tag{4.9}$$

Consequently, following six independent elements are defined

$$\begin{aligned} &J_x = \sum_i m_i (y_i^2 + z_i^2), \quad J_y = \sum_i m_i (z_i^2 + x_i^2), \quad J_z = \sum_i m_i (x_i^2 + y_i^2) \\ &J_{xy} = \sum_i m_i x_i y_i, \quad J_{yz} = \sum_i m_i y_i z_i, \quad J_{zx} = \sum_i m_i z_i x_i, \end{aligned} \tag{4.10}$$

where J_x, J_y and J_z are called *moments of inertia with respect to the axes Ox, Oy* and *Oz*, respectively, and J_{xy}, J_{yz}, J_{zx} are called *products of inertia.*

The *polar moment of inertia* J_O and three *moments of inertia with respect to* the orthogonal *planes Oxy, Oyz* and *Ozx*, respectively, can be computed starting from above six elements, if these quantities are already known:

$$\begin{aligned} J_O &= \sum_i m_i \tilde{r}_i^2 = \frac{1}{2}(J_x + J_y + J_z) \\ J_{xOy} &= \sum_i m_i z_i^2 = \frac{1}{2}(J_x + J_y - J_z) \\ J_{yOz} &= \sum_i m_i x_i^2 = \frac{1}{2}(J_y + J_z - J_x) \\ J_{zOx} &= \sum_i m_i y_i^2 = \frac{1}{2}(J_z + J_x - J_y). \end{aligned} \tag{4.11}$$

Remarks

1. From the above formulations it is seen that J_x, J_y, J_z will always be positive since they involve the product of distance squared and mass, while the products of inertia can be positive, negative or zero. When one or more of the axes of coordinates are axes of symmetry for the mass point system, the products of inertia are zero.

2. The radius of gyration of a system of particles of total mass m, with respect to an axis Δ, has units of length and is defined by the formula $i_\Delta = \sqrt{J_\Delta / m}$. In a similar manner, we can define the following radii of gyration: $i_x = \sqrt{J_x/m}$, $i_y = \sqrt{J_y/m}$, $i_z = \sqrt{J_z/m}$, $i_O = \sqrt{J_O/m}$.
3. The moments of inertia, the products of inertia and the tensor of inertia for a rigid body of mass m, which is defined as the integral $\hat{J}_O = \int_D \tilde{r}\tilde{r}^T dm$ with respect to the local-fixed system $Oxyz$ (Fig. 4.2), are all constant once they are computed in this coordinate frame, where $\tilde{r}$ is the skew-symmetric matrix associated with the column matrix $\vec{r} = [x \quad y \quad z]^T$ of the position vector of an arbitrary point A on the body.

4.2.2 Generalized Huygens-Steiner Theorem

Components of the inertia tensor of a rigid body are defined in a coordinate system $Oxyz$, whose origin O and whose axes are attached to this body. In this frame, let $\vec{r}_C$ and $\vec{r}$ be the position vectors of the mass centre C and the arbitrary point A on the body having elementary mass dm (Fig. 4.4). The vector $\vec{\rho}$ and the skew-symmetric matrix $\tilde{\rho}$, that define the location of A with respect to another frame $\Omega\xi\eta\varsigma$, can be written as shown

$$\vec{\rho} = \vec{\rho}_O + a^T\vec{r}, \quad \tilde{\rho} = \tilde{\rho}_O + a^T\tilde{r}a, \tag{4.12}$$

where a is the transformation matrix between these two coordinate systems.

The moments of inertia depend on the pole with respect to which computations are made. If the inertia tensor $\hat{J}_O$ for a rigid body is known about the point O, it is convenient to determine the inertia tensor $\hat{J}_\Omega$ of the same body with respect to the origin of a new frame $\Omega\xi\eta\varsigma$, as follows

$$\hat{J}_\Omega = \int_D \tilde{\rho}\tilde{\rho}^T dm = \tilde{\rho}_O\tilde{\rho}_O^T \int_D dm + a^T \int_D \tilde{r}\tilde{r}^T dma + \tilde{\rho}_O^T a^T \int_D \tilde{r} dma + a^T \int_D \tilde{r}^T dma\tilde{\rho}_O. \tag{4.13}$$

Bearing in mind that the last two integrals define the skew-symmetric tensors of static moments $\tilde{S}_O = m\tilde{r}_C$, $\tilde{S}_O^T = m\tilde{r}_C^T$ and that $\int_D dm = m$ is the mass of the body, the final result is therefore

$$\hat{J}_\Omega = a^T\hat{J}_O a + m\tilde{\rho}_O\tilde{\rho}_O^T + m\tilde{\rho}_O^T a^T \tilde{r}_C a + m a^T \tilde{r}_C^T a \tilde{\rho}_O. \tag{4.14}$$

This is the matrix form of the *generalized Huygens-Steiner theorem* on the variation of tensor of inertia for a rigid body [3].

Remarks

1. Transformation matrix a proves to be unit matrix I and the relation (4.14) is written in a simplified form, if the axes of the two coordinate systems are parallel

$$\hat{J}_\Omega = \hat{J}_O + m\{\tilde{r}_C^2 - \tilde{\rho}_C^2\}. \tag{4.15}$$

2. If O and C are two coincident points, such that $\vec{r}_C = \vec{0}$, from (4.15) we obtain a matrix form of the known transfer theorem or *parallel axes theorem*:

$$\hat{J}_\Omega = \hat{J}_C + m\tilde{\rho}_C\tilde{\rho}_C^T, \tag{4.16}$$

which states that the mass moments of inertia, for example, defined with respect to the axes of a non-centroidal coordinate system $\Omega\xi\eta\varsigma$ are equal to the mass moments of inertia defined with respect to the axes of the centroidal coordinate system $Cxyz$ plus the product of the mass m and the square of the distances between the respective axes. These moments of inertia, as well as the products of inertia, can be expressed from (4.16) in terms of those defined with respect to the centroidal coordinate system as

$$\begin{aligned} &J_\xi = J_x + m(\eta_C^2 + \varsigma_C^2), \quad J_\eta = J_y + m(\varsigma_C^2 + \xi_C^2), \quad J_\varsigma = J_z + m(\xi_C^2 + \eta_C^2) \\ &J_{\xi\eta} = J_{xy} + m\xi_C\eta_C, \quad J_{\eta\varsigma} = J_{yz} + m\eta_C\varsigma_C, \quad J_{\varsigma\xi} = J_{zx} + m\varsigma_C\xi_C, \end{aligned} \tag{4.17}$$

where $J_x, J_y, J_z, J_{xy}, J_{yz}$ and J_{zx} are the moments and products of inertia with respect to the centre of mass C and $\xi_C, \eta_C, \varsigma_C$ are the components of the vector $\overrightarrow{\Omega C} = \vec{\rho}_C$ that define the location of the centre of the mass with respect to the reference point Ω. Since the moments of inertia are always positive, it can be seen from the preceding equations that any translation of a coordinate system away from the centre of mass always leads to an increase of the moments of inertia. The products of inertia, however, may increase or decrease, depending upon the direction of the translation. The smallest moments of inertia attain their minimum values at the mass centre of the rigid body.

3. If O and C are again two coincident points, the general matrix relation (4.14) becomes

$$\hat{J}_\Omega = a^T\hat{J}_C a + m\tilde{\rho}_C\tilde{\rho}_C^T. \tag{4.18}$$

4. If Ω and O are two coincident points, so that $\vec{\rho}_O = \vec{0}$, from the general Eq. (4.14) we obtain the matrix formula of variation of the inertia tensor relative to the concurrent axes:

$$\vec{J}_O = a\hat{J}_\Omega a^T, \tag{4.19}$$

with the following usual expressions

$$\begin{aligned}
J_x &= J_\xi \alpha_{11}^2 + J_\eta \alpha_{12}^2 + J_\varsigma \alpha_{13}^2 - 2J_{\xi\eta}\alpha_{11}\alpha_{12} - 2J_{\eta\varsigma}\alpha_{12}\alpha_{13} - 2J_{\varsigma\xi}\alpha_{13}\alpha_{11} \\
J_y &= J_\eta \alpha_{21}^2 + J_\eta \alpha_{22}^2 + J_\varsigma \alpha_{23}^2 - 2J_{\xi\eta}\alpha_{21}\alpha_{22} - 2J_{\eta\varsigma}\alpha_{22}\alpha_{23} - 2J_{\varsigma\xi}\alpha_{23}\alpha_{21} \\
J_z &= J_\eta \alpha_{31}^2 + J_\eta \alpha_{32}^2 + J_\varsigma \alpha_{33}^2 - 2J_{\xi\eta}\alpha_{31}\alpha_{32} - 2J_{\eta\varsigma}\alpha_{32}\alpha_{33} - 2J_{\varsigma\xi}\alpha_{33}\alpha_{31} \\
J_{xy} &= -J_\xi \alpha_{11}\alpha_{21} - J_\eta \alpha_{12}\alpha_{22} - J_\varsigma \alpha_{13}\alpha_{23} \\
&\quad + J_{\xi\eta}(\alpha_{11}\alpha_{22} + \alpha_{12}\alpha_{21}) + J_{\eta\varsigma}(\alpha_{12}\alpha_{23} + \alpha_{13}\alpha_{22}) + J_{\varsigma\xi}(\alpha_{13}\alpha_{21} + \alpha_{11}\alpha_{23}) \\
J_{yz} &= -J_\xi \alpha_{21}\alpha_{31} - J_\eta \alpha_{22}\alpha_{32} - J_\varsigma \alpha_{23}\alpha_{33} \\
&\quad + J_{\xi\eta}(\alpha_{21}\alpha_{32} + \alpha_{22}\alpha_{31}) + J_{\eta\varsigma}(\alpha_{22}\alpha_{33} + \alpha_{23}\alpha_{32}) + J_{\varsigma\xi}(\alpha_{23}\alpha_{31} + \alpha_{21}\alpha_{33}) \\
J_{zx} &= -J_\xi \alpha_{31}\alpha_{11} - J_\eta \alpha_{32}\alpha_{12} - J_\varsigma \alpha_{33}\alpha_{13} \\
&\quad + J_{\xi\eta}(\alpha_{31}\alpha_{12} + \alpha_{32}\alpha_{11}) + J_{\eta\varsigma}(\alpha_{32}\alpha_{13} + \alpha_{33}\alpha_{12}) + J_{\varsigma\xi}(\alpha_{33}\alpha_{11} + \alpha_{31}\alpha_{13}).
\end{aligned} \tag{4.20}$$

5. Taking into account the Lagrange identity

$$\{\tilde{r}\vec{u}\}^T\{\tilde{r}\vec{u}\} + (\vec{u}^T\vec{r})^2 = \vec{r}^T\vec{r}, \tag{4.21}$$

the moment of inertia of a rigid body about an arbitrary *axis* Δ, oriented by the unit vector $\vec{u}$ (Fig. 4.5), as

$$J_\Delta = \int_D \delta^2 dm = \int_D \{\vec{r}^T\vec{r} - (\vec{u}^T\vec{r})^2\}dm, \tag{4.22}$$

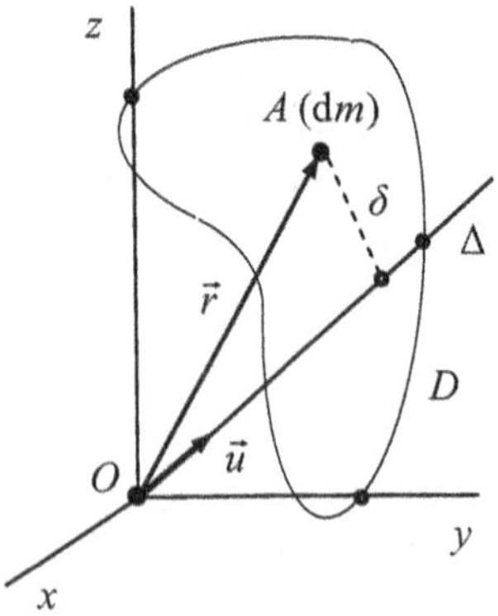

Fig. 4.5 Axial moment of inertia J_Δ of a rigid body

can, ultimately, be expressed in a quadratic form as follows

$$J_\Delta = \int_D \{\tilde{r}\vec{u}\}^T\{\tilde{r}\vec{u}\}dm = \vec{u}^T \int_D \tilde{r}\tilde{r}^T dm\, \vec{u} = \vec{u}^T \hat{J}_O \vec{u}. \tag{4.23}$$

6. Starting from another known identity

$$\{\tilde{r}\vec{u}_1\}^T\{\tilde{r}\vec{u}_2\} = -\{\tilde{r}^T\vec{u}_1\}\{\tilde{r}^T\vec{u}_2\}, \tag{4.24}$$

the *product of inertia* $J_{\Delta_1\Delta_2}$ of a rigid body relative to two orthogonal planes $\Delta_1\Delta_3$ and $\Delta_2\Delta_3$ of a Cartesian system $\Delta_1\Delta_2\Delta_3$, as

$$J_{\Delta_1\Delta_2} = \int_D \{\vec{u}_1^T\vec{r}\}\{\vec{u}_2^T\vec{r}\}dm, \tag{4.25}$$

can (also) be written as follows

$$J_{\Delta_1\Delta_2} = -\vec{u}_1^T \int_D \tilde{r}\tilde{r}^T dm\, \vec{u}_2 = -\vec{u}_1^T \hat{J}_O \vec{u}_2, \tag{4.26}$$

where $\vec{u}_1$ and $\vec{u}_2$ are the unit vectors of the axes Δ_1 and Δ_2. Consequently, the inertia moments and the products of inertia (4.11) are—easily—expressed as follows

$$\begin{aligned} &J_x = \vec{i}^T\hat{J}_O\vec{i}, \quad J_y = \vec{j}^T\hat{J}_O\vec{j}, \quad J_z = \vec{k}^T\hat{J}_O\vec{k} \\ &J_{xy} = -\vec{i}^T\hat{J}_O\vec{j}, \quad J_{yz} = -\vec{j}^T\hat{J}_O\vec{k}, \quad J_{zx} = -\vec{k}^T\hat{J}_O\vec{i}. \end{aligned} \tag{4.27}$$

4.2.3 Principal Axes and Principal Moments of Inertia

Consider a rigid body, placed in a Cartesian coordinate system *Oxyz*, and an arbitrary axis Δ passing through the origin O (Fig. 4.5). If we denote by $\vec{u} = [\alpha \quad \beta \quad \gamma]^T$ the unit vector of this axis, the moment of inertia of the body with respect to Δ is expressed by (4.23), as

$$J_\Delta = \vec{u}^T \hat{J}_O \vec{u}. \tag{4.28}$$

We will now determine the orientation of this axis about which the value of the moment of inertia J_Δ becomes maximum and minimum. This is possible if one applies the following derivatives as necessary conditions

$$\frac{\partial\Phi}{\partial\alpha} = 0, \quad \frac{\partial\Phi}{\partial\beta} = 0, \quad \frac{\partial\Phi}{\partial\gamma} = 0, \tag{4.29}$$

on the following scalar function

$$\Phi(\vec{u}) = \vec{u}^T \hat{J}_O \vec{u} + \lambda(1 - \vec{u}^T I \vec{u}), \tag{4.30}$$

where I represents the 3×3 unit matrix and λ is a Lagrange multiplier.

The system of three homogeneous linear equations obtained from (4.29) can be concentrated as follows

$$I\{\hat{J}_O - \lambda I\}\vec{u} = 0. \tag{4.31}$$

Since these equations cannot have trivial solutions $\alpha = 0, \beta = 0, \gamma = 0$, it is necessary that the determinant must vanishes as $\det\{\hat{J}_O - \lambda I\} = 0$. The roots $\lambda_1 = J_1, \lambda_2 = J_2$ and $\lambda_3 = J_3$ of the algebraic equations, given by (4.31) are three positive eigen-values representing the *principal moments of inertia*, and the three eigen-vectors $\vec{u}_1, \vec{u}_2$ and $\vec{u}_3$ are precisely the unit vectors of *principal axes of inertia* [4].

Indeed, if $\lambda_1 = J_1$ is one of these roots, and $\vec{u}_1 = [\alpha_1 \quad \beta_1 \quad \gamma_1]^T$ is the unit vector of first direction, then by pre-multiplying (4.31) by $\vec{u}_1^T$, it follows that

$$\vec{u}_1^T \hat{J}_O \vec{u}_1 - \lambda_1 \vec{u}_1^T I \vec{u}_1 = 0, \tag{4.32}$$

and so

$$J_1 = \lambda_1 = J_x\alpha_1^2 + J_y\beta_1^2 + J_z\gamma_1^2 - 2J_{xy}\alpha_1\beta_1 - 2J_{yz}\beta_1\gamma_1 - 2J_{zx}\gamma_1\alpha_1. \tag{4.33}$$

In the following, by pre-multiplying same Eq. (4.31), written for the first principal direction of inertia, by $\vec{u}_2^T$, it results that:

$$\vec{u}_2^T \hat{J}_O \vec{u}_1 - \lambda_1 \vec{u}_2^T I \vec{u}_1 = 0 \tag{4.34}$$

and finally $J_{21} = -\lambda_1 \vec{u}_2^T \vec{u}_1$. If the index 1 is replaced by 2 and the same time the index 2 is replaced by 1, we get $J_{12} = -\lambda_2 \vec{u}_1^T \vec{u}_2$. By subtracting, the two above equations lead, ultimately, to the solutions

$$J_{12} = J_{21} = 0, \quad \vec{u}_1^T \vec{u}_2 = 0. \tag{4.35}$$

Therefore, if the orientation of the *Oxyz* body-fixed axes is changed, the moments and the products of inertia will change in value. There is one unique orientation of the frame for which the products of inertia vanish and the three directions Δ_1, Δ_2 and Δ_3 are orthogonal. For this orientation, the inertia matrix takes the diagonal form

$$\hat{J}_O = \begin{bmatrix} J_1 & 0 & 0 \\ 0 & J_2 & 0 \\ 0 & 0 & J_3 \end{bmatrix}, \tag{4.36}$$

where J_1, J_2, J_3 are called the *principal moments of inertia*. The axes $\Delta_1, \Delta_2, \Delta_3$, whose unit vectors are $\vec{u}_1, \vec{u}_2, \vec{u}_3$, respectively, are called the *principal axes of inertia*.

4.2.4 Inertia Moments of Planar Systems

Considering a lamina body of mass m, which lies in the Oxy plane (Fig. 4.6), the *planar state of inertia* is defined by two moments of inertia about the axes Ox, Oy and one product of inertia, as

$$J_x = \int_D y^2 dm, \quad J_y = \int_D x^2 dm, \quad J_{xy} = \int_D xydm. \tag{4.37}$$

The inertia moment with respect to Oz is equal to the polar moment of inertia: $J_O = J_z = J_x + J_y$.

The parallel axes theorem is similar to (4.16), where the following expressions should be introduced

$$\vec{\rho}_C = \begin{bmatrix} \xi_C \\ \eta_C \\ 0 \end{bmatrix}, \quad \hat{J}_C = \begin{bmatrix} J_x & -J_{xy} & 0 \\ -J_{yx} & J_y & 0 \\ 0 & 0 & J_z \end{bmatrix}; \tag{4.38}$$

it results that

$$J_\xi = J_x + m\eta_C^2, \quad J_\eta = J_y + m\xi_C^2, \quad J_{\xi\eta} = J_{xy} + m\xi_C\eta_C \tag{4.39}$$

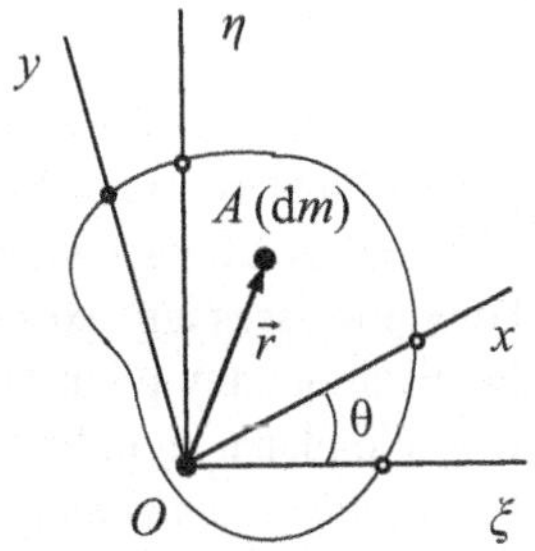

Fig. 4.6 Inertia moments of a planar rigid body

Starting from the formula of variation of inertia tensor relative to the concurrent axes $\hat{J}_O = a\hat{J}_\Omega a^T$ and using the *reduced form* of the matrices

$$\hat{J}_\Omega = \begin{bmatrix} J_\xi & -J_{\xi\eta} \\ -J_{\eta\xi} & J_\eta \end{bmatrix}, \quad a = \begin{bmatrix} \cos\theta & \sin\theta \\ -\sin\theta & \cos\theta \end{bmatrix}, \tag{4.40}$$

we obtain three usual relations

$$\begin{aligned} J_x &= J_\xi \cos^2\theta + J_\eta \sin^2\theta - J_{\xi\eta}\sin 2\theta \\ J_y &= J_\xi \sin^2\theta + J_\eta \cos^2\theta + J_{\xi\eta}\sin 2\theta \\ J_{xy} &= \frac{1}{2}(J_\xi - J_\eta)\sin 2\theta + J_{\xi\eta}\cos 2\theta. \end{aligned} \tag{4.41}$$

If the first and second equations are added together, we can show that the polar moment of inertia about the origin O is independent of the orientation of moving planar frame

$$J_O = J_x + J_y = J_\xi + J_\eta. \tag{4.42}$$

The development of the reduced determinant

$$\det\{\hat{J}_O - \lambda I\} = \begin{vmatrix} J_x - \lambda & -J_{xy} \\ -J_{yx} & J_y - \lambda \end{vmatrix} = 0 \tag{4.43}$$

leads to the principal moments of inertia $\lambda_1 = J_1$ and $\lambda_2 = J_2$, as follows

$$\begin{aligned} J_1 &= \frac{1}{2}(J_x + J_y) + \frac{1}{2}\sqrt{(J_x - J_y)^2 + 4J_{xy}^2} = J_{\max} \\ J_2 &= \frac{1}{2}(J_x + J_y) - \frac{1}{2}\sqrt{(J_x - J_y)^2 + 4J_{xy}^2} = J_{\min}. \end{aligned} \tag{4.44}$$

The angle θ_1 that defines the orientation of the principal axes of inertia may be found by differentiating the first of Eq. (4.41) with respect to θ and setting the result at zero; it results in two solutions

$$\tan 2\theta_1 = \frac{2J_{xy}}{J_y - J_x}, \quad \theta_2 = \theta_1 + \frac{\pi}{2}. \tag{4.45}$$

We note that if a planar rigid body possesses an axis of symmetry, through a point O, this axis must be a principal axis of the body about O. On the other hand, a principal axis does not need to be an axis of symmetry and whether or not a planar body possesses any axes of symmetry it will unfailingly have two principal axes of inertia about any point O. If the point O is chosen to coincide with the mass centre C of the planar rigid body, the two principal axes about its centre are referred to as the *principal centroidal axes* of the body.

4.2.5 General Matrix of Inertia of a Rigid Body

The static moments of the skew-symmetric tensor $\tilde{S}_O$ (4.2), the moments of inertia and the products of inertia of a rigid body from the symmetric tensor $\hat{J}_O$ (4.10) can be assembled, if we use the dual vectors.

Considering a rigid body attached to $Oxyz$ frame (Fig. 4.4), two matrices are associated with the position vector $\vec{r}$ of the point A:

$$\vec{R} = \begin{bmatrix} \vec{r} \\ \vec{0} \end{bmatrix}, \quad R = \begin{bmatrix} I & 0 \\ \tilde{r} & 0 \end{bmatrix}. \tag{4.46}$$

The distribution of mass inside the spatial rigid body can be evaluated by a symmetric *general tensor of inertia* with respect to the origin O:

$$M_O = \int_D RR^T dm = \begin{bmatrix} \hat{m} & m\tilde{r}_C^T \\ m\tilde{r}_C & \hat{J}_O \end{bmatrix}, \tag{4.47}$$

where $\hat{m} = mI$ is the diagonal matrix of the mass, $m\tilde{r}_C = \tilde{S}_O$ is the skew-symmetric matrix of static moments, and $\hat{J}_O$ is the symmetric matrix of inertia moments.

Knowing completely the spatial state of inertia of a rigid body through M_O, evaluated with respect to the coordinate system $Oxyz$, we propose to determine a new general tensor of inertia M_Ω of the same body relative to another frame $\Omega\xi\eta\varsigma$. Using the general transformation matrix A (2.26), we can define the matrix relations

$$R^\circ = \Re_O^\circ + A^T R A, \quad R_C^\circ = \Re_O^\circ + A^T R_C A, \tag{4.48}$$

where the following matrices are associated with the position vector $\vec{\rho}_O$ of the origin O

$$\vec{R}_O^\circ = \begin{bmatrix} \vec{\rho}_O \\ \vec{0} \end{bmatrix}, \quad \Re_O^\circ = \begin{bmatrix} 0 & 0 \\ \tilde{\rho}_O & 0 \end{bmatrix}. \tag{4.49}$$

Substituting in the expression of the general matrix of inertia M_Ω yields

$$\begin{aligned} M_\Omega = \int_D R^\circ R^{\circ T} dm = \Re_O^\circ \Re_O^{\circ T} \int_D dm + A^T \int_D RR^T dm\, A \\ + \Re_O^\circ A^T \int_D R^T dm\, A + A^T \int_D R dm\, A\, \Re_O^{\circ T}, \end{aligned} \tag{4.50}$$

where

$$\int_D R dm = \begin{bmatrix} \hat{m} & 0 \\ m\tilde{r}_C & 0 \end{bmatrix} = mR_C, \quad \int_D RR^T dm = M_O. \tag{4.51}$$

After some developments, ultimately, we obtain a new expression of the *generalized Huygens-Steiner theorem* on the variation of general tensor of inertia for a rigid body [3]:

$$M_\Omega = A^T M_O A + m\Re_O^\circ \Re_O^{\circ T} + m\Re_O^\circ A^T R_C^T A + mA^T R_C A \Re_O^{\circ T}. \tag{4.52}$$

Starting from this general formula, we immediately find the same equation of variation of the skew-symmetric tensor of static moments (4.5) and same equation of variation of the symmetric inertia tensor of moments of inertia (4.14):

$$\begin{aligned} \tilde{S}_\Omega &= a^T \tilde{S}_O a + m\tilde{\rho}_O \\ \hat{J}_\Omega &= a^T \hat{J}_O a + m\tilde{\rho}_O \tilde{\rho}_O^T + m\tilde{\rho}_O a^T \tilde{r}_C^T a + ma^T \tilde{r}_C a \tilde{\rho}_O^T. \end{aligned} \tag{4.53}$$

If Ω and O are two coincident points, the general equation (4.52) leads to the formula of variation of the general tensor of inertia with respect to the concurrent axes

$$M_O = AM_\Omega A^T, \tag{4.54}$$

with its two known components

$$\tilde{S}_O = a\tilde{S}_\Omega a^T, \quad \hat{J}_O = a\hat{J}_\Omega a^T. \tag{4.55}$$

4.3 Kinetic Impulse of a System of Particles

A system of mass points $A_i, \ldots, A_j, \ldots, A_n$ $(i, j = 1, 2, \ldots, n; i \neq j)$ is moving in the space of a Cartesian coordinate system $\Omega\xi\eta\varsigma$, which is centred at a fixed point Ω. Let $m_i, \ldots, m_j, \ldots, m_n$ be the masses of the points and $\vec{\rho}_i, \vec{v}_i, \ldots, \vec{\rho}_j, \vec{v}_j, \ldots, \vec{\rho}_n, \vec{v}_n$ their position vectors and velocities, respectively, with respect to this fixed frame of reference (Fig. 4.7).

The total *linear momentum* or *kinetic impulse* $\vec{h}$ of a system of particles is defined as the sum of kinetic impulses of various particles:

$$\vec{h} = \sum_i m_i \vec{v}_i. \tag{4.56}$$

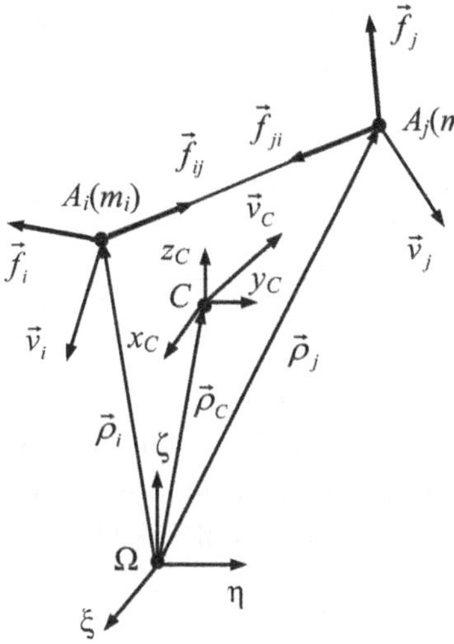

Fig. 4.7 System of mass particles

Since the location of the mass centre C of the system is determined from (4.1), where m is the total mass of all particles, then taking the derivatives, we have

$$\vec{h} = m\vec{v}_C, \tag{4.57}$$

which states that the total kinetic impulse of a system of particles is equivalent to the kinetic impulse of a particle of mass m, moving with the velocity $\vec{v}_C$ of the centre C. The vector $\vec{h}$ has the same direction as $\vec{v}_C$ and a magnitude mv_C of unit kg m/s.

The kinetic impulse of a *rigid body* (Fig. 4.4), is obtained by replacing the sum by the integral of the three-dimensional domain D occupied by the body

$$\vec{h}^{\circ} = \int_D \dot{\vec{\rho}} dm = m\dot{\vec{\rho}}_C. \tag{4.58}$$

4.4 Kinetic Moment of a Rigid Body

The *angular momentum* or *kinetic moment* of the particle A_i about a fixed point Ω is defined as the moment about the same point Ω of the kinetic impulse $m_i\vec{v}_i$ (Fig. 4.7). Denoting by $\vec{\rho}_i$ the position vector of A_i, the total *angular momentum* or *kinetic moment* $\vec{k}_\Omega$ about Ω of the system of particles is determined by the vector summing up the kinetic moments of all particles, as

$$\vec{k}_\Omega = \sum_i \vec{\rho}_i \wedge m_i \vec{v}_i. \tag{4.59}$$

Integrating over the entire domain D occupied by a *rigid body* (Fig. 4.4), we will use the projections on the *fixed frame* $\Omega\xi\eta\zeta$ to obtain a matrix formula of *kinetic moment*

$$\vec{k}^{\circ}_\Omega = \int_D \tilde{\rho}\dot{\vec{\rho}}dm, \tag{4.60}$$

where $\tilde{\rho}$ is the skew-symmetric matrix operator associated with the position vector $\vec{\rho}$ of an arbitrary point A on the body. The expression of $\vec{k}^{\circ}_\Omega$ has specific forms for particular motions of a rigid body.

4.4.1 *Translation*

Since the constitutive points of a rigid body subjected to a translation (Fig. 2.2) have the same velocity $\dot{\vec{\rho}} = \vec{v} = \dot{\vec{\rho}}_C = \vec{v}_C$, at a given instant, the kinetic moment becomes

$$\vec{k}^{\circ}_\Omega = m\tilde{\rho}_C\vec{v}_C. \tag{4.61}$$

4.4.2 *Rotation About a Fixed Axis*

Consider a rigid body of mass m which rotates about a fixed axis Δ (Fig. 4.8). Let P be a point on the body, of elementary mass dm, and $\vec{r}$ its position vector with respect to a body-frame $Oxyz$ centred at fixed point O on Δ. Taking into account the transformation matrix a between a fixed frame and the rotating frame attached at the body, we can express the position vector $\vec{\rho}$ and the velocity $\dot{\vec{\rho}}$ of same point P, measured from the fixed coordinate system:

$$\vec{\rho} = a^T\vec{r}, \quad \tilde{\rho} = a^T\tilde{r}a, \quad \dot{\vec{\rho}} = a^T\tilde{\omega}\vec{r} = a^T\tilde{r}^T\vec{\omega}. \tag{4.62}$$

Substituting into the general equation (4.60) and recognizing the array of inertia tensor of the body, we have successively the following expression for the kinetic moment about O:

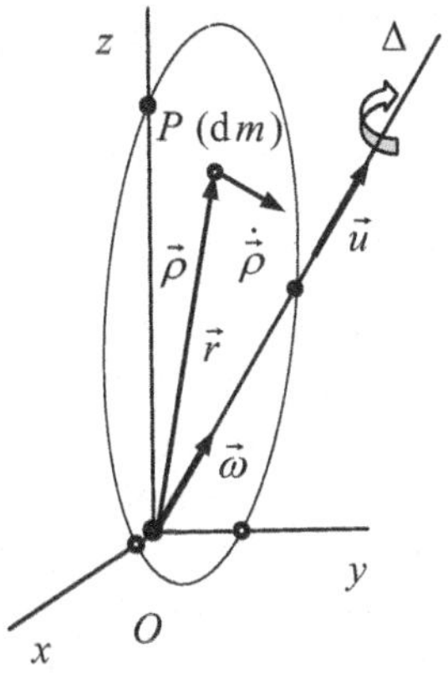

Fig. 4.8 Rotation of the rigid body

$$\vec{k}_O^{\circ} = \int_D \tilde{\rho}\dot{\vec{\rho}}dm = a^T \int_D \tilde{r}aa^T\tilde{\omega}\,\vec{r}dm = a^T \int_D \tilde{r}\tilde{r}^T dm\vec{\omega} = a^T \hat{J}_O\vec{\omega}. \tag{4.63}$$

Using the projections on the *moving frame Oxyz*, we find the vector

$$\vec{k}_O = a\vec{k}_O^{\circ} = \hat{J}_O\vec{\omega} \tag{4.64}$$

with three usual rectangular components

$$\begin{aligned} k_x &= J_x\omega_x - J_{xy}\omega_y - J_{xz}\omega_z \\ k_y &= -J_{yx}\omega_x + J_y\omega_y - J_{yz}\omega_z \\ k_z &= -J_{zx}\omega_x - J_{zy}\omega_y + J_z\omega_z. \end{aligned} \tag{4.65}$$

Remarks

1. Vectors $\vec{k}_O$ and $\vec{\omega}$ will, in general, have different directions.
2. If Δ and Oz are two coincident axes, the components (4.65) yield

$$k_x = -J_{xz}\omega, \quad k_y = -J_{yz}\omega, \quad k_z = J_z\omega. \tag{4.66}$$

3. The kinetic moment $\vec{k}_O$ of a rigid body and its angular velocity $\vec{\omega} = \omega\vec{u}$ will have the same direction if and only if $\vec{\omega}$ is directed along an axis of symmetry Oz, so $\vec{k}_O = J_z\vec{\omega}$.
4. In the case of a non-symmetrical lamina body lying in the Oxy plane and rotating about Oz, the kinetic moment is represented by the vector $\vec{k}_O = J_z\vec{\omega}$.

4.4.3 Spherical Motion of a Rigid Body

In the case of a rigid body constrained to rotate in three-dimensional space about a fixed point O (Fig. 2.5), the kinetic moment about O can be computed using same expressions (4.64), (4.65), where here $\vec{\omega}$ is (here) the instantaneous angular velocity of the body, so that $\vec{k}_O = \hat{J}_O\vec{\omega}$.

The kinetic moment corresponding to a given angular velocity $\vec{\omega}$ is independent of the choice of the coordinate axes. Equation $\vec{k}_O = \hat{J}_O\vec{\omega}$ may be simplified if the coordinate axes of $Oxyz$ are oriented such that they become principal axes of inertia for the body at the point O. When these axes are used, the products of inertia vanish, and the principal moments of inertia about the axes are represented as $J_x = J_1, J_y = J_2 . J_z = J_3$, the array of tensor of inertia then takes a diagonal form (4.36) and the components of kinetic moment reduce to $k_x = J_1\omega,\ k_y = J_2\omega$, $k_z = J_3\omega$. We add that if three principal moments of inertia are equal, $J_1 = J_2 = J_3 = J$, the components k_x, k_y, k_z of the kinetic moment about O are proportional to the components $\omega_x, \omega_y, \omega_z$ of the angular velocity, and the vectors $\vec{k}_O$ and $\vec{\omega}$ are collinear at any instant [5].

4.4.4 General Motion of a Rigid Body

Consider a rigid body attached at the moving frame $Oxyz$ (Fig. 4.9), having mass m and its centre of mass at C. We will determine the kinetic moment in the most general case of three-dimensional motion of a rigid body. Using two frames $\Omega\xi\eta\varsigma$ and $Oxyz$, the position vectors $\vec{\rho}_O$ and $\vec{r}$ are drawn from the origin of coordinates Ω to another origin O and from O to an arbitrary particle P of elementary mass dm.

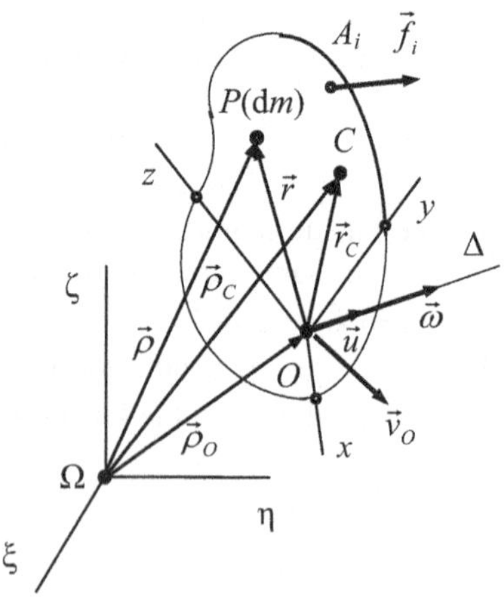

Fig. 4.9 General motion of the rigid body

Using the transformation matrix a, the position vectors $\vec{\rho}, \vec{\rho}_C$ and the velocities $\dot{\vec{\rho}}, \dot{\vec{\rho}}_C$ for the point P and the mass centre C of the body, respectively, are expressed as follows

$$\begin{aligned} &\vec{\rho} = \vec{\rho}_O + a^T\vec{r}, \quad \tilde{\rho} = \tilde{\rho}_O + a^T\tilde{r}a, \quad \dot{\vec{\rho}} = \dot{\vec{\rho}}_O + a^T\tilde{\omega}\vec{r} \\ &\vec{\rho}_C = \vec{\rho}_O + a^T\vec{r}_C, \quad \tilde{\rho}_C = \tilde{\rho}_O + a^T\tilde{r}_C a, \quad \dot{\vec{\rho}}_C = \dot{\vec{\rho}}_O + a^T\tilde{\omega}\vec{r}_C. \end{aligned} \tag{4.67}$$

The kinetic moment as measured from the fixed reference system will be computed relative to the pole Ω, as

$$\vec{k}^{\circ}_{\Omega} = \int_D \tilde{\rho}\dot{\vec{\rho}}dm = m\tilde{\rho}_O\dot{\vec{\rho}}_C + ma^T\tilde{r}_C a\dot{\vec{\rho}}_O + a^T\hat{J}_O\vec{\omega}. \tag{4.68}$$

Pre-multiplying by a, we find the final expression of the kinetic moment, when the components are now measured and used in the *body-fixed frame*

$$\vec{k}_{\Omega} = a\vec{k}^{\circ}_{\Omega} = ma\tilde{\rho}_O a^T\vec{v}_C + m\tilde{r}_C\vec{v}_O + \hat{J}_O\vec{\omega}. \tag{4.69}$$

The later term, $\vec{k}_O = \hat{J}_O\vec{\omega}$, is a vector which represents precisely the kinetic moment with respect to the point O of the same rigid body subjected to a relative spherical motion about the origin O of an imaginary translating frame, which can be computed using the relations (4.64) and (4.65).

Remarks

1. The kinetic moment of a rigid body, which has a cylindrical motion about and along a fixed axis Δ is computed using the same formula (4.69).
2. If O and the mass centre C are two coincident points, the vector $\vec{k}_{\Omega}$ becomes

$$\vec{k}_{\Omega} = ma\tilde{\rho}_C a^T\vec{v}_C + \hat{J}_C\vec{\omega}, \tag{4.70}$$

where $m\vec{v}_C = \vec{h}$ is the linear momentum or kinetic impulse of the body measured in the moving frame. This significant relation is known as the *Koenig theorem* on the kinetic moment of a *rigid body* subjected to a general spatial motion. For a deformable system of particles, the later term is replaced by the total kinetic moment $\vec{k}_C$ of this system with respect to the origin C of a coordinate frame $Cx_Cy_Cz_C$ which translates with the velocity $\vec{v}_C$ (Fig. 4.7).

Once the kinetic impulse $\vec{h} = m\vec{v}_C$ and the kinetic moment $\vec{k}_C$ have been determined, the total kinetic moment $\vec{k}_{\Omega}$ of a system of particles about a fixed point Ω can be obtained by adding the moment about Ω of the vector $\vec{h}$ and the kinetic moment $\vec{k}_C$ about the centre C of the same system subjected to a relative motion with respect to a central translating frame.

4.4.5 Planar Motion of a Rigid Body

Considering a fixed frame $\Omega\xi\eta\varsigma$, the kinetic moment $\vec{k}_\Omega$ will be computed for a body subjected to a general planar motion. Here, we will assume the body is non-symmetric with respect to the $\Omega\xi\eta$ reference plane (Fig. 2.6).

Using the rotation matrix a_φ (2.32), the expression of the kinetic moment is of similar form to that derived for a rigid body having a general three-dimensional motion:

$$\vec{k}_\Omega = m a_\varphi \tilde{\rho}_O a_\varphi^T \vec{v}_C + m \tilde{r}_C \vec{v}_O + \hat{J}_O \vec{\omega}, \tag{4.71}$$

where we denote

$$a_\varphi = \begin{bmatrix} \cos\varphi & \sin\varphi & 0 \\ -\sin\varphi & \cos\varphi & 0 \\ 0 & 0 & 1 \end{bmatrix}, \quad \vec{v}_O = a_\varphi \begin{bmatrix} \dot{\xi}_O \\ \dot{\eta}_O \\ 0 \end{bmatrix}, \quad \vec{\omega} = \dot{\varphi} \begin{bmatrix} 0 \\ 0 \\ 1 \end{bmatrix}. \tag{4.72}$$

Equation (4.71) reduces to a simpler and usual formula applied to planar motions, if point O coincides with the mass centre C:

$$\vec{k}_\Omega = m a_\varphi \tilde{\rho}_C a_\varphi^T \vec{v}_C + \hat{J}_C \vec{\omega}, \tag{4.73}$$

where $m\vec{v}_C = \vec{h}$ is the kinetic impulse of a planar rigid measured in the moving frame. This result states that when the kinetic moment of a body subjected to a planar motion is computed about a fixed point Ω, this is equivalent to merging the moment of kinetic impulse $\vec{h}$ about Ω and the central kinetic moment $\hat{J}_C \vec{\omega}$.

Assuming a *lamina body* in a planar motion or a rigid body which is symmetrical with respect to the reference plane, the later term of (4.73) represents a vector $\vec{k}_C = J_z \vec{\omega}$, having the same direction as $\vec{\omega}$, where J_z represents the moment of inertia about a central axis Cz, perpendicular on the fixed plan $\Omega\xi\eta$.

4.4.6 Kinetic Wrench of the Rigid Body

Consider again the rigid body of mass m, attached at O into the frame $Oxyz$ and subjected to a general three-dimensional motion with respect to the fixed coordinate system $\Omega\xi\eta\varsigma$ (Fig. 4.9). The absolute kinematics of the body is known at any instant through the *twist of velocities*, relative to point O, expressed by the dual vector [3].

$$\vec{V}_O = \begin{bmatrix} \vec{v}_O \\ \vec{\omega} \end{bmatrix}. \tag{4.74}$$

Since the body is considered as being made up of a large number of particles P, the field of elementary kinetic impulses $\dot{\vec{\rho}}dm$ can be reduced at the pole Ω to two vectors: the total kinetic impulse $\vec{h}^\circ = \int_D \dot{\vec{\rho}}dm$ and the total kinetic moment $\vec{k}^\circ_\Omega = \int_D \tilde{\rho}\dot{\vec{\rho}}dm$.

Using the above notations (4.46), (4.48), (4.49) and pursuing a general definition

$$\vec{K}^\circ_\Omega = \int_D R^\circ \dot{\vec{R}}^\circ dm, \tag{4.75}$$

these two significant vectors $\vec{h}^\circ$ and $\vec{k}^\circ_\Omega$ can be merged as a *dual kinetic moment*, called the *kinetic wrench* of the rigid body with respect to fixed point Ω, as

$$\vec{K}^\circ_\Omega = \begin{bmatrix} \vec{h}^\circ \\ \vec{k}^\circ_\Omega \end{bmatrix}. \tag{4.76}$$

Substituting $\vec{R}^\circ = \vec{R}^\circ_O + A^T\vec{R}$, $R^\circ = \Re^\circ_O + A^T RA$, $\dot{\vec{R}}^\circ = \dot{\vec{R}}^\circ_O + A^T\tilde{\Omega}\vec{R}$ into (4.75), we obtain the following formula of the kinetic wrench measured in the *local* body-frame

$$\vec{K}_\Omega = A\vec{K}^\circ_\Omega = mA\Re^\circ_O A^T\vec{V}_C + M_O\vec{V}_O, \tag{4.77}$$

for the general motion and the cylindrical motion of a rigid body, where M_O is the general inertia tensor (4.47) of the body with respect to O and A represents the general transformation matrix (2.26).

Expanding into expression (4.77), we find the known components (4.58), (4.69) of the kinetic wrench in the local body-frame

$$\vec{h} = m\vec{v}_C, \quad \vec{k}_\Omega = ma\tilde{\rho}_O a^T\vec{v}_C + m\tilde{r}_C\vec{v}_O + \hat{J}_O\vec{\omega}. \tag{4.78}$$

Starting from (4.77), we will now consider in particular some types of motion.

1. ***Translation***. If a rigid body of mass m (Fig. 2.2) is subjected to either rectilinear or curvilinear translation of velocity $\vec{v}_C$, the dual vector (4.77), computed about Ω, becomes

$$\vec{K}_\Omega = \begin{bmatrix} m\vec{v}_C \\ m\tilde{\rho}_C\vec{v}_C \end{bmatrix}. \tag{4.79}$$

2. ***Rotation about a fixed point or a fixed axis***. In the case of spherical motion of a rigid body of mass m and tensor of inertia $\hat{J}_O$ (Fig. 2.5), or a rigid body rotating about a fixed axis Δ (Fig. 4.8), the dual kinetic wrench about O is

$$\vec{K}_O = M_O \vec{V}_O = \begin{bmatrix} m\tilde{\omega}\vec{r}_C \\ \hat{J}_O\vec{\omega} \end{bmatrix}. \tag{4.80}$$

3. ***Planar motion***. It is convenient to consider that the point O and the centre C are coincident and the Cz axis is perpendicular to the reference plane, if the body is subjected to general planar motion (Fig. 2.6). Hence, the vector $\vec{K}_\Omega$ is given by the expression

$$\vec{K}_\Omega = \begin{bmatrix} m\vec{v}_C \\ ma_\varphi\tilde{\rho}_C a_\varphi^T\vec{v}_C + \hat{J}_C\vec{\omega} \end{bmatrix}. \tag{4.81}$$

4.5 Kinetic Energy of a Rigid Body

The kinetic energy E of a system of particles A_i of masses m_i and velocities $\vec{v}_i$ (Fig. 4.7) is defined as the sum of the kinetic energies of the various particles which constitute the system

$$E = \frac{1}{2}\sum_i m_i \vec{v}_i^2. \tag{4.82}$$

In the case of a rigid body or a deformable continuous system, an arbitrary particle P of elementary mass dm has at any given instant a velocity $\vec{v}^\circ = a^T\vec{v} = \dot{\vec{\rho}}$. The kinetic energy of the entire body is determined integrating the energies for each particle

$$E = \frac{1}{2}\int_D \dot{\vec{\rho}}^2 dm = \frac{1}{2}\int_D \dot{\vec{\rho}}^T\dot{\vec{\rho}}\, dm. \tag{4.83}$$

This expression can be written in specific forms for different types of motion accomplished by the rigid body.

4.5.1 Translation

When a rigid body of mass m is subjected to either rectilinear or curvilinear translation, the points have the same velocity $\dot{\vec{\rho}} = \vec{v}_C$ at a given instant. From (4.83), the kinetic energy of the body is therefore

$$E = \frac{1}{2}\vec{v}_C^2 \int_D dm = \frac{1}{2} m\,\vec{v}_C^2 = \frac{1}{2} m\,\vec{v}_C^T \vec{v}_C, \tag{4.84}$$

where $\vec{v}_C$ is the velocity of mass centre C at the instant considered (Fig. 2.2).

4.5.2 Rotation About a Fixed Axis

Here, the rigid body is rotating with angular velocity $\vec{\omega} = \omega\,\vec{u}$ about the fixed axis Δ of unit vector $\vec{u}$, which passes through the point O (Fig. 4.8). The velocity of a point P of mass dm located by the position vector $\vec{r}$ is

$$\dot{\vec{\rho}} = \dot{a}^T \vec{r} = a^T \tilde{\omega} \vec{r} = a^T \tilde{r}^T \vec{\omega}. \tag{4.85}$$

Substituting into (4.83) and factoring in the inertia tensor of the body, we obtain successively

$$E = \frac{1}{2}\int_D \dot{\vec{\rho}}^T \dot{\vec{\rho}}\, dm = \frac{1}{2}\int_D \vec{\omega}^T \tilde{r} a a^T \tilde{r}^T \vec{\omega}\, dm = \frac{1}{2}\vec{\omega}^T \int_D \tilde{r}\tilde{r}^T dm \vec{\omega} = \frac{1}{2}\vec{\omega}^T \vec{k}_O$$
$$= \frac{1}{2}\vec{\omega}^T \hat{J}_O \vec{\omega}. \tag{4.86}$$

Identifying the moments and the products of inertia relative to axes attached at O, the energy expression becomes

$$E = \frac{1}{2}\omega^2 \vec{u}^T \hat{J}_O \vec{u} = \frac{1}{2} J_\Delta \omega^2$$
$$= \frac{1}{2} J_x \omega_x^2 + \frac{1}{2} J_y \omega_y^2 + \frac{1}{2} J_z \omega_z^2 - J_{xy}\omega_x\omega_y - J_{yz}\omega_y\omega_z - J_{zx}\omega_z\omega_x. \tag{4.87}$$

In the particular case of a rigid body rotating about a fixed axis $\Delta \equiv Oz$, the kinetic energy can be given in a simplified form $E = \frac{1}{2} J_z \omega^2$.

4.5.3 Spherical Motion of a Rigid Body

If O is a fixed point in the rigid body (Fig. 2.5), the kinetic energy can be computed using the same expressions (4.86), (4.87), where $\vec{\omega}$ is now the instantaneous angular velocity of the rigid body

$$E = \frac{1}{2}\vec{\omega}^T \hat{J}_O \vec{\omega}. \tag{4.88}$$

Supposing that the axes Ox, Oy, Oz are chosen to be oriented along the principal directions of inertia for the body, the above equation becomes

$$E = \frac{1}{2}J_1\omega_x^2 + \frac{1}{2}J_2\omega_y^2 + \frac{1}{2}J_3\omega_z^2, \tag{4.89}$$

where J_1, J_2, J_3 are the principal moments of inertia relative to point O.

4.5.4 General Motion of a Rigid Body

Consider a rigid body, of mass m and centre of mass C, having a three-dimensional motion (Fig. 4.9). The absolute velocity $\dot{\vec{\rho}}$ of the particle P of elementary mass dm is expressed as a sum of the velocity $\dot{\vec{\rho}}_O$ of the reference point O of the body and the velocity of P about O, thus

$$\dot{\vec{\rho}} = \dot{\vec{\rho}}_O + a^T \tilde{\omega} \vec{r} = a^T \{\vec{v}_O + \tilde{r}^T \vec{\omega}\}. \tag{4.90}$$

The kinetic energy for the entire body is obtained successively from the general equation (4.83), as follows

$$\begin{aligned} E &= \frac{1}{2}\int_D \dot{\vec{\rho}}^T \dot{\vec{\rho}}\, dm = \frac{1}{2}\int_D \{\vec{v}_O^T + \vec{\omega}^T \tilde{r}\} a a^T \{\vec{v}_O + \tilde{r}^T \vec{\omega}\} dm \\ &= \frac{1}{2} m \vec{v}_O^T \vec{v}_O + m \vec{v}_O^T \tilde{\omega} \vec{r}_C + \frac{1}{2}\vec{\omega}^T \hat{J}_O \vec{\omega}, \end{aligned} \tag{4.91}$$

where we identify $\int_D dm = m$, $\int_D \vec{r} dm = m\vec{r}_C = \vec{S}_O$, $\int_D \tilde{r}\tilde{r}^T dm = \hat{J}_O$.

The later term $E_O = \frac{1}{2}\vec{\omega}^T \hat{J}_O \vec{\omega}$ represents precisely the kinetic energy of the body subjected to a relative spherical motion about the origin O of an imaginary translating frame, which is expressed by the relation (4.88).

Simplification occurs, however, if the reference point O is the centre of mass C:

$$E = \frac{1}{2} m \vec{v}_C^T \vec{v}_C + \frac{1}{2} \vec{\omega}^T \hat{J}_C \vec{\omega}. \tag{4.92}$$

This relation is known as the *Koenig theorem* on the kinetic energy of a rigid body subjected to a general spatial motion. In the case of a deformable system of particles, the later term is replaced by the total kinetic energy E_C of this system with respect to a coordinate frame $Cx_Cy_Cz_C$ which is fixed at C and translates with the velocity $\vec{v}_C$.

The kinetic energy of a system of particles in a fixed frame can be separate into two parts: kinetic energy of the centre C, having a mass equal to the mass of entire system, and the kinetic energy of same system of particles in its relative motion about the mass centre. Particularly, in the case of a cylindrical motion of a rigid body about and along a fixed axis Δ, of unit vector $\vec{u}$, with the velocities $\vec{v}_O = v_O \vec{u}, \vec{\omega} = \omega \vec{u}$, the kinetic energy is computed, starting from the same formula (4.91), as follows

$$E = \frac{1}{2} m v_O^2 + \frac{1}{2} J_\Delta \omega^2. \tag{4.93}$$

4.5.5 Planar Motion of the Rigid Body

Assuming that the body is non-symmetric with respect to the Oxy reference plane (Fig. 2.6), the kinetic energy can be expressed using same formula (4.91) as

$$E = \frac{1}{2} m \vec{v}_O^T \vec{v}_O + m \vec{v}_O^T \tilde{\omega}_\varphi \vec{r}_C + \frac{1}{2} \vec{\omega}_\varphi^T \hat{J}_O \vec{\omega}_\varphi, \tag{4.94}$$

where the last term yields:

$$E_O = \frac{1}{2} \dot{\varphi}^2 \vec{k}^T \hat{J}_O \vec{k} = \frac{1}{2} J_z \dot{\varphi}^2. \tag{4.95}$$

As a special case, if O coincides with the mass centre C of the body, then $\tilde{r}_C = 0$ and therefore we have

$$E = \frac{1}{2} m v_C^2 + \frac{1}{2} J_z \omega^2. \tag{4.96}$$

Here J_z is the moment of inertia for the body about an axis which is perpendicular to the plane of motion and passes permanently through the mass centre. The terms on the right-hand side are always positive, since the velocities are squared.

4.5.6 *General Expression of Kinetic Energy of a Rigid Body*

Using the notations (4.48), (4.49) and pursuing the definition

$$E = \frac{1}{2}\int_D \overset{\circ}{\vec{R}}{}^T \overset{\circ}{\vec{R}}\, dm, \tag{4.97}$$

the kinetic energy of a rigid subjected to a general three-dimensional motion (Fig. 4.9) can be expressed in a simplified form.

Substituting $\overset{\circ}{\vec{R}} = \overset{\circ}{\vec{R}}_O + A^T\tilde{\Omega}\vec{R}$, we obtain, successively, the following initial formula of the kinetic energy for the general motion of a rigid body:

$$E = \frac{1}{2}m\overset{\circ}{\vec{R}}{}_O^T \overset{\circ}{\vec{R}}_O + m\overset{\circ}{\vec{R}}{}_O^T A^T\tilde{\Omega}\vec{R}_C + \frac{1}{2}\vec{\Omega}^T M_O\vec{\Omega}, \tag{4.98}$$

Since $\vec{V}_O = A\overset{\circ}{\vec{R}}_O + \vec{\Omega}, \vec{\Omega}^T = [\vec{0}^T \quad \vec{\omega}^T]$, the above equation can be written in a definitive general form [3].

$$E = \frac{1}{2}\vec{V}_O^T M_O \vec{V}_O, \tag{4.99}$$

which is equivalent to the known (4.91) expression

$$E = \frac{1}{2}m\,\vec{v}_O^T\vec{v}_O + m\,\vec{v}_O^T\tilde{\omega}\,\vec{r}_C + \frac{1}{2}\vec{\omega}^T\hat{J}_O\vec{\omega}. \tag{4.100}$$

4.6 Power and Work of the Forces Acting on a System of Particles

If $\vec{\rho}_i$ denotes the position vector of an arbitrary point A_i, of velocity $\vec{v}_i$, the elementary *displacement* is represented by the differential $d\vec{\rho}_i = \vec{v}_i dt$. Assuming that a force $\vec{f}_i$ is acting on this particle, the *power* developed at this instant by the force is defined as the scalar product of this force and the velocity $\vec{v}_i$ of the point (Fig. 4.7)

$$P_i = \vec{v}_i^T\vec{f}_i = \vec{f}_i^T\vec{v}_i. \tag{4.101}$$

The elementary *work* done by the force $\vec{f}_i$ corresponding to the displacement $d\vec{\rho}_i$, during the interval of time dt, is defined as the quantity

$$dW_i = P_i dt = \vec{v}_i^T \vec{f}_i dt = \vec{f}_i^T d\vec{\rho}_i \tag{4.102}$$

The unit of power used in SI is *watt* (1 W = 1 kg m^2 s^{-3}), while the unit of work is *watt-second* (1 W s = 1 kg m^2 s^{-2}).

The *work* of a variable force $\vec{f}_i$, defined through its rectangular components X_i, Y_i, Z_i, during a *finite* displacement of the actuated particle A_i between two successive positions A and B, is obtained by integrating the above relation along the path described by the particle

$$W_{AB} = \int_A^B f_i^T \vec{v}_i dt = \int_A^B (X_i \dot{x}_i + Y_i \dot{y}_i + Z_i \dot{z}_i) dt. \tag{4.103}$$

When the work done by a force during the moving of a particle from one point to another is *independent of the path* followed by the particle, then this force is called a *conservative force.*

It is important to point out that the in general the analytical expression of the work is not an exact differential. But, in the special cases where the expression of the work is an exact differential, the force vector is said to be *conservative* since it can be obtained using a potential function. Non-conservative forces, however, cannot be derived from a potential function, and hence their work is not equal to the variation of any particular function.

The weight of a particle or a rigid body and the force of an elastic spring are two examples of *conservative forces* often encountered in mechanics.

For example, the work done by the *weight* $\vec{G} = m\vec{g}$ of a particle which is moving up along the path from the position A to position B is independent of the path since it depends only on the displacement of the centre of gravity and is equal to the product of mg and the vertical displacement, thus $W_{AB} = mg(z_A - z_B)$. We note that the work of the weight is positive when the particle is displaced downward. The work of the force $F = -kx$, developed into a linear elastic *spring* of *stiffness k*, attached to a fixed point and acting on a particle or at a point of a body, during a finite rectilinear displacement from $A(x_A)$ to $B(x_B)$, is also independent of the path of the point and depends only on the extension or compression of the spring, as $W_{AB} = \frac{1}{2}k(x_A^2 - x_B^2)$. We add that this expression is negative when $x_B > x_A$, that being when the particle is moving so as to further elongate the spring.

The *potential function* or *potential energy V* of a particle is a measure of the amount of work of a *conservative force* when the particle moves from a given position to a fixed datum.

The work done by any *conservative force* $\vec{f} = -\overrightarrow{grad}\, V$ in moving a particle from point A to point B is measured by the difference of *potential function V*, as follows

$$W_{AB} = -\int_A^B \overrightarrow{grad}\, V d\vec{\rho} = -\int_A^B dV = V_A - V_B. \tag{4.104}$$

For example, the *elastic potential energy* which a spring imparts to an attached particle or a rigid body when the spring is elongated or compressed from an initial free position $(x = 0)$ to a final position is $V = \frac{1}{2}kx^2$. Also, the *gravitational potential energy* of a particle or a rigid body of weight $G = mg$ is determined from knowing the height of the centre of gravity C above or below a horizontal fixed datum $(z = 0)$, as $V = mgz_C$. In contrast to these forces, *frictional forces* are *non-conservative*.

The total power developed by the external forces $\vec{f}_i$ and internal forces $\vec{f}_{ij}$, acting upon a mechanical system of particles, is defined as the sum of the powers of various forces, corresponding to the displacements $d\vec{\rho}_i = \vec{v}_i dt$ of the particles of the system (Fig. 4.7):

$$P = P_{ext} + P_{\text{int}}, \tag{4.105}$$

where $P_{ext} = \sum_i \vec{v}_i^T \vec{f}_i$ is the power of external forces, $P_{\text{int}} = \sum_i \sum_j \vec{v}_{ij}^T \vec{f}_{ij}$ is the power of internal forces and $\vec{v}_{ij} = \vec{v}_i - \vec{v}_j$ is the relative velocity of two points A_i and A_j of this system.

The total work of all forces acting on the various particles, whether these forces are external or internal, can be separate, as follows

$$dW_{ext} = P_{ext} dt, \quad dW_{\text{int}} = P_{\text{int}} dt. \tag{4.106}$$

A number of forces frequently encountered in applications of kinetics perform no work. They are forces applied to fixed points or acting in a direction perpendicular. The cases where the work done by some internal forces vanish are: reaction at a frictionless pin when the body supported rotates about the pin, reaction at a frictionless surface when the body in contact moves along the surface, system of smooth, bodies connected by inextensible cords or strings, bodies enmeshed with other bodies or internal reaction and friction force at a roller moving without sliding along a fixed or moving surface. In all these cases, the work of forces acting at the points of contact are eliminated, since they occur in equal but opposite collinear pairs and each pair of forces moves through an equal distance when the system undergoes a displacement [2].

If the *internal force field* is a conservative one and if $\vec{f}_{ij} = -\overrightarrow{grad}\, V_{ij}$, then a *potential function* or a total *potential energy* V of the system of particles can be defined by integration, starting from the relation

$$dV = -\sum_i \sum_j \vec{v}_{ij}^T \vec{f}_{ij} dt. \tag{4.107}$$

When the internal forces acting upon a mechanical system are all conservative, the sum $E+V$ is called the total *mechanical energy* of a system of particles. The unit of mechanical energy used in SI is *joule* (1 J = 1 kg m^2 s^{-2}).

4.7 Power and Work of the Forces Acting on a Rigid Body

Consider the various external forces $\vec{f}_i$ acting on a rigid body subjected to a general three-dimensional motion in the space of the coordinate system $\Omega\xi\eta\varsigma$ (Fig. 4.9).

The kinematics of the body at a given instant is known through the characteristic vectors $\vec{v}_O$ and $\vec{\omega}$ measured in the local-body frame $Oxyz$. Using the transformation matrix a, the velocity of an arbitrary point A_i on the body, located by the position vector $\overrightarrow{OA_i} = \vec{r}_i$, is expressed in the fixed frame $\Omega\xi\eta\varsigma$ as $\dot{\vec{\rho}}_i = a^T\{\vec{v}_O + \tilde{r}_i^T\vec{\omega}\}$.

The power developed by all external forces $\vec{f}_i^{\circ} = a^T\vec{f}_i$, acting in the points A_i at a given instant, is

$$\begin{aligned} P_{ext} &= \sum_i \dot{\vec{\rho}}_i^T \vec{f}_i^{\circ} = \sum_i \{\vec{v}_O^T + \vec{\omega}^T\tilde{r}_i\} a a^T \vec{f}_i = \vec{v}_O^T \sum_i \vec{f}_i \\ &+ \vec{\omega}^T \sum_i \tilde{r}_i \vec{f}_i = \vec{v}_O^T \vec{f} + \vec{\omega}^T \vec{m}_O. \end{aligned} \tag{4.108}$$

Since $\vec{f}$ and $\vec{m}_O$ are the components of the *wrench* of external forces expressed by the *dual moment*

$$\vec{M}_O = \sum_i R_i \vec{F}_i = \sum_i \begin{bmatrix} I & 0 \\ \tilde{r}_i & 0 \end{bmatrix} \begin{bmatrix} \vec{f}_i \\ 0 \end{bmatrix} = \begin{bmatrix} \vec{f} \\ \vec{m}_O \end{bmatrix}, \tag{4.109}$$

and $\vec{V}_O^T = \begin{bmatrix} \vec{v}_O^T & \vec{\omega}^T \end{bmatrix}$ is the line matrix of the *dual velocity* given by the *twist* of velocity field of the rigid body, the power becomes

$$P_{ext} = \vec{V}_O^T \vec{M}_O. \tag{4.110}$$

If the rigid body is subjected to *translation*, the power developed by the external forces is $P_{ext} = \vec{v}_O^T \vec{f}$.

In a particular case where the rigid body is *rotating* with an angular velocity $\vec{\omega} = \omega\vec{u}$ about a fixed axis Δ of unit vector $\vec{u}$ passing through O, the power yields successively

$$P_{ext} = \vec{\omega}^T \vec{m}_O = \omega \vec{u}^T \vec{m}_O = \omega\, m_\Delta, \tag{4.111}$$

where m_Δ is precisely the resultant moment of external forces with respect to the axis of rotation. The total work of the couple of moment m_Δ during a finite rotation of the rigid body is obtained by integrating both members of $dW_{ext} = P_{ext}dt$ from the initial value θ_1 of the angle of rotation to the final value θ_2, as follows:

$$W_{ext} = \int_1^2 m_\Delta \omega\, dt = \int_1^2 m_\Delta d\theta. \tag{4.112}$$

When the couple of moment m_Δ has a constant magnitude, the above formula reduces at $W_{ext} = m_\Delta(\theta_2 - \theta_1)$, where the work is positive, provided m_Δ and $\theta_2 - \theta_1$ have the same sense of direction.

References

1. Beer, F.P., Johnston, Jr., E.R.: Vector Mechanics for Engineers, Dynamics. McGraw-Hill Ryerson (1988)
2. Hibbeler, R.C.: Engineering Mechanics, Dynamics. Prentice-Hall, Inc. (1997)
3. Staicu, S.: Mecanica Teoretica. Edit. Didactica & Pedagogica, Bucharest (1998)
4. Voinea, R., Stroe, I., Predoi, M.: Technical Mechanics. Edit. Politehnica Press, Bucharest (2010)
5. Nikravesh, P.: Computer-Aided Analysis of Mechanical Systems. Prentice-Hall International, Inc. (1988)

Chapter 5
Dynamics of the Rigid Body

When rigid bodies are accelerated and the magnitude or the direction of their velocities changes, it is necessary to use Newton's fundamental laws of dynamics to relate the motions of the bodies to the forces acting upon them. When the resultant of many forces acting on a mass particle is not zero, that particle will have the acceleration proportional to the magnitude of the resultant and the direction of this force.

5.1 Fundamental System of Differential Equations of Motion for a System of Particles

Consider a system of n particles isolated within an enclosed region, in space, as shown schematically in Fig. 4.7. Concentrated at the point A_i, the mass m_i is located by the position vector $\vec{\rho}_i$ directed from the origin Ω of an inertial frame $\Omega\xi\eta\zeta$ towards this particle.

There are two types of forces acting on each particle, including an externally-applied active or reactive force, represented symbolically by $\vec{f}_i$, and internal forces $\vec{f}_{ij}$ of interaction between A_i and the other particles belonging to same system, where $i < j$ and $\vec{f}_{ii} = \vec{0}$. The external force represents, for example, the effect of gravitational, electrical, magnetic or contact forces between the particle A_i and adjacent bodies or particles not included within the system.

The forces of interaction between particles in a system occur in equal magnitude but in opposite collinear pairs and must satisfy Newton's third law of action and reaction: the force on A_i due to A_j must be exactly equal to the negative of the force acting on A_j due to A_i

S. Staicu, *Dynamics of Parallel Robots*, Parallel Robots: Theory and Applications,
https://doi.org/10.1007/978-3-319-99522-9_5

$$\vec{f}_{ij} + \vec{f}_{ji} = \vec{0}, \quad \vec{\rho}_i \wedge \vec{f}_{ij} + \vec{\rho}_j \wedge \vec{f}_{ji} = \vec{0} \quad (i,j = 1,2,\ldots,n; i < j). \tag{5.1}$$

To derive the equations of motion for both unconstrained and constrained systems of particles, only two of Newton's laws of motion for a single particle are needed as postulates: the second law of total force acting on a particle and the third law of action and reaction.

Let $\vec{\gamma}_i = \ddot{\vec{\rho}}_i$ be the acceleration of A_i with respect to the conventionally-fixed frame of reference. The equations of motion for the system can be written as a collection of equations of motion for all particles taken individually. Thus, for each of the particles A_i we apply second Newton's law of motion

$$m_i \ddot{\vec{\rho}}_i = \vec{f}_i + \sum_j \vec{f}_{ij} \quad (i,j = 1,2,\ldots,n; i < j). \tag{5.2}$$

Describing the motion of the system of n particles, these n vector relations constitute the *fundamental system of differential equations*. If the external forces $\vec{f}_i$ acting on every particle are known and the forces of interaction $\vec{f}_{ij}$ between particles will generally depend on the position vectors $\vec{\rho}_i$ and the velocities $\vec{v}_i = \dot{\vec{\rho}}_i$, the system of differential equations can be integrated, giving first the velocities $\dot{\vec{\rho}}_i$ and, finally, the position vectors $\vec{\rho}_i$.

5.2 Theorem of Kinetic Impulse

Here, it is demonstrated that Newton's Second Law can be expressed in an alternative form relating the rate of change of the total linear momentum or kinetic impulse to the resultant of external forces acting on the particles of a mechanical system.

Considering a system of mass particles (Fig. 4.7), the expression (4.56) of linear momentum or kinetic impulse $\vec{h}$ can be differentiated with respect to time t to obtain

$$\dot{\vec{h}} = \sum_i m_i \ddot{\vec{\rho}}_i. \tag{5.3}$$

Returning now to the fundamental system, its vector equation (5.2) can be summed together as

$$\sum_i m_i \ddot{\vec{\rho}}_i = \sum_i \vec{f}_i + \sum_i \sum_j \vec{f}_{ij}, \tag{5.4}$$

which is valid for the entire mechanical system of particles. The double sum contains both $\vec{f}_{ij}, \vec{f}_{ji}$ and, consequently, the system of internal forces is equipollent to zero and it is found that $\sum_i \sum_j \vec{f}_{ij} = \vec{0}$. Note, however, that it does not follow that the internal forces have no effect on the particle under consideration. Hence, only the sum of external forces will remain, and therefore the equation of motion for the system of particles becomes the *theorem of kinetic impulse*

$$\dot{\vec{h}} = \sum_i \vec{f}_i. \tag{5.5}$$

The derivative with respect to time of the total kinetic impulse $\vec{h}$ of a system of mass particles or a *rigid body* is equal to the resultant of all external forces acting on the system.

Since the total m of the system, as the sum of the individual masses, is constant, the relation (5.5) leads to the *theorem of mass centre motion*

$$m\vec{\gamma}_C = \sum_i \vec{f}_i, \tag{5.6}$$

where $\vec{\gamma}_C$ is the acceleration of the mass centre C. This result states that the centre of mass of a system of particles or a *rigid body* moves as if it were a particle of entire mass m of the system under the action of a force equal to the resultant vector of the external forces.

The equation of motion (5.6) will be useful for solving problems involving force, velocity and time. Integrating between the limits t' and t'', we have

$$m\vec{v}_C'' - m\vec{v}_C' = \sum_i \int_{t'}^{t''} \vec{f}_i dt, \tag{5.7}$$

where the vector $\vec{I}_i = \int_{t'}^{t''} \vec{f}_i dt$ of unit N s is the total *impulse of the external force*, which measures the effect of $\vec{f}_i$ during the interval of time $\Delta t = t'' - t'$ when this force acts.

If no external force acts on the particles of a system, during the interval of time Δt, the right-hand member of Eq. (5.7) will be equal to zero and the equation reduces to $\vec{h} = \overrightarrow{const.}$, which express that the total kinetic impulse of the system of particles is *conserved*. This is the *law of conservation* of kinetic impulse. Moreover, the mass centre C of this mechanical system or *rigid body* moves in a straight line with a constant velocity $\vec{v}_C = \overrightarrow{const}$.

5.3 Theorem of Kinetic Moment

Consider the same system of particles A_i (Fig. 4.7), subjected to external forces $\vec{f}_i$ and internal forces $\vec{f}_{ij}$. Differentiating with respect to time t the expression (4.59) of angular momentum or kinetic moment $\vec{k}_\Omega$ of the system of mass particles, we obtain

$$\dot{\vec{k}}_\Omega = \sum_i \vec{\rho}_i \wedge m_i \ddot{\vec{\rho}}_i. \tag{5.8}$$

Pre-multiplying each of the differential equations of the fundamental system (5.2) by the position vector $\vec{\rho}_i$ and adding the results, we have

$$\sum_i \vec{\rho}_i \wedge m_i \ddot{\vec{\rho}}_i = \sum_i \vec{\rho}_i \wedge \vec{f}_i + \sum_i \sum_j \vec{\rho}_i \wedge \vec{f}_{ij}. \tag{5.9}$$

Because the resultant moment about Ω of the internal forces is equipollent to zero, $\sum_i \sum_j \vec{\rho}_i \wedge \vec{f}_{ij} = \vec{0}$, only the torque of the external forces will remain, and, consequently, the equation of motion for the entire system becomes the *theorem of kinetic moment*

$$\dot{\vec{k}}_\Omega = \sum_i \vec{\rho}_i \wedge \vec{f}_i. \tag{5.10}$$

It will be further that the derivative with respect to time of the total kinetic moment $\vec{k}_\Omega$ of a system of mass particles or a *rigid body*, computed with respect to a fixed point Ω, is equal to the resultant torque about the same point, of all external forces acting on the system.

Integrating between the limits t' and t'', we obtain a new expression for the variation of kinetic moment during the interval of time $\Delta t = t'' - t'$:

$$\vec{k}''_\Omega - \vec{k}'_\Omega = \sum_i \int_{t'}^{t''} \vec{\rho}_i \wedge \vec{f}_i dt. \tag{5.11}$$

In some applications, such as problems involving central forces, the torque about a fixed point Ω of each of the external forces can be zero without any of the forces being zero. In such cases, the total kinetic moment of the mechanical system about Ω is *conserved*, $\vec{k}_\Omega = const.$

5.4 Theorem of Kinetic Moment with Respect to a Translating Frame

Sometimes, it is convenient to consider the motion of a system of particles or a *rigid body* with respect to a moving frame which translates with respect to a fixed frame of reference.

It will be seen that the fundamental relation (5.10) change when the frame is replaced by an arbitrary translating frame of reference *Oxyz*, and that the theorem of kinetic moment has same general form as for the motion of the mechanical system relative to a central translating frame $Cx_Cy_Cz_C$ (Fig. 4.7).

Starting from the general relation (4.69), now given in a vector form $\vec{k}_\Omega = \vec{\rho}_O \wedge m\vec{v}_C + \vec{r}_C \wedge m\vec{v}_O + \vec{k}_O$, we express, and subsequently differentiate with respect to time the kinetic moment $\vec{k}_O = \sum_i \vec{r}_i \wedge m_i\vec{v}_{ir}$ of a system of particles relative to O; it follows that

$$\dot{\vec{k}}_O = \dot{\vec{k}}_\Omega - \vec{v}_{Cr} \wedge m\vec{v}_O - \vec{r}_C \wedge m\vec{\gamma}_O - \vec{v}_O \wedge m\vec{v}_C - \vec{\rho}_O \wedge m\vec{\gamma}_C. \tag{5.12}$$

Taking account of both the theorem of mass centre motion (5.6) and the fundamental theorem of the kinetic moment (5.10) and considering the vector relations $\vec{\rho}_i = \vec{\rho}_O + \vec{r}_i$, $\vec{v}_C = \vec{v}_O + \vec{v}_{Cr}$, the above relation (5.12) becomes

$$\dot{\vec{k}}_O = \sum_i \vec{r}_i \wedge \vec{f}_i - \vec{r}_C \wedge m\vec{\gamma}_O. \tag{5.13}$$

If the point O and the mass centre C are two coincident points, Eq. (5.13) leads to the following useful expression of the theorem of kinetic moment for the relative motion of a system of particles or a *rigid body* with respect to a central translating frame:

$$\dot{\vec{k}}_C = \sum_i \vec{r}_i \wedge \vec{f}_i, \tag{5.14}$$

which express that the derivative of the kinetic moment about mass centre C of a system of particles or a *rigid body* is equal to the resultant torque about C of all external forces.

If the resultant torque about the mass centre of the external forces is zero, it follows from (5.14) tat the total kinetic moment of the mechanical system about C is *conserved*, $\vec{k}_C = const.$

5.5 Theorem of Kinetic Energy

Consider once again the system of particles A_i of masses m_i and velocities $\vec{v}_i = \dot{\vec{\rho}}_i$ (Fig. 4.7), subjected to a system of external forces $\vec{f}_i$ and internal forces $\vec{f}_{ij}$.

Differentiating with respect to time t the matrix expression $E = \frac{1}{2}\sum_i m_i \vec{v}_i^T \vec{v}_i$ of total kinetic energy for a system of particles (4.82), we obtain

$$\dot{E} = \frac{1}{2}\frac{d}{dt}\sum_i m_i \vec{v}_i^T \vec{v}_i = \sum_i m_i \dot{\vec{\rho}}_i^T \ddot{\vec{\rho}}_i. \tag{5.15}$$

Pre-multiplying each of the differential equations of fundamental system (5.2) by the line matrix of the velocity $\vec{v}_i{}^T = \dot{\vec{\rho}}_i^T$ and then adding these results, we get

$$\sum_i m_i \vec{v}_i^T \ddot{\vec{\rho}}_i = \sum_i \vec{v}_i^T \vec{f}_i + \sum_i \sum_j \vec{v}_i^T \vec{f}_{ij} = P_{ext} + P_{\text{int}}. \tag{5.16}$$

The lines (5.15) and (5.16) lead, through comparison, to the equation of motion

$$\dot{E} = P_{ext} + P_{\text{int}}. \tag{5.17}$$

Therefore, the derivative with respect to time of the total kinetic energy of a system of particles is equal to the sum of powers developed by all external and internal forces acting on the mechanical system.

Using the elementary total work of the external and internal forces (4.106), the last expression of the theorem can be written in a differential form

$$dE = dW_{ext} + dW_{\text{int}}. \tag{5.18}$$

Note that although the internal forces on adjacent particles A_i and A_j occur in an equal but opposite collinear pair, the total work done by each of these forces will, in general, not cancel out since the absolute velocities of any two particles will be different.

Integrating in time between the limits t' and t'', corresponding to a *finite* displacement of the system of particles between two successive positions, we obtain a *useful* form of the theorem of kinetic energy

$$E'' - E' = W_{ext} + W_{\text{int}}, \tag{5.19}$$

which states that the variation of total kinetic energy is equal to the work done by all external and internal forces acting on the particles of the system.

5.6 Conservation of Mechanical Energy

The *potential energy* V of a system of particles provides a measure of the work which can be done be all internal conservative forces acting on it.

When an isolated system of particles moves only under the action of internal conservative forces, the theorem of kinetic energy (5.18) can be modified in an integrated form

$$E_m = E + V = const. \tag{5.20}$$

which expresses precisely that the mechanical energy E_m as a sum of kinetic energy and of the potential energy of the system is *conserved.* This is the *law of conservation* of mechanical energy.

We note that while the weight of a particle and the elastic force exerted by a spring are two conservative forces, a *friction force* is a non-conservative force and the work done by this dissipative force cannot be expressed as a change in potential energy.

5.7 Theorem of Kinetic Energy with Respect to a Translating Frame

The general equation of motion (5.17) change when the fixed system of reference $\Omega\xi\eta\zeta$ is replaced by an arbitrary translating frame $Oxyz$, and the theorem of variation of kinetic energy will have the same general form as for the motion of the mechanical system relative to a central translating frame $Cx_Cy_Cz_C$ (Fig. 4.7).

Starting from the relation (4.91), now given for a *system of particles* $E = E_O + \frac{1}{2}m\,\vec{v}_O^T\vec{v}_O + m\,\vec{v}_{Cr}^T\,\vec{v}_O$, we express and afterwards will differentiate with respect to time the kinetic energy $E_O = \frac{1}{2}\sum_i m_i\vec{v}_{ir}^T\vec{v}_{ir}$ for the relative motion of the same system with respect to a translating frame $Oxyz$; it follows that

$$\dot{E}_O = \dot{E} - m\,\vec{v}_O^T\dot{\vec{v}}_O - m\vec{v}_O^T\dot{\vec{v}}_{Cr} - m\vec{v}_{Cr}^T\dot{\vec{v}}_O. \tag{5.21}$$

Taking account of both the theorem of kinetic energy (5.17) and the theorem of mass centre motion (5.6), and considering the vector relations $\vec{v}_i = \vec{v}_O + \vec{v}_{ir}$, $\vec{v}_C = \vec{v}_O + \vec{v}_{Cr}$, the above relation (5.21) becomes

$$\dot{E}_O = P_{O\text{ext}} + P_{O\text{int}} - m\vec{v}_{Cr}^T\vec{\gamma}_O, \tag{5.22}$$

where the following powers are developed by the external forces $\vec{f}_i$ and internal forces $\vec{f}_{ij}$, acting on the mechanical system of particles during its relative motion about the point O:

$$P_{Oext} = \sum_i \vec{v}_{ir}^T \vec{f}_i, \quad P_{Oint} = \sum_i \sum_j \vec{v}_{ir}^T \vec{f}_{ij}. \tag{5.23}$$

If the point O and the mass centre C are two coincident points, the above Eq. (5.22) leads to the theorem of kinetic energy for the relative motion of a system of particles or a *rigid body* with respect to a central translating frame:

$$\dot{E}_C = P_{Cext} + P_{Cint}. \tag{5.24}$$

5.8 Equations of Motion in Dynamics of the Rigid Body

The rigid body can be regarded as a collection of a very large number of particles of total mass m and a centre of mass at C, which has not contact with any other bodies (Fig. 4.9).

Subjected to several external forces $\vec{f}_i^\circ$, the body has a general three-dimensional motion in the space of the coordinate system $\Omega\xi\eta\varsigma$: *translation* along any direction and *rotation* about any instantaneous axis. The forces and moments acting on a rigid body can be due to the action of the springs, dampers or gravity among them. Conventionally, the forces $\vec{f}_i^\circ$ are replaced by an equivalent *wrench* consisting of one resultant force $\sum_i \vec{f}_i^\circ$ and one torque $\sum_i \tilde{\rho}_i \vec{f}_i^\circ$ as a sum of the moments of the individual forces with respect to fixed point Ω. The forces $\vec{f}_i^\circ = a^T \vec{f}_i$ and the components of the wrench must be transformed into the coordinate system in which the equations of motion are derived, using the orthogonal matrix of transformation a (2.8).

The fundamental theorems of kinetic impulse (5.5) and kinetic moment (5.10) for a system of particles will now be applied to the analysis of general motion of rigid body and of systems of rigid bodies.

Considering first the motion of the mass centre C of the rigid body with respect to fixed frame of reference $\Omega\xi\eta\zeta$, we recall the expression (4.58) of the kinetic impulse of a rigid body

$$\vec{h}^\circ = m\dot{\vec{\rho}}_C = a^T \vec{h} = a^T m \vec{v}_C = a^T m(\vec{v}_O + \tilde{\omega}\vec{r}_C). \tag{5.25}$$

and we write the matrix form for the theorem of mass centre motion into the usual frame $Oxyz$ (Fig. 4.9)

$$m(\dot{\tilde{\omega}} + \tilde{\omega}^2)\vec{r}_C = \sum_i \vec{f}_i - m(\dot{\vec{v}}_O + \tilde{\omega}\vec{v}_O) \tag{5.26}$$

Turning now to the motion of the body relative to its mass centre C, we recall the expression of the relative kinetic moment $\vec{k}_O^\circ = a^T\vec{k}_O = a^T\hat{J}_O\vec{\omega}$ in the fixed frame $\Omega\xi\eta\zeta$ and we compute the derivative with respect to time

$$\dot{\vec{k}}_O^\circ = a^T\tilde{\omega}\hat{J}_O\vec{\omega} + a^T\hat{J}_O\dot{\vec{\omega}}. \tag{5.27}$$

Using the projections on the axes of the usual moving frame $Oxyz$ (Fig. 4.9), the theorem of kinetic moment (5.13) for the relative motion about the point O of the body becomes

$$\hat{J}_O\dot{\vec{\omega}} + \tilde{\omega}\hat{J}_O\vec{\omega} = \sum_i \tilde{r}_i\vec{f}_i - m\tilde{r}_C(\dot{\vec{v}}_O + \tilde{\omega}\vec{v}_O) \tag{5.28}$$

If O and C are two coincident points ($\vec{r}_C = \vec{0}$), the theorem of mass centre motion (5.26) and the above matrix equation (5.28) become

$$\begin{aligned} m(\dot{\vec{v}}_C + \tilde{\omega}\vec{v}_C) &= \sum_i \vec{f}_i \\ \hat{J}_C\dot{\vec{\omega}} + \tilde{\omega}\hat{J}_C\vec{\omega} &= \sum_i \tilde{r}_i\vec{f}_i. \end{aligned} \tag{5.29}$$

This is the *fundamental Newton-Euler equations* that govern the motion of the rigid body. These equations can easily be used in problems involving the planar or spherical motions of several connected rigid bodies.

Fundamental theorems in the dynamics of rigid body (5.26) and (5.28) can be merged if these equations are pre-multiplied by the line matrices of the characteristic vectors $\vec{v}_O^T$ and $\vec{\omega}_O^T$, respectively, and afterwards are added to the results

$$m\vec{v}_O^T(\dot{\vec{v}}_O + \dot{\tilde{\omega}}\vec{r}_C) + \vec{\omega}^T(\hat{J}_O\dot{\vec{\omega}} + m\tilde{r}_C\dot{\vec{v}}_O) = \vec{v}_O^T\sum_i \vec{f}_i + \vec{\omega}^T\sum_i \tilde{r}_i\vec{f}_i. \tag{5.30}$$

This scalar equation represents precisely the theorem of kinetic energy for the general motion of a rigid body, which could be written in a compact form, using the notations (2.24), (4.47) and (4.108)

$$\vec{V}_O^T M_O\dot{\vec{V}}_O = \vec{V}_O^T\vec{M}_O. \tag{5.31}$$

If the points O and C are coincident ($\vec{r}_C = \vec{0}$), we have found the following simplified relation

$$m\vec{v}_C^T\dot{\vec{v}}_C + \vec{\omega}^T\hat{J}_C\dot{\vec{\omega}} = \vec{v}_C{}^T\sum_i \vec{f}_i + \vec{\omega}^T\sum_i \tilde{r}_i\vec{f}_i. \tag{5.32}$$

5.8.1 Planar Motion of a Rigid Body

If all links of a mechanical system undergo motion in one plane or in parallel planes, the system is said to be experiencing *planar motion*. The planar motion of a rigid body of mass m could be considered as a combination of a translation within a fixed reference plane $\Omega\xi\eta$ and a rotation about an axis perpendicular to that plane. The path of motion of any particle of the body is a plane curve parallel to the reference plane (Fig. 5.1).

It is necessary to define coordinates that specify at all times the location of the body: two absolute coordinates $\xi_C, \eta_C, \zeta_C = const.$ of the origin C of a central body-fixed coordinate system $Cxyz$ and one angle of rotation φ of this system relative to the inertial frame $\Omega\xi\eta\varsigma$, where the axes Cz and $\Omega\zeta$ remain permanently parallel.

Having three non-collinear points B_j of position vectors $\overrightarrow{CB_j} = \vec{r}_j^B = \begin{bmatrix} x_j^B & y_j^B & z_B^j \end{bmatrix}^T$ $(j = 1, 2, 3)$, which are moving in the plane $\Omega\xi\eta$, a non-symmetric body with respect to this plane is acted upon by external forces $\vec{f}_i$ in its points A_i of the position vectors $\vec{r}_i = \begin{bmatrix} x_i & y_i & z_i \end{bmatrix}^T$. Being constrained in three coplanar points, the body with two nonzero central products of inertia J_{xz} and J_{yz} would translate only in the plane and rotate only about Cz.

Since the planar motion may be viewed within the $\Omega\xi\eta\varsigma$ system of reference, all the active forces and the reactions $\vec{N}_j^B = \begin{bmatrix} 0 & 0 & N_j^B \end{bmatrix}^T$ can then be projected onto the axes of the moving frame $Cxyz$.

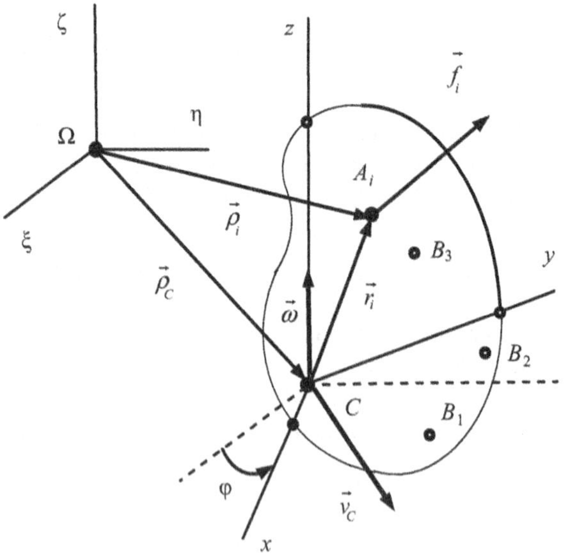

Fig. 5.1 Rigid body in a planar motion

The equations of motion for a constrained body can be simplified if the motion is planar. The translational equation of motion for the mass centre of the rigid body, given by (5.26), can be written as

$$m(\dot{\vec{v}}_C + \tilde{\omega}\vec{v}_C) = \sum_i \vec{f}_i + \sum_j \vec{N}_j^B, \tag{5.33}$$

where the vector of translational velocity $\vec{v}_C$, the angular velocity $\vec{\omega}$ of the body and the matrix of transformation a_φ (2.32) can be written, respectively, as

$$\vec{v}_C = \begin{bmatrix} v_{Cx} & v_{Cy} & 0 \end{bmatrix}^T, \quad \vec{\omega} = \begin{bmatrix} 0 & 0 & \dot{\varphi} \end{bmatrix}^T, \quad a_\varphi = rot(z, \varphi) \tag{5.34}$$

We will now determine the effects caused by the torque of external active forces and reactions computed about the moving mass centre C, using the theorem of kinetic moment (5.29)

$$\hat{J}_C\dot{\vec{\omega}} + \tilde{\omega}\hat{J}_C\vec{\omega} = \sum_i \tilde{r}_i \vec{f}_i + \sum_j \tilde{r}_j^B \vec{N}_j^B, \tag{5.35}$$

where the total resultant moment of forces has three nonzero components in the directions x, y and z.

Six independent scalar equations may be generated from (5.33) and (5.35), to describe the planar motion of the rigid body and evaluate the intensities N_j^B of the three reactions

$$\begin{aligned}
& m(\dot{v}_{Cx} - \dot{\varphi} v_{Cy}) = \sum_i X_i \\
& m(\dot{v}_{Cy} + \dot{\varphi} v_{Cx}) = \sum_i Y_i \\
& 0 = \sum_i Z_i + \sum_j N_j^B \\
& -J_{xz}\ddot{\varphi} + J_{yz}\dot{\varphi}^2 = \sum_i M_{xi} + \sum_j x_j^B N_j^B \\
& -J_{yz}\ddot{\varphi} - J_{xz}\dot{\varphi}^2 = \sum_i M_{yi} - \sum_j y_j^B N_j^B \\
& J_z\ddot{\varphi} = \sum_i M_{zi}
\end{aligned} \tag{5.36}$$

Using the following transformation $\vec{v}_C = a_\varphi[\dot{\xi}_C \quad \dot{\eta}_C \quad 0]^T$, three independent differential equations from (5.36) yield by integration three unknown variables $\xi_C(t)$, $\eta_C(i)$ and $\varphi(t)$.

5.8.2 Spherical Motion of a Rigid Body

The *spherical motion* accomplished by a rigid body with a fixed point O is equivalent to a rotation of this body with angular velocity $\vec{\omega}$ about an instantaneous axis which passes through O. The motion of the body is completely characterized by the motion of a sphere of centre O detached from the same body (Fig. 5.2).

To define the orientation of the body relative to the fixed $O\xi\eta\varsigma$ frame, a second system of reference $Oxyz$ will be attached to this body, so that their axes Ox, Oy and Oz are oriented in the direction of the principal rotating body's axes of inertia. It is more convenient to use this later moving frame, as this ensures that the tensor of inertia (4.10) and the axial moments of inertia of the body, computed in an $Oxyz$ frame, will maintain the same values over the course of the motion.

Denote by the vectors $\sum_i \vec{f}_i$ and $\sum_i \tilde{r}_i \vec{f}_i$, the resultant force and, respectively, the resultant torque of the external forces acting on the body and denote by $\vec{R}_O(X_O, Y_O, Z_O)$ the reaction acting in the point O, as the centre of a spherical joint.

Using the Euler angles ψ, θ, φ as orientation coordinates, the angular velocity $\vec{\omega}$, the angular acceleration $\dot{\vec{\omega}}$ and the velocity $\vec{v}_C = \tilde{\omega}\vec{r}_C$ of mass centre C can be expressed in term of these variables and their first and second derivatives. Written into the moving frame $Oxyz$, the rate of change of kinetic impulse (5.26) and kinetic

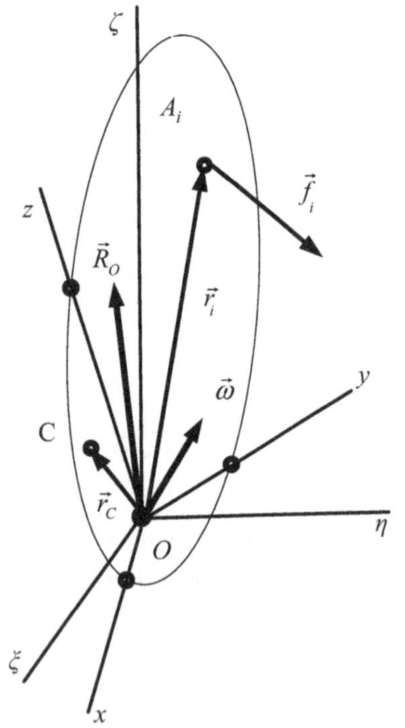

Fig. 5.2 Rigid body in a spherical motion

moment (5.29) leads to the *Newton-Euler formalism* of matrix equations of a rigid body, which is constrained to rotate about a fixed point O:

$$\begin{aligned} m(\dot{\tilde{\omega}} + \tilde{\omega}^2)\vec{r}_C &= \sum_i \vec{f}_i + \vec{R}_O \\ \hat{J}_O\dot{\vec{\omega}} + \tilde{\omega}\hat{J}_O\vec{\omega} &= \sum_i \tilde{r}_i\vec{f}_i, \end{aligned} \tag{5.37}$$

where the later equation will not contain the unknown reaction at O.

Starting from the second equation (5.37), we develop its components in the directions of the moving axes x, y and z, to obtain the following three scalar differential relations, called *Euler's equations*

$$\begin{aligned} J_x\dot{\omega}_x + (J_z - J_y)\omega_y\omega_z &= \sum_i M_{xi} \\ J_y\dot{\omega}_y + (J_x - J_z)\omega_z\omega_x &= \sum_i M_{yi} \\ J_z\dot{\omega}_z + (J_y - J_x)\omega_x\omega_y &= \sum_i M_{zi}. \end{aligned} \tag{5.38}$$

These equations can be analytically or numerically integrated for any initial conditions, in order to determine the joint velocities $\omega_x, \omega_y, \omega_z$ and to obtain three unknown functions $\psi(t), \theta(t), \varphi(t)$ [1]–[3].

If the external forces $\vec{f}_i$ are equipollent to a single resultant vector passing through the fixed point, all component of the resultant torque from the right hand of the relations (5.38) are *zero* and, consequently, the general matrix equations (5.28), (5.30) will lead to the *Euler-Poinsot case*, as

$$\hat{J}_O\dot{\vec{\omega}} + \tilde{\omega}\hat{J}_O\vec{\omega} = 0, \quad \vec{\omega}^T\hat{J}_O\dot{\vec{\omega}} = \frac{d}{dt}\left(\frac{1}{2}\vec{\omega}^T\hat{J}_O\vec{\omega}\right) = 0. \tag{5.39}$$

We remark, immediately, that pre-multiplying the first above relation by the line matrix $\vec{\omega}^T$, we obtain the second relation as an integral of three differential equations. A new integral of the same relations can be obtained if we pre-multiply the first matrix relation (5.39) by the line matrix $2\vec{\omega}^T\hat{J}_O$, as follows

$$2\vec{\omega}^T\hat{J}_O^2\dot{\vec{\omega}} = \frac{d}{dt}(\vec{\omega}^T\hat{J}_O^2\vec{\omega}) = 0. \tag{5.40}$$

Finally, we establish a new system of three independent equations containing the relations of the conservation of the kinetic energy E (5.39), the conservation of the intensity k_O of the kinetic moment (5.40) and the first differential equation selected from (5.38), for example:

$$
\begin{aligned}
&J_x\omega_x^2 + J_y\omega_y^2 + J_z\omega_z^2 = 2E \\
&J_x^2\omega_x^2 + J_y^2\omega_y^2 + J_z^2\omega_z^2 = k_O^2 \\
&J_x\dot{\omega}_x + (J_z - J_y)\omega_y\omega_x = 0,
\end{aligned}
\tag{5.41}
$$

where the components of angular velocity $\vec{\omega}$ are expressed in the body-fixed frame (2.48) in terms of Euler angles and their time derivatives

$$
\vec{\omega} = \begin{bmatrix} \cos\theta\cos\varphi & \sin\varphi & 0 \\ -\cos\theta\sin\varphi & \cos\varphi & 0 \\ \sin\theta & 0 & 1 \end{bmatrix} \begin{bmatrix} \dot{\psi} \\ \dot{\theta} \\ \dot{\varphi} \end{bmatrix}.
\tag{5.42}
$$

The system of Eq. (5.41) allows us to express ω_y and ω_z as functions of ω_x

$$
\omega_y = \sqrt{a^2 - b^2\omega_x^2}, \quad \omega_z = \sqrt{c^2 - d^2\omega_x^2}
\tag{5.43}
$$

with the notations

$$
\begin{aligned}
a^2 &= \frac{2EJ_z - k_O^2}{J_y(J_z - J_y)}, \quad b^2 = \frac{J_x(J_z - J_x)}{J_y(J_z - J_y)} \\
c^2 &= \frac{k_O^2 - 2EJ_y}{J_z(J_z - J_y)}, \quad d^2 = \frac{J_x(J_x - J_y)}{J_z(J_z - J_y)}.
\end{aligned}
\tag{5.44}
$$

These four expressions are all positive, if the moments of inertia satisfy the inequalities

$$
J_y < J_x < J_z, \quad J_y < \frac{k_O^2}{2E} < J_z.
\tag{5.45}
$$

Because the square roots of the terms from (5.44) will provide real values, the domain of the function ω_x is easily established as

$$
-\frac{c}{d} < \omega_x(t) < \frac{c}{d}.
\tag{5.46}
$$

From third relation (5.41) we obtain a differential equation

$$
\frac{d\omega_x}{\sqrt{(a^2 - b^2\omega_x^2)(c^2 - d^2\omega_x^2)}} = \frac{J_y - J_z}{J_x}dt.
\tag{5.47}
$$

Using following change of variable $\omega_x = \frac{c}{d}u$ and adding the notations

$$
k = \frac{bc}{ad}, \quad \lambda = \frac{J_y - J_z}{J_z}ad,
\tag{5.48}
$$

the integration of the differential equation (5.47) leads to *incomplete elliptic integrals* of first kind of module k

$$t = \frac{1}{\lambda} \int_{u_0}^{u} \frac{du}{\sqrt{(1-u^2)(1-k^2u^2)}} = F(\alpha, k) - F(\alpha_0, k), \tag{5.49}$$

where $\alpha = \arcsin\left(\frac{d}{c}\omega_x\right)$ and $\alpha_0 = \arcsin\left(\frac{d}{c}\omega_x^\circ\right)$ are two arguments. The function F can be obtained from tables or through numerical integration. The other two angular velocities can immediately be obtained from (5.43).

5.8.3 Spherical Motion of a Rigid Body Under Its Own Weight

Assume that the unique force acting on the rigid body is its own weight $G = mg$, applied along the vertical axis $O\zeta$ at the mass centre C of position vector $\vec{r}_C = \begin{bmatrix} x_C & y_C & z_C \end{bmatrix}^T$ (Fig. 5.3).

Using the matrix of transformation a (2.38), the theorem of kinetic moment (5.37) about the fixed point O can be written in a matrix form as

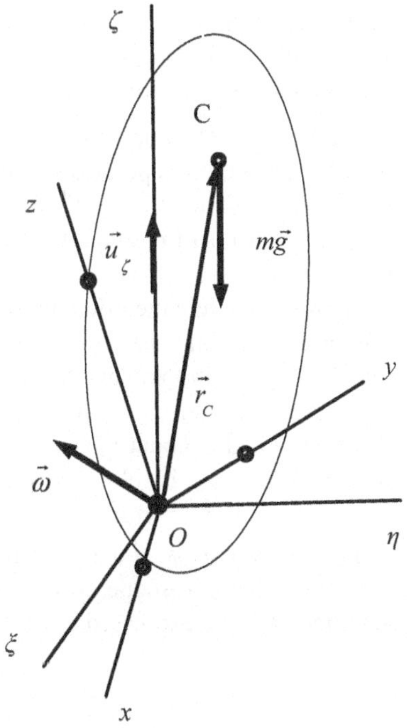

Fig. 5.3 Rigid body in a spherical motion under own weight

$$\hat{J}_O\dot{\vec{\omega}} + \tilde{\omega}\hat{J}_O\vec{\omega} = -mg\tilde{r}_C a\vec{u}_\zeta, \quad \vec{u}_\zeta = [0 \quad 0 \quad 1]^T. \tag{5.50}$$

Pre-multiplying (5.50) by $(au)^T$, we obtain a relation characterizing the *conservation* of the *kinetic moment* about vertical axis $O\zeta$:

$$\vec{u}_\zeta^T a^T \hat{J}_O\vec{\omega} = const. \tag{5.51}$$

If we pre-multiply by $\vec{\omega}^T$ the same relation (5.50), we express a new relation

$$\frac{d}{dt}(\vec{\omega}^T \hat{J}_O\vec{\omega}) = -2mg\vec{u}_\zeta^T a^T \tilde{\omega}\vec{r}_C, \tag{5.52}$$

from which we obtain the *low of conservation* of the *mechanical energy*

$$\vec{\omega}^T \hat{J}_O\vec{\omega} + 2mg\zeta_C = const., \quad \zeta_C = \vec{u}_\zeta^T a^T \vec{r}_C. \tag{5.53}$$

5.8.4 Gyroscope

A special rigid body of three degrees of freedom, commonly known in mechanics, is the *gyroscope*. When mounted in a Cardan suspension (Fig. 5.4), the gyroscope can assume any orientation, but its mass centre must remain fixed at O. The study of this device motion is one of the more interesting in spatial dynamics. The gyroscope is an axially symmetric rigid body, consisting of a rotor **1** that spins about its axis Oz, which is mounted on an inner circular ring **2**, so that the mass centre remains in a fixed position O, hence $\vec{\rho}_C = \vec{0}$. This first ring is free to rotate about another axis Ox' which is mounted perpendicular to Oz on an outer circular ring **3**, rotating freely about a fixed vertical axis $O\zeta$. Because of the symmetry of the rotor about its Oz axis, the products of inertia are equal to zero, while the axial moments of inertia are $J_x = J_y < J_z$.

In order to define the orientation of the rigid body, let us select the fixed frame of reference $O\xi\eta\zeta$ and the rotating system of axes $Oxyz$ attached to the gyroscope, with the Oz axis in the plane of inner circular ring **2**. The position of the rigid body at any given instant is defined by means of three successive rotations of Euler's angles ψ, θ and φ about the axes $O\zeta$, Ox' and Oz, respectively. The general matrix of transformation is obtained here through three basic matrices of rotation $R_\psi = rot(\zeta, \psi)$, $R_\theta = rot(x', \theta)$, $R_\varphi = rot(z, \varphi)$, as $R = R_\varphi R_\theta R_\psi$.

The derivatives $\dot{\psi}$, $\dot{\theta}$ and $\dot{\varphi}$ define respectively the rate of *precession*, *nutation* and *spin* of the gyroscope at the instant considered. The angular velocity $\vec{\omega}$ of the rigid body will be expressed as the sum of three angular velocities corresponding to

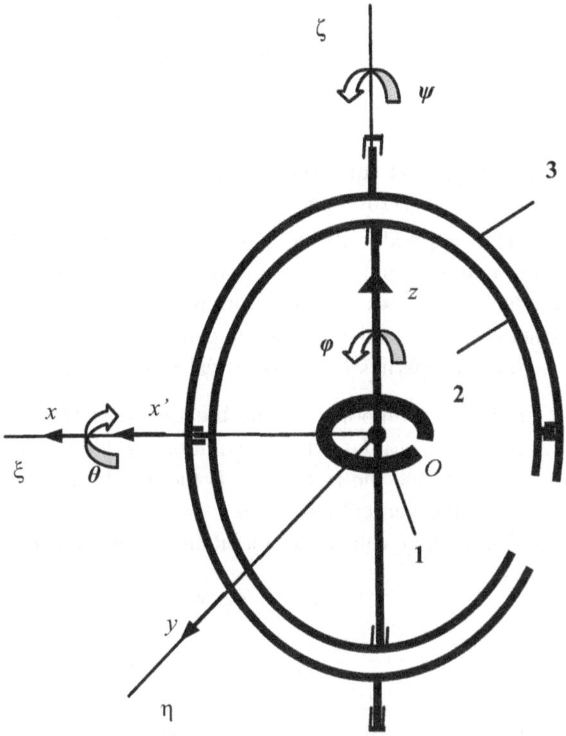

Fig. 5.4 Initial configuration of the gyroscope

three concurrent rotations, starting from the skew-symmetric matrix $\tilde{\omega} = R\dot{R}^T$, which is associated with the vector

$$\vec{\omega} = \dot{\psi} R_\varphi R_\theta \vec{u}_3 + \dot{\theta} R_\varphi \vec{u}_1 + \dot{\varphi} \vec{u}_3, \tag{5.54}$$

with its components evaluated in the $Oxyz$ mobile frame

$$\begin{aligned} \omega_x &= \dot{\psi} \sin\theta \sin\varphi + \dot{\theta} \cos\varphi \\ \omega_y &= \dot{\psi} \sin\theta \cos\varphi - \dot{\theta} \sin\varphi \\ \omega_z &= \dot{\psi} \cos\theta + \dot{\varphi} \end{aligned} \tag{5.55}$$

We will analyze the motion about its mass centre of the gyroscope under no force except its own weight, but with some initial conditions imposed on the angular velocities. The matrix equation $\hat{J}_O \dot{\vec{\omega}} + \tilde{\omega} \hat{J}_O \vec{\omega} = \vec{0}$ or the following analytical relations define the spherical motion of the gyroscope subjected to its weight when the mass of two circular rings is ignored

$$\begin{aligned} &J_x\dot{\omega}_x + (J_z - J_x)\omega_x\omega_z = 0 \\ &J_x\dot{\omega}_y - (J_z - J_x)\omega_x\omega_z = 0 \\ &J_z\dot{\omega}_z = 0. \end{aligned} \tag{5.56}$$

From the last equation is derived $\omega_z = \dot{\psi}\cos\theta + \dot{\varphi} = \omega_{z0} = const.$

When the gyroscope is mounted in gimbal rings, it becomes free of external moments applied to its base, and, consequently, its kinetic moment $\vec{k}_O$ maintain the same intensity and the same fixed orientation along the axis $O\zeta$ when the base is rotated. Two laws of conservation (5.41) relative to the kinetic moment and the kinetic energy of the gyroscope can be considered, if necessary.

Some properties of the spherical motion will be expressed.

1. From the component along the spin axis Oz of the kinetic moment we obtain $\cos\theta = J_z\omega_{z0}/k_O$, when the *nutation angle* θ formed by $O\zeta$ and Oz axes remains *constant* and the rigid body is moving in a special case of a *steady precession* about the vertical fixed axis.
2. The angular velocity $\vec{\omega}$ maintains a constant intensity ω since

$$J_x(\omega_x^2 + \omega_y^2) = 2E - J_z\omega_{z0}^2. \tag{5.57}$$

3. The angle δ formed by $\vec{\omega}$ and $O\zeta$ axis remains *constant* because $k_O\omega\cos\delta = 2E = const.$
4. Developing the expression of the kinetic moment $k_O = \vec{u}^T R^T \hat{J}_O\vec{\omega}$, $\vec{u} = [0 \quad 0 \quad 1]^T$, we remark that the rate of precession $\dot{\psi}$ and the rate of spin $\dot{\varphi}$ remain constant:

$$\dot{\psi} = \frac{k_O}{J_x} = const., \quad \dot{\varphi} = \left(\frac{1}{J_z} - \frac{1}{J_x}\right)k_O\cos\theta = const. \tag{5.58}$$

5. The motion of the gyro about its mass centre can be represented by the motion of a *body cone* called *polodic cone* which rolls without slipping on a fixed *space cone* called a *herpolodic cone* which is described with a constant angular velocity $\dot{\psi}$ by the *instantaneous axis of rotation*. In the case of *steady precession*, the two cones are circular, since the angles $\theta - \delta$ and δ that the vector $\vec{\omega}$ forms, respectively, with the axis of symmetry of the body and with the precession axis are constant.
6. From two analytical equations of motion (5.56) we obtain the differential equation

$$\ddot{\omega}_x + p^2\omega_x = 0, \quad p = \omega_{z0}\left(\frac{J_z}{J_x} - 1\right) \tag{5.59}$$

with the solutions

$$\begin{aligned} \omega_x &= \sqrt{\omega_{x0}^2 + \omega_{y0}^2}\sin(pt + \phi_0) \\ \omega_y &= -\sqrt{\omega_{x0}^2 + \omega_{y0}^2}\cos(pt + \phi_0), \quad \phi_0 = a\tan(-\omega_{x0}/\omega_{y0}) \end{aligned} \tag{5.60}$$

where ω_{x0}, ω_{y0} are two components of the angular velocity of perturbation.

For small perturbations ω_{x0}, ω_{y0}, the rotation axis of the gyroscope remains in the proximity of its initial direction and oscillates at the large frequency $v = 1/p$ in a *stable* motion. This property of stability is used for gyroscopes which must maintain the same orientation of their axes of rotation as was imposed at take-off, during the arbitrary motion of a ship or an aircraft.

5.8.5 Gyroscopic Moment

Consider a new gyroscope represented by a symmetrical rigid body of mass m which is centred at a fixed point O on its axis of symmetry Oz (Fig. 5.5). The mass centre of C the axis-symmetrical body is supposed to be on the Oz axis at a distance $OC = r_C$. We denote by J_z the moment of inertia of the rotor about the spin axis and by $J_x = J_y < J_z$ its moments of inertia about the transverse axes through O. The gyroscope spins about the axis of symmetry at an angular velocity which is considerably greater than the precession angular velocity during its rotation about a vertical fixed axis.

Fig. 5.5 Steady precession of the gyroscope

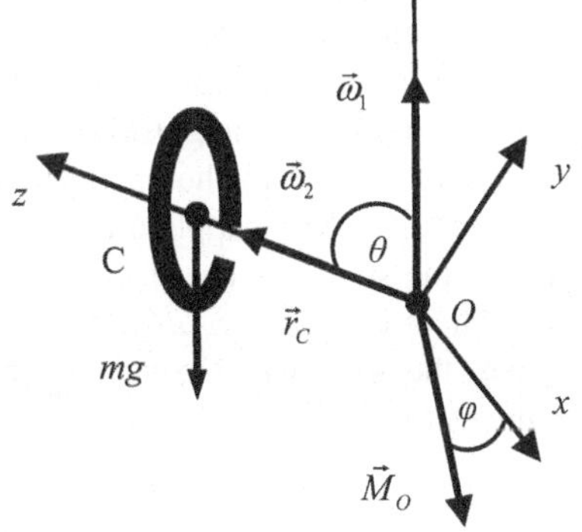

Knowing the weight $\vec{G} = m\vec{g}$ as external force applied to the rigid body, we will investigate the particular case of gyroscopic motion in a *steady precession* about the vertical fixed axis $O\zeta$, in which, for some initial conditions, the nutation angle θ, the rate of precession $\dot{\psi} = \omega_1$ and the rate of spin $\dot{\varphi} = \omega_2$ all remain constant.

We first note that the angular velocity $\vec{\omega}$ (5.55) of the rotor reduces to the components

$$\omega_x = \omega_1 \sin\theta \sin\varphi, \quad \omega_y = \omega_1 \sin\theta \cos\varphi, \quad \omega_z = \omega_1 \cos\theta + \omega_2. \tag{5.61}$$

Derivatives with respect to time of (5.61) give the components of angular acceleration in the $Oxyz$ mobile frame

$$\varepsilon_x = \omega_1\omega_2 \sin\theta \cos\varphi, \quad \varepsilon_y = -\omega_1\omega_2 \sin\theta \sin\varphi, \quad \varepsilon_z = 0. \tag{5.62}$$

Because of the symmetry of the rotor about the axis Oz axis the product of inertia is equal to zero. In this case, the equations of motion (5.38) reduce to

$$\begin{aligned} [J_z\omega_2 + (J_z - J_x)\omega_1 \cos\theta]\omega_1 \sin\theta \cos\varphi &= \sum M_{xi} \\ -[J_z\omega_2 + (J_z - J_x)\omega_1 \cos\theta]\omega_1 \sin\theta \sin\varphi &= \sum M_{yi} \\ 0 &= \sum M_{zi}, \end{aligned} \tag{5.63}$$

which yields the components of the moment $\vec{M}_O$ of the required external forces applied to the gyroscope

$$\begin{aligned} &\sum M_{xi} = M_O \cos\varphi, \quad \sum M_{yi} = -M_O \sin\varphi, \quad \sum M_{zi} = 0 \\ &\vec{M}_O = \left(1 + \frac{J_z - J_x}{J_z}\frac{\omega_1}{\omega_2}\cos\theta\right) J_z \vec{\omega}_1 \wedge \vec{\omega}_2. \end{aligned} \tag{5.64}$$

Three vectors $\vec{\omega}_1$, $\vec{\omega}_2$ and $\vec{M}_O$ act along their respective positive axes and therefore $\vec{M}_O$ is perpendicular to the plane of two angular velocities. The vector product of the angular velocities should be correctly chosen such that the *gyroscopic moment* $\vec{M}_g = -\vec{M}_O$ can counterbalance the moment $mg\tilde{r}_C^T R \vec{u}_\zeta$ (5.50) of the rotor's weight about O.

Since the mass centre of the rotor is fixed in the frame $Oxyz$, the forces which must be applied to the gyroscope to maintain its steady precession reduce to a pair of moment equal to the right hand of relation (5.64). We note that this pair of moments should be applied about an axis perpendicular to the precession axis $O\zeta$ and to the spin axis Oz of the gyroscope. If the rigid body rotates with very high angular velocity $\omega_2 \gg \omega_1$ or if we consider the particular case of interest $\theta = \pi/2$ where the precession axis and the spin axis are at right angle, then Eq. (5.64) reduces to

$$M_O = J_z|\vec{\omega}_1 \wedge \vec{\omega}_2| = J_z\omega_1\omega_2. \tag{5.65}$$

If we apply to the gyro a couple M_O about an horizontal axis perpendicular to its axis of spin, the body will precess about an axis perpendicular to both the spin axis and the couple axis in a sense such that the vectors representing the precession, the spin and the couple, respectively, form a right-handed triad. The *gyroscopic effect* must be taken into account, for example, in the design of bearings supporting rotating shafts subjected to forced precession.

References

1. Voinea, R., Stroe, I., Predoi, M.: Technical Mechanics. Edit. Politehnica Press, Bucharest (2010)
2. Staicu, S.: Mecanica teoretica. Edit. Didactica & Pedagogica, Bucharest (1998)
3. Wittenberg, J.: Dynamics of System of Rigid Bodies. B. G. Teubner, Stuttgart (1977)

Chapter 6
Analytical Mechanics

Starting from fundamental principles, Newtonian mechanics establish general theorems on the motion of a mass points system, computing all external and internal reaction forces. Founded by Lagrange, analytical mechanics introduce new methods, writing general equations of motion of constrained systems which can eliminate the reaction forces.

6.1 Principle of Virtual Work

A *possible* elementary displacement $\delta \vec{r}_i$ of a constrained particle under the action of an arbitrary fictitious force is called *virtual displacement*. This displacement is, moreover, *independent of time* and *consistent* with the kinematics constraints imposed by the motion of the mass point. Virtual displacements are *imaginary* in the sense that they are assumed to occur while the time is held fixed.

The virtual work of a force vector is defined to be the scalar product of virtual change of the position vector of the point of application of the force and the force vector.

Considering a system of particles under only smooth constraints (Fig. 4.7), a general expression of the *principle of virtual work* is as follows: for a mass points constrained system, without any internal or external friction, the total virtual work of the reaction forces, during any virtual displacement, is zero [1, 2].

Consider an imaginary or *virtual* general movement of a *rigid body* (Fig. 4.9), which indicates an instantaneous displacement $\delta \vec{\rho}_O$ accompanied by an instantaneous rotation $\delta\theta$ about the axis (Δ) of unit vector $\vec{u}$. Pursuing the matrix relations (4.107) and (4.110), the *virtual work* of an external force $\vec{f}_i$, which is located by $\vec{r}_i = \overrightarrow{\Omega A_i}$ in A_i and undergoes these virtual displacements, is a scalar quantity

S. Staicu, *Dynamics of Parallel Robots*, Parallel Robots: Theory and Applications,
https://doi.org/10.1007/978-3-319-99522-9_6

$$\delta W_i = \delta \vec{\rho}_O^T \vec{f}_i + \delta \theta \vec{u}^T \tilde{r}_i \vec{f}_i. \tag{6.1}$$

The *principle of virtual work* states that if a constrained system of connected rigid bodies T_j, which is in *equilibrium* under various external forces $\vec{f}_i^j$, is given an arbitrary *virtual displacement* from that the position of equilibrium, the total *virtual work* δW done by the forces during this displacement is *zero*, in the assumption of neglecting any external or in internal friction

$$\delta W = \sum_i \sum_j \delta \vec{\rho}_O^{jT} \vec{f}_i^j + \sum_i \sum_j \delta \theta_j \vec{u}^T \tilde{r}_i^j \vec{f}_i^j = 0. \tag{6.2}$$

This *necessary* condition is also *sufficient* for a system of rigid bodies whose connections are *frictionless*, which are subjected to a spatial force system: if the total virtual work done by the forces acting on the mechanical system is zero for any independent virtual displacement, the system is in equilibrium.

Application of the principle of virtual work leads to a number of equations equal to the number of system degrees of freedom, where the joint reaction forces are automatically eliminated.

For a system having h *degrees of freedom*, corresponding to h independent generalized coordinates, it is possible to write successively h independent scalar virtual work equations, such as one for every virtual displacement taken along the variation of each of generalized coordinates, when $h-1$ remaining independent coordinates are held fixed.

The principle of virtual work is generally used to determine the equilibrium configuration for a series of multiply connected rigid bodies and is particularly effective when applied to the solution of problems involving the equilibrium of machines and mechanisms consisting of several connected members.

The procedure of applying the principle of virtual work is called the *method of virtual displacements*.

In the study of problems of equilibrium or motion of certain mechanical systems, it is more advantageous to replace the determination of the virtual displacement $\delta \vec{r}_i$ by the computing of the instantaneous virtual velocity $\vec{v}_{iv}$ of the point A_i where the force $\vec{f}_i$ is acting, followed by the evaluation of the *virtual power* $P_{iv} = \vec{v}_{iv}^T \vec{f}_i$ of this force. Thus, for a constrained system of connected rigid bodies T_j acted upon by the external forces $\vec{f}_i^j$, we express the *principle of virtual power* or the *principle of virtual velocities* as follows

$$P_v = \sum_i \sum_j \vec{v}_{Ov}^{jT} \vec{f}_i^j + \sum_i \sum_j \vec{\omega}_v^{jT} \tilde{r}_i^j \vec{f}_i^j = 0, \tag{6.3}$$

or in the usual form (4.110), $P_v = \sum_j \vec{V}_{Ov}^{jT} \vec{M}_O^j = 0$, using the *twist* $\vec{V}_{Ov}^{jT} = \begin{bmatrix} \vec{v}_{Ov}^{jT} & \vec{\omega}_v^{jT} \end{bmatrix}$ of virtual velocity field and the *wrench* $\vec{M}_O^j$ (4.109) of vectors $\vec{f}_i^j$.

Configuration of equilibrium of a constrained mechanical system, without any external or internal friction, acted upon only by its own weight corresponds to a maximum position, a minimum position or a stationary position of the mass centre C of the system, $\delta\zeta_C = 0$, where ζ_C is the altitude of the point C with respect to a fixed frame $\Omega\xi\eta\zeta$ (Fig. 4.7). This particular form of expressing the condition of equilibrium is called *Torricelli's principle*.

6.2 D'Alembert's Principle

Under the action of externally applied forces $\vec{f}_i$, the particles A_i of masses m_i within a constrained mechanical system receive the accelerations $\vec{\gamma}_i$ (Fig. 4.7).

The study of the dynamics of the particles A_i is based on the *fundamental system of differential equations* (5.2)

$$m_i\vec{\gamma}_i = \vec{f}_i + \vec{R}_i \quad (i, j = 1, 2, \ldots, n;\ i<j), \tag{6.4}$$

where $\vec{R}_i$ is the resultant of all internal and external forces acting upon A_i.

Writing these relations in a new equivalent form

$$\vec{R}_i + \vec{f}_i + \vec{f}_i^{in} = \vec{0}, \tag{6.5}$$

the following vectors are considered as an additional force applied on the particles A_i to maintain the equilibrium of the system

$$\vec{f}_i^{in} = -m_i\vec{\gamma}_i. \tag{6.6}$$

This force which is *fictitiously* added to the system of internal and external forces, is called *inertia force*. Dynamically, inertia is the property of a body that causes it to resist any effort to change its motion.

If we apply the *inertia forces* $(-m_i\vec{\gamma}_i)$ on a constrained mechanical system subjected to external forces $\vec{f}_i$, then the system will be in *equilibrium*. These new conditions stating a *dynamic equilibrium* constitute the *d'Alembert principle*. Consequently, the wrench of the external actions and reactions acting on a rigid body and the wrench of inertia forces are two vector systems in equilibrium. The study of dynamics is reduced to the static equilibrium.

The principle of virtual power can easily be generalized to study the dynamics of a constrained mechanical system of particles, without any internal or external friction, using the conditions of equilibrium implemented by d'Alembert's principle in a mathematical statement, which contain the total virtual power developed by all active forces and inertia forces

$$P_v = \sum_i \vec{v}_{iv}^T \vec{f}_i + \sum_i \vec{v}_{iv}^T(-m_i\vec{\gamma}_i) = 0. \tag{6.7}$$

6.2.1 Wrench of Inertia Forces

Inertia forces, in general, depend on the mass and shape of the body as well as its velocity and acceleration. The equipollency of inertia forces of a system of particles (Fig. 4.7) of mass m can be treated in the same way as the externally-applied forces.

Any system of inertia forces acting on a mass system can be replaced by an equipollent of two vectors that consists of one resultant vector $\vec{f}_{in}$, called *inertia force*, and a resultant moment $\vec{m}_{\Omega in}$ of inertia forces with respect to the origin Ω of the fixed frame $\Omega\xi\eta\zeta$, called *couple of inertia,* as follows

$$\begin{aligned} \vec{f}_{in} &= \sum_i (-m_i\vec{\gamma}_i) = -\dot{\vec{h}} = -m\vec{\gamma}_C \\ \vec{m}_{\Omega in} &= \sum_i \vec{\rho}_i \wedge (-m_i\vec{\gamma}_i) = -\dot{\vec{k}}_\Omega, \end{aligned} \tag{6.8}$$

where $\vec{h}$ and $\vec{k}_\Omega$ represent respectively the kinetic impulse and the kinetic moment of the system of particles.

6.2.2 Inertia Forces Applied to a Rigid Body

Study of the equipollency of inertia forces in the case of the rigid body need to consider the type of motion accomplished by this body [3].

1. ***Translation***. If a rigid body of mass m (Fig. 2.2), is subjected to either rectilinear or curvilinear translation of acceleration $\vec{\gamma}_C$, the inertia forces vector field is represented by one *inertia force* $\vec{f}_{in} = -m\vec{\gamma}_C$, acting in the mass centre C.
2. ***Rotation about a fixed point***. Starting from (4.80), in the case of spherical motion of a rigid body of mass m and tensor of inertia $\hat{J}_O$ (Fig. 5.2), a dual vector with respect to fixed point O for the inertia forces is computed

$$\vec{M}^\circ_{Oin} = -\dot{\vec{K}}^\circ_O = -\frac{d}{dt}(A^T M_O \vec{V}_O). \tag{6.9}$$

Rearranging, we obtain the *wrench* of inertia forces in the local body-frame

$$\vec{M}_{Oin} = A\vec{M}^\circ_{Oin} = -M_O\dot{\vec{V}}_O - \tilde{\Omega}M_O\vec{V}_O, \tag{6.10}$$

with its components

$$\begin{aligned} \vec{f}_{in} &= a\vec{f}^\circ_{in} = -m(\tilde{\omega}^2 + \dot{\tilde{\omega}})\vec{r}_C \\ \vec{m}_{Oin} &= a\,\vec{m}^\circ_{Oin} = -\hat{J}_O\dot{\vec{\omega}} - \tilde{\omega}\hat{J}_O\vec{\omega}. \end{aligned} \tag{6.11}$$

When a rigid body is rotating about a *fixed axis* Δ (Fig. 4.8), passing through the point O, we can consider the same wrench of inertia forces of above components (6.11).

3. ***Planar motion***. To simplify the calculus, consider that the point O and the centre C are coincident and the Cz axis is perpendicular to the reference plane, if the body of mass m and central tensor of inertia $\hat{J}_C$ is subjected to a general planar motion (Fig. 5.1). Starting from (6.9), the *wrench* of inertia forces about fixed point Ω can be determined

$$\begin{bmatrix} \vec{f}^{\circ}_{in} \\ \vec{m}^{\circ}_{\Omega in} \end{bmatrix} = -\dot{\vec{K}}^{\circ}_{\Omega} = -\frac{d}{dt}(A^T_{\varphi}\vec{K}_{\Omega}). \tag{6.12}$$

Replacing the vector $\vec{K}_{\Omega}$ (4.81) and rearranging, we obtain the components

$$\vec{f}_{in} = a_{\varphi}\vec{f}^{\circ}_{in} = -m(\dot{\vec{v}}_C + \tilde{\omega}\,\vec{v}_C)$$

$$\vec{m}_{\Omega in} = a_{\varphi}\vec{m}^{\circ}_{\Omega in} = a_{\varphi}\tilde{\rho}_C a^T_{\varphi}\vec{f}_{in} - \hat{J}_C\dot{\vec{\omega}} - \tilde{\omega}\,\hat{J}_C\vec{\omega}. \tag{6.13}$$

Using a known relation of transfer $\vec{m}_{\Omega in} = m_{Cin} + a_{\varphi}\tilde{\rho}_C a^T_{\varphi}\vec{f}_{in}$, we can express the matrix wrench of inertia forces with respect to mass centre $\vec{M}_{Cin} = -M_C\dot{\vec{V}}_C - \tilde{\Omega}M_C\vec{V}_C$ from which, finally, we obtain two useful components

$$\vec{f}_{in} = -m(\dot{\vec{v}}_C + \tilde{\omega}\,\vec{v}_C), \quad \vec{m}_{Cin} = -\hat{J}_C\dot{\vec{\omega}} - \tilde{\omega}\hat{J}_C\vec{\omega}. \tag{6.14}$$

The above relations (6.14) and (6.15) can be *formally* used to establish the wrench of inertia forces with respect to mass centre C of a rigid body having a *general motion.*

If the points C and O are different, the wrench of inertia forces with respect to the mobile point O is expressed in the local body-frame

$$\vec{f}_{in} = -m\{\vec{\gamma}_O + (\tilde{\omega}^2 + \dot{\tilde{\omega}})\vec{r}_C\}, \quad \vec{m}_{Oin} = -\{m\tilde{r}_C\vec{\gamma}_O + \hat{J}_O\dot{\vec{\omega}} + \tilde{\omega}\hat{J}_O\vec{\omega}\}. \tag{6.15}$$

Remark

Let a multi-body mechanical system, that consists of T_j bodies of masses m_j and central tensors of inertia $\hat{J}^j_C$, be acted upon by a system of external forces $\vec{f}^j_i$ and subjected to internal and external constraints, without any friction. The virtual power developed by the reaction joint forces, arising as the result of the connectivity between different bodies, is zero. Taking into account of inertia forces, the conditions of the *dynamics equilibrium* for a moving system of bodies can be developed using the principle of virtual power, as follows:

$$P_v = \sum_j \vec{V}^{jT}_{Cv}\vec{M}^j_C + \sum_j \vec{V}^{jT}_{Cv}\vec{M}^j_{Cin} = 0, \tag{6.16}$$

where two wrenches $\vec{M}^j_C$ and $\vec{M}^j_{Cin}$ of externally-applied forces and inertia forces, respectively, such as the twist of virtual velocity field $\vec{V}^{jT}_{Cv} = \begin{bmatrix} \vec{v}^{jT}_{Cv} & \vec{\omega}^{jT}_v \end{bmatrix}$, are computed in the mass centre C_j. Practically, the compact matrix equation (6.16) can be expanded in an equivalent form to the virtual work of the forces $\vec{f}^j_i$ and the inertia forces, as

$$\begin{aligned} &\sum_i \sum_j \vec{v}^{jT}_{Cv}\vec{f}^j_i + \sum_i \sum_j \vec{\omega}^{jT}_v \tilde{r}^j_i \vec{f}^j_i \\ &\quad - \sum_j m_j \vec{v}^{jT}_{Cv}(\dot{\vec{v}}^j_C + \tilde{\omega}_j \vec{v}^j_C) - \sum_j \vec{\omega}^{jT}_v (\hat{J}^j_C \dot{\vec{\omega}}_j + \tilde{\omega}_j \hat{J}^j_C \vec{\omega}_j) = 0 \end{aligned} \tag{6.17}$$

6.3 Lagrange Equations

A synthesis of the principle of virtual work and d'Alembert's principle allows us to formulate the dynamics equations of a mechanical system using any set of h independent scalar parameters.

Lagrange's equations are applicable to mechanical systems with either holonomic or non-holonomic constraints. The constraint equations and their first derivatives must be factored into the equations of motion to produce a number of equations that accords with the number of unknowns. In this regard, these equations are more suitable for modelling the dynamics of parallel manipulators, which are complicated in cases where multiple closed-loop chains exist.

A holonomic system of n interconnected particles, without any external friction, is now considered. The global position vector $\vec{r}_i$ of the arbitrary particle A_i, of mass m_i, can be written in terms of *generalized coordinates* $q_1, q_2, \ldots, q_k, \ldots, q_h$, as

$$\vec{r}_i = \vec{r}_i[q_1(t), q_2(t), \ldots, q_k(t), \ldots, q_h(t)]. \tag{6.18}$$

If the system of particles, under action of external and internal forces $\vec{f}_i$ and also of inertia forces $(-m_i\vec{\gamma}_i)$ is in equilibrium, the total virtual work of both systems of forces is zero

$$\sum_i \vec{f}_i \delta\vec{r}_i + \sum_i (-m_i\vec{\gamma}_i)\delta\vec{r}_i = 0, \tag{6.19}$$

where $\delta\vec{r}_i$ is the virtual displacement $\delta\vec{r}_i = \sum_k \frac{\partial \vec{r}_i}{\partial q_k}\delta q_k$.

Substituting $\delta\vec{r}_i$ in (6.19), the virtual work of forces $\vec{f}_i$ can be written as

$$\sum_i \vec{f}_i \delta \vec{r}_i = \sum_k \left(\sum_i \vec{f}_i \frac{\partial \vec{r}_i}{\partial q_k} \right) \delta q_k = \sum_k \Phi_k \delta q_k, \tag{6.20}$$

where

$$\Phi_k = \sum_i \vec{f}_i \frac{\partial \vec{r}_i}{\partial q_k} \quad (k = 1, 2, \ldots, h) \tag{6.21}$$

is the vector of *generalized forces* associated with the system of generalized coordinates q_k.

In the body of virtual work of inertia forces $\sum_k \left(\sum_i -m_i \vec{\gamma}_i \frac{\partial \vec{r}_i}{\partial q_k} \right) \delta q_k$, we develop successively the following term

$$\sum_i -m_i \vec{\gamma}_i \frac{\partial \vec{r}_i}{\partial q_k} = -\frac{d}{dt} \sum_i m_i \vec{v}_i \frac{\partial \vec{r}_i}{\partial q_k} + \sum_i m_i \vec{v}_i \frac{d}{dt} \left(\frac{\partial \vec{r}_i}{\partial q_k} \right), \tag{6.22}$$

where the velocities $\vec{v}_i$ of A_i will be obtained by differentiating the position vector (6.18) with respect to time

$$\vec{v}_i = \sum_k \frac{\partial \vec{r}_i}{\partial q_k} \dot{q}_k + \frac{\partial \vec{r}_i}{\partial t}. \tag{6.23}$$

Considering that $\vec{v}_i$ is a vector function of $2h$ independent variables—generalized coordinates $\vec{q} = (q_1, q_2, \ldots, q_h)$ and *generalized velocities* $\dot{\vec{q}} = (\dot{q}_1, \dot{q}_2, \ldots, \dot{q}_h)$—we can compute the following derivative

$$\frac{\partial \vec{v}_i}{\partial q_k} = \sum_{j=1}^{k} \frac{\partial^2 \vec{r}_i}{\partial q_j \partial q_k} \dot{q}_j + \frac{\partial^2 \vec{r}_i}{\partial t \partial q_k}, \quad \frac{\partial \vec{v}_i}{\partial \dot{q}_k} = \frac{\partial \vec{r}_i}{\partial q_k}. \tag{6.24}$$

A comparison of (6.24) and the following derivative with respect to time

$$\frac{d}{dt} \left(\frac{\partial \vec{r}_i}{\partial q_k} \right) = \sum_{j=1}^{k} \frac{\partial^2 \vec{r}_i}{\partial q_k \partial q_j} \dot{q}_j + \frac{\partial^2 \vec{r}_i}{\partial q_k \partial t}, \tag{6.25}$$

allows us to establish a significant identity

$$\frac{d}{dt} \left(\frac{\partial \vec{r}_i}{\partial q_k} \right) = \frac{\partial \vec{v}_i}{\partial q_k}. \tag{6.26}$$

Replacing (6.24) and (6.26) in (6.22), we obtain

$$\sum_i -m_i \vec{\gamma}_i \frac{\partial \vec{r}_i}{\partial q_k} = -\frac{d}{dt}\left(\frac{\partial E}{\partial \dot{q}_k}\right) + \frac{\partial E}{\partial q_k}, \tag{6.27}$$

where E is precisely the system's kinetic energy, as defined by

$$E(q_1, q_3, \ldots, q_k, \ldots, q_h, \dot{q}_1, \dot{q}_2, \ldots, \dot{q}_h, t) = \frac{1}{2}\sum_i m_i \vec{v}_i^2 . \tag{6.28}$$

Using the fundamental equations (6.19), one concludes that the system equations of motion can, finally, by written

$$\frac{d}{dt}\left(\frac{\partial E}{\partial \dot{q}_k}\right) - \frac{\partial E}{\partial q_k} = \Phi_k \quad (k = 1, 2, \ldots, h). \tag{6.29}$$

These differential relations are called *Lagrange's equations* of the second kind for holonomic mechanical systems.

6.3.1 Conservative System of Forces

If all external and internal forces are *conservative*, their vectors $\vec{f}_i = \vec{\nabla} U_i$ are expressed by the derivatives of the potential functions U_i, called *force functions* and, consequently, the generalized forces become

$$\Phi_k = \sum_i \frac{\partial \vec{r}_i}{\partial q_k} \vec{\nabla} U_i = \frac{\partial U}{\partial q_k}, \tag{6.30}$$

where the potential function $U(q_1, q_2, \ldots, q_h) = \sum_i U_i$, which does not depend on the generalized velocities, is the *total force function*.

Denoting by $L = E + U$ the *Lagrange kinetic function* or *Lagrangian*, above Eq. (6.29) assume the definitive form

$$\frac{d}{dt}\left(\frac{\partial L}{\partial \dot{q}_k}\right) - \frac{\partial L}{\partial q_k} = 0 \quad (k = 1, 2, \ldots, h). \tag{6.31}$$

6.3.2 Non-holonomic Mechanical Systems

If the system of particles is acted upon by conservative forces, under v *non-holonomic* constraints, then the virtual displacements $\delta q_k (k = 1, 2, \ldots, h)$ are replaced by the virtual velocities $\dot{q}_k^v (k = 1, 2, \ldots, h)$.

Based on the kinematics relations of constraint, we obtain the following matrix equation on the virtual velocities

$$C_{vh}\dot{\vec{q}}_v = 0, \tag{6.32}$$

where

$$C_{vh} = \begin{bmatrix} c_{11} & c_{12} & \dots & c_{1h} \\ c_{21} & c_{22} & \dots & c_{2h} \\ \dots & \dots & \dots & \dots \\ c_{v1} & c_{v2} & \dots & c_{vh} \end{bmatrix}, \quad \dot{\vec{q}}_v = \begin{bmatrix} \dot{q}_1^v \\ \dot{q}_2^v \\ \dots \\ \dot{q}_h^v \end{bmatrix}. \tag{6.33}$$

Using the vector $\vec{\lambda} = [\lambda_1 \quad \lambda_2 \quad \dots \quad \lambda_s \quad \dots \quad \lambda_v]^T$ for the Lagrange multipliers, a new form of Lagrange's equations is established:

$$\frac{d}{dt}\left(\frac{\partial L}{\partial \dot{q}_k}\right) - \frac{\partial L}{\partial q_k} = \sum_{s=1}^{v} \lambda_s c_{sk} \quad (k = 1, 2, \dots, h). \tag{6.34}$$

These relations are added at the constraint conditions (6.32) to determine a system of $h + v$ differential equations, from which one obtains, successively, the multipliers λ_s and the generalized coordinates q_k.

6.3.3 Laws of Conservation in Analytical Mechanics

A generalized coordinate that does not appear in the Lagrangian L, during an interval of time, is called *cyclic* or ignorable. For a cyclic coordinate q_l we have $\frac{\partial L}{\partial q_l} = 0$ and, consequently, its corresponding Lagrange equation from (6.31) can be replaced by a differential equation equivalent to the relation

$$p_l = \frac{\partial L}{\partial \dot{q}_l} = \frac{\partial E}{\partial \dot{q}_l} = const. \tag{6.35}$$

This result implies that the *generalized kinetic impulse* p_l associated with the cyclic coordinate q_l is *conserved.* This generalized kinetic impulse could be a linear momentum or an angular momentum.

Consider that all the forces acting on the mechanical system are conservative and that the constraints are holonomic and scleronomic, thus $\frac{\partial L}{\partial t} = 0$. We will transform, successively, the Lagrange's equations (6.31), multiplying each of these relations by $\dot{q}_k$ and afterwards adding the results, as

$$\sum_{k=1}^{h} \dot{q}_k \frac{d}{dt}\left(\frac{\partial L}{\partial \dot{q}_k}\right) - \sum_{k=1}^{h} \dot{q}_k \frac{\partial L}{\partial q_k} = \frac{d}{dt}\left(\sum_{k=1}^{h} \dot{q}_k \frac{\partial L}{\partial \dot{q}_k} - L\right) = 0. \tag{6.36}$$

This implies that in the case of conservative system, the scalar function H is a constant of the system

$$H(q_1, q_2, \ldots, q_h, p_1, p_2, \ldots, p_h, t) = \sum_{k=1}^{h} p_k \dot{q}_k - L = const. \tag{6.37}$$

The analytical function H thus defined is called *Hamilton's function* or *Hamiltonian*.

We also note that since $\sum_{k=1}^{h} \dot{q}_k \frac{\partial L}{\partial \dot{q}_k} = 2E$, the Hamiltonian H takes the following form

$$H = 2E - (E + U) = E + V = const., \tag{6.38}$$

which implies that, for a conservative system, the Hamiltonian's function is precisely the mechanical energy, being the sum of the kinetic energy and potential energy of the system, and it remain constant throughout the period of system motion. This represents the *law* of *conservation* of *mechanical energy*.

6.4 Canonical Hamiltonian Equations

If a mechanical system has h degrees of freedom, Lagrange's equations define h second-order differential relations expressed in terms of the generalized coordinates and their derivatives [2].

The Hamiltonian formulation leads to a set of $2h$ first-order differential relations defined in terms of generalized coordinates q_k and generalized kinetic impulse p_k as defined above (6.35).

Written in a simplified form

$$\dot{p}_k = \frac{\partial L}{\partial q_k} \quad (k = 1, 2, \ldots, h), \tag{6.39}$$

Lagrange's equations (6.31) can be considered as a linear system of relations in $\dot{q}_1, \dot{q}_2, \ldots, \dot{q}_h$, giving the following solutions

$$\dot{q}_l = \dot{q}_l(q_1, q_2, \ldots, q_h, p_1, p_2, \ldots, p_h, t) \ldots (l = 1, 2, \ldots, h) \tag{6.40}$$

Moreover, two significant analytical functions, the Lagrangian L and the Hamiltonian H, can be expressed as two functions of q_k, p_k, $(k = 1, 2, \ldots, l)$ and the time t.

Taking into account of (6.37), from the following derivatives of H with respect to q_k and p_k

$$\begin{aligned}\frac{\partial H}{\partial q_k} &= \sum_{l=1}^{k} p_l \frac{\partial \dot{q}_l}{\partial q_k} - \frac{\partial L}{\partial q_k} - \sum_{l=1}^{h} \frac{\partial L}{\partial \dot{q}_l}\frac{\partial \dot{q}_l}{\partial q_k} \\ \frac{\partial H}{\partial p_k} &= \dot{q}_k + \sum_{l=1}^{k} p_l \frac{\partial \dot{q}_l}{\partial p_k} - \sum_{l=1}^{h} \frac{\partial L}{\partial \dot{q}_l}\frac{\partial \dot{q}_l}{\partial p_k},\end{aligned} \tag{6.41}$$

we have

$$\frac{\partial H}{\partial p_k} = \dot{q}_k, \quad \frac{\partial H}{\partial q_k} = -\dot{p}_k \quad (k = 1, 2, \ldots, h). \tag{6.42}$$

Replacing the h second-order differential equations derived from Lagrange's equations, new $2h$ first-order differential relations (6.42) are called *canonical Hamiltonian equations*.

The principle of virtual work, Lagrange's equations and the canonical Hamiltonian equations can be used to define the same set of differential equations. The answer to the question of which method is best for formulating the dynamics equations depends on the application.

References

1. Shabana, A.: Computational Dynamics. Wiley, New York (1994)
2. Voinea, R., Stroe, I., Predoi, M.: Technical Mechanics. Edit. Politehnica Press, Bucharest (2010)
3. Staicu, S.: Mecanica Teoretica. Edit. Didactica & Pedagogica, Bucharest (1998)

Chapter 7
Dynamics of Constrained Robotic Systems

Serial manipulators are mechanisms which consist of a series of single-DOF active revolute or prismatic joints connecting the fixed base to the end-effectors. These robots have good operating characteristics, such as a large workspace and high flexibility, but have some disadvantages, such as low precision, low stiffness and high vibrations and deflections.

To overcome the aforementioned drawbacks of the serial architectures, there have been considerable developments in the field of manipulating structures, now known as the *parallel manipulators*. Constrained systems differ from free systems in that the freedom of motion of one or more position variables is limited by mechanical constraints. Parallel robots are closed chains consisting of a fixed base and a moving platform that are connected to each other by a set of serial legs. These legs, which typically have the same kinematical structure, are connected to both the platforms at points that are distributed in a geometrically symmetric fashion. In engineering, holonomic constraints are achieved by means of inflexible guides, joints, levers, bearings, rods and other connections. For an industrial robot, for example, each degree of freedom is normally assigned one rigid body and a drive motor.

As the mechanical complexity of new robotic devices increases, it is becoming more important to have good formal procedures for representing kinematics and dynamics relationships. Based on the matrix mechanics of a rigid body, recursive relations about the kinematics and the dynamics of the constrained robotic systems schematized by kinematical chains are developed in this chapter. By using a set of successive mobile frames, we will first analyze the geometrical properties and the kinematics of the vector system of velocities and accelerations for each element of the robot. Expressed for every independent loop of a parallel mechanism, useful *conditions of connectivity* regarding the relative velocities and the relative accelerations will be determined for the direct or the inverse kinematics problem.

With known positions, velocities and accelerations for a given trajectory of the end-effectors, the required actuation forces or torques, the internal reaction joint forces and, finally, the dimensions of the robot can easily be determined by solving

S. Staicu, *Dynamics of Parallel Robots*, Parallel Robots: Theory and Applications,
https://doi.org/10.1007/978-3-319-99522-9_7

the inverse dynamic problem. Dynamics modelling of parallel robots has been a subject of great research interest, because the dynamics models are also needed for control purposes. The dynamics problem is solved in two efficient ways: using an approach based on the general Principle of virtual work and sometimes applying sometimes the formalism of Lagrange equations of second kind. Final matrix relations written in a recursive compact form express only the *explicit dynamics equations* of the constrained robotic systems. When establishing the expressions for the active forces and the active torques in an inverse dynamics problem, these equations are in fact useful for real-time control of the robot's evolution.

This methodology might be serviceable for the kinematics analysis and the nonlinear dynamics of a multi-body system consisting of interconnected rigid or, possibly, deformable bodies. Examples of mechanical structures that can be modeled as constrained robotic systems are: mechanisms, planar and spatial parallel robots, mobile robots, parallel kinematics machines, gear trains for robotics and robotic hybrid architectures.

7.1 Geometric Model of the Robot

Most robotic control commands are simply executed in the joint space, while the robotic motions are specified in the task space. Therefore, first it is necessary to solve the inverse kinematics problem in order to find the corresponding sets of joint angles, given the desired position of the end-effectors. In order to define the motion characteristics of the robots, it is necessary to represent first the $3D$ pose of the moving platform with respect to a fixed coordinate frame. This naturally leads to the need for a systematic representation of the position and location of all rigid bodies in space. Relations between the geometrical parameters of the robot and the final motion of the moving platform will be derived and analyzed.

The study of parallel robots leads to a great number of loops but also to difficulty in modelling and simulation, since the model is highly dependent on the architecture and there is no general method to determine the models as there is for serial manipulators.

Generally, a serial leg of a robot consists of a base link T_0 and n moving links connected by N joints. The kinematic joints constrain the virtual position of the mechanism and transmit the relative motion from one body to another. To describe the geometry of the links, starting from T_0, we number the movable links sequentially from T_1 to T_n and the joints from O_1 to O_N. Thus, apart from the base link, each link has two joints; the link T_k has the joint O_k at its proximal end and another joint O_{k+1} at the distal end.

Defined as the *reference frame*, a fixed frame $O_0x_0y_0z_0$ is attached to the base (T_0) at a convenient location. We suppose that the moving bodies $T_k(k = 1, 2, \ldots, n)$ are permanently attached to the set of mobile frames $O_kx_ky_kz_k$. In order to illustrate the development of recursive kinematics relations, we consider an arbitrary body T_k which is connected to the adjoined body T_{k-1} and which has a

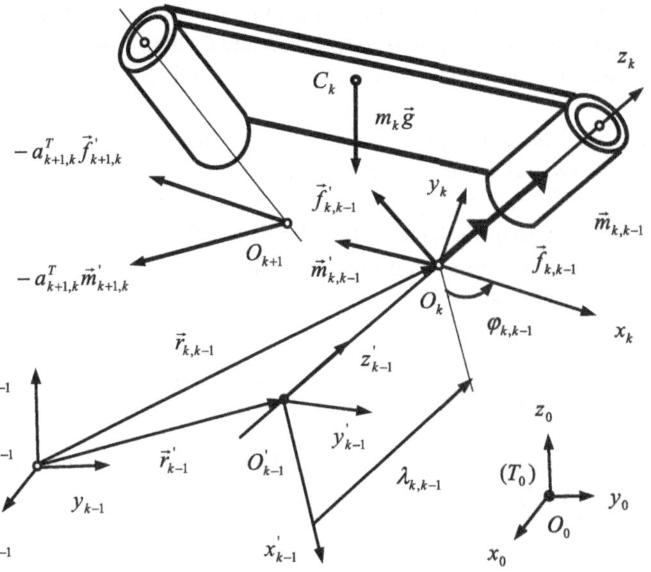

Fig. 7.1 Geometry and forces exerted on the link T_k

relative cylindrical motion (Fig. 7.1). The two-degrees-of-freedom cylindrical joint allows relative translation along and relative rotation about the joint axis z_k. A combination of two displacements is called *screw displacement* [1].

Independent geometric parameters describing the relative position of the body T_k with respect to its adjacent link T_{k-1} will be uniquely determined: the translation displacement $O'_{k-1}O_k = \lambda_{k,k-1}$ of the origin O_k and the joint angle $\varphi_{k,k-1}$ of rotation about the positive z_k axis, according to the right-hand rule. These variables in fact represent the distance of translation and the angle of rotation needed to bring a link from its reference configuration to the next configuration and they should not be confused with the Denavit-Hartenberg parameters [2]. However, their relation to the D-H parameters can be derived from these two coordinates by creating relative displacements with respect to the reference configuration and by performing some simple matrix operations [3].

Linked to rigid body T_k, the mobile frame $O_kx_ky_kz_k$ make either a translation, rotation or simultaneously a complex motion of translation-rotation along the common axis z_k of the connecting joint between two adjacent bodies T_k and T_{k-1}. Any vector or tensor will be expressed using their components about the axes of the attached frame $O_kx_ky_kz_k$. For a revolute joint, the displacement $\lambda_{k,k-1}$ is constant and the angle $\varphi_{k,k-1}$ is a variable that measures the relative rotation of link T_k with respect to the near link T_{k-1}, while for a prismatic joint the angle $\varphi_{k,k-1}$ is constant and the displacement $\lambda_{k,k-1}$ is a variable that measures the relative translation between two links. We note that a universal joint can be geometrically modelled by two intersecting revolute joints and that the modelling of a spherical joint can be performed by considering a set of three successive rotations about three concurrent axes.

In what follows, we will apply the concept of successive screw displacements to geometrical analysis of the closed-loop kinematical chains and we note that a *joint variable* is the displacement required to move a link from its initial location to actual position. However, a variable associated to an active joint is denoted as an *input variable* of the active joint. To simplify the kinematics analysis, a customary *associated spatial graph* of the mechanism is constructed. In the graph representation, the links are denoted by vertices and the joints by edges, with the addition that the vertex denoting the fixed link is the base link 0 and that the corresponding vertices of all moving links k are numbered from 1 to n (Fig. 7.2).

Two independent coordinates $\lambda_{k,k-1}$ and $\varphi_{k,k-1}$ determine uniquely the position of the vector

$$\vec{r}_{k,k-1} = \vec{r}\,'_{k-1} + \lambda_{k,k-1} a^{T}_{k-1} \vec{u}_3, \quad a^{T}_{k-1} = a^{\psi T}_{k-1} a^{\theta T}_{k-1}, \quad \vec{u}_3 = [0 \quad 0 \quad 1]^T \tag{7.1}$$

and generate the *variable* orthogonal transformation 3×3 matrix $a^{\varphi}_{k,k-1}$ for the relative rotation of the link T_k about z_k axis, where

$$\vec{r}_{k,k-1} = \begin{bmatrix} x_{k,k-1} \\ y_{k,k-1} \\ z_{k,k-1} \end{bmatrix}, \quad \overrightarrow{O_{k-1}O'_{k-1}} = \vec{r}\,'_{k-1} = \begin{bmatrix} x'_{k-1} \\ y'_{k-1} \\ z'_{k-1} \end{bmatrix}$$
$$a^{\varphi}_{k,k-1} = rot(z, \varphi_{k,k-1}) = \begin{bmatrix} \cos\varphi_{k,k-1} & \sin\varphi_{k,k-1} & 0 \\ -\sin\varphi_{k,k-1} & \cos\varphi_{k,k-1} & 0 \\ 0 & 0 & 1 \end{bmatrix}. \tag{7.2}$$

The above *constant* matrix product $a_{k-1} = a^{\theta}_{k-1} a^{\psi}_{k-1}$ of two orthogonal matrices a^{θ}_{k-1} and a^{ψ}_{k-1} allows us to give the position of the future screw relative displacement axis $O'_{k-1}O_k$ and the pose of a frame $O'_{k-1}x'_{k-1}y'_{k-1}z'_{k-1}$, which is fixed in the body of same rigid body T_{k-1} [4]:

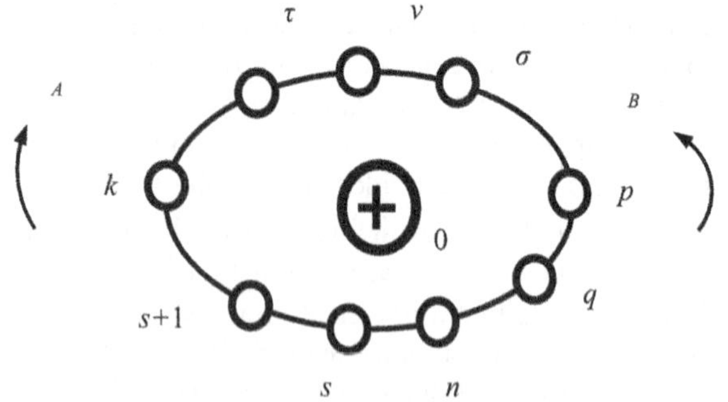

Fig. 7.2 Associated graph of the mechanism

$$
\begin{aligned}
a_{k-1}^{\theta} &= rot(y, \theta_{k-1}) = \begin{bmatrix} \cos\theta_{k-1} & 0 & -\sin\theta_{k-1} \\ 0 & 1 & 0 \\ \sin\theta_{k-1} & 0 & \cos\theta_{k-1} \end{bmatrix}, \\
a_{k-1}^{\psi} &= rot(z, \psi_{k-1}) = \begin{bmatrix} \cos\psi_{k-1} & \sin\psi_{k-1} & 0 \\ -\sin\psi_{k-1} & \cos\psi_{k-1} & 0 \\ 0 & 0 & 1 \end{bmatrix}.
\end{aligned}
\tag{7.3}
$$

Now, the position of the permanent axis z_k of the relative cylindrical motion is completely determined by the constant vector $\vec{r}_{k-1}^{\prime}$ and the constant matrix a_{k-1}. After one translation and two orthogonal rotations followed by one relative rotation or translation about z_x axis, a geometrical change of coordinates between the T_{k-1} and T_k frame is completely given by the relations (7.1)–(7.3). If every link is connected to at least two other links, finally, a general kinematical chain can form one or more independent closed-loops.

Some matrix equations, relating the location of a link T_σ and the position of the end-effector to the independent joint variables, should be expressed in a geometrical modelling of the robotic systems. The recursive form of the position vector for the origin O_σ

$$
\vec{r}_{\sigma s} = \vec{r}_{\sigma-1,s} + a_{\sigma-1,s}^{T}\vec{r}_{\sigma,\sigma-1}, \quad \vec{r}_{ss} = \vec{0}, \quad a_{ss} = I, \quad \sigma \geq s+1 \tag{7.4}
$$

and following transformation matrices [3]

$$
a_{\sigma s} = \prod_{j=1}^{\sigma-s} a_{\sigma-j+1,\sigma-j}, \quad a_{\sigma s}^{T} = \prod_{j=s+1}^{\sigma} a_{j,j-1}^{T}, \quad a_{j,j-1} = a_{j,j-1}^{\varphi} a_{j-1} \tag{7.5}
$$

give immediately yield the relative position of the link T_σ in the frame linked to the body T_s. When the change of coordinates is taken in succession starting from the base T_s and ending to the link T_σ, the corresponding matrices are certainly multiplied. In the *forward kinematics*, the joint variables are given and the problem is to find where the end-effector is with respect to the fixed base, while in the *inverse kinematics*, the pose of the end-effector is given and the problem is to compute the joint displacements needed to bring the end-effector to the desired position.

If the reference point of the moving platform T_n is located precisely at its mass centre C_n, the position of $T_n \equiv O_n x_n y_n z_n$ with respect to the fixed reference frame $O_0 x_0 y_0 z_0$ can be described along every leg by the position vector $\vec{r}_0^{C_n}$ of the origin C_n and the *absolute* rotation matrix a_{n0}, as follows

$$
\begin{aligned}
\vec{r}_0^{C_n} &= \vec{r}_{10} + \sum_{k=1}^{n-1} a_{k0}^{T}\vec{r}_{k+1,k} + a_{n0}^{T}\vec{r}_n^{C_n}, \quad a_{n0} = \prod_{k=1}^{n} a_{n-k+1,n-k} \\
\vec{r}_n^{C_n} &= \overrightarrow{O_n C_n} = \begin{bmatrix} x_n^{C_n} & y_n^{C_n} & z_n^{C_n} \end{bmatrix}^{T}.
\end{aligned}
\tag{7.6}
$$

These two loop-closure equations constitute the *geometrical model* of a parallel robot. Generally, the number of the relations that can be established for a mechanism is dependent on the number of independent closed-loops.

7.2 Velocities and Accelerations

The concepts developed for the position representation will form the basis for linear velocities and angular velocity and acceleration representation. A systematic approach is now derived to analyze the kinematics of spatial parallel mechanisms.

Kinematic analysis refers to the study of the geometry of motion in a robot without considering the forces and torques that cause the motion. The relationship between the velocity of the moving platform and the joint velocities can ultimately be written as a set of linear equations $J_1\dot{q} = J_2\dot{X}$, where $\dot{q}$ is the column matrix of joint velocities, $\dot{X}$ represents the generalised velocities, and J_1, J_2 are the inverse and direct Jacobian matrices.

The relative cylindrical motion of the link T_k is determined by the *twist* of two vectors evaluated in the frame $O_k x_k y_k z_k$: the linear relative velocity $\vec{v}_{k,k-1}$ of O_k and the relative angular velocity vector $\vec{\omega}_{k,k-1}$, thus

$$\vec{v}_{k,k-1} = \dot{\lambda}_{k,k-1}\vec{u}_3, \quad \vec{\omega}_{k,k-1} = \dot{\varphi}_{k,k-1}\vec{u}_3. \tag{7.7}$$

We note that a 3×3 useful skew-symmetric matrix *associated* with the angular velocity $\vec{\omega}_{k,k-1}$ is derived

$$\tilde{\omega}_{k,k-1} = a_{k,k-1}\dot{a}^T_{k,k-1} = \dot{\varphi}_{k,k-1}\tilde{u}_3, \quad \tilde{u}_3 = \begin{bmatrix} 0 & -1 & 0 \\ 1 & 0 & 0 \\ 0 & 0 & 0 \end{bmatrix}, \quad \det\tilde{\omega}_{k,k-1} = 0. \tag{7.8}$$

To describe the kinematic state of each link T_σ with respect to the frame T_s, we compute the angular velocity $\vec{\omega}_{\sigma s}$ and the linear velocity $\vec{v}_{\sigma s}$ of the reference point O_σ in terms of characteristic vectors of the preceding body. Starting from the first moving link T_s and ending at the last link T_σ, both vectors can easily be expressed in a recursive manner (3.25)

$$\vec{v}_{\sigma s} = a_{\sigma,\sigma-1}\{\vec{v}_{\sigma-1,s} + \tilde{\omega}_{\sigma-1,s}\vec{r}_{\sigma,\sigma-1}\} + \dot{\lambda}_{\sigma,\sigma-1}\vec{u}_3, \quad \vec{\omega}_{\sigma s} = a_{\sigma,\sigma-1}\vec{\omega}_{\sigma-1,s} + \dot{\varphi}_{\sigma,\sigma-1}\vec{u}_3 \tag{7.9}$$

The kinematic *conditions of connectivity* will be given through some constraint relations written in a forward computation between the relative velocities of each independent closed loop of the mechanism. Starting from the base (T_0) and moving towards to the moving platform T_n, the absolute velocities $\vec{v}^A_{n0}$ and $\vec{\omega}^A_{n0}$ are computed along first leg A, for example. These vectors should coincide with the two

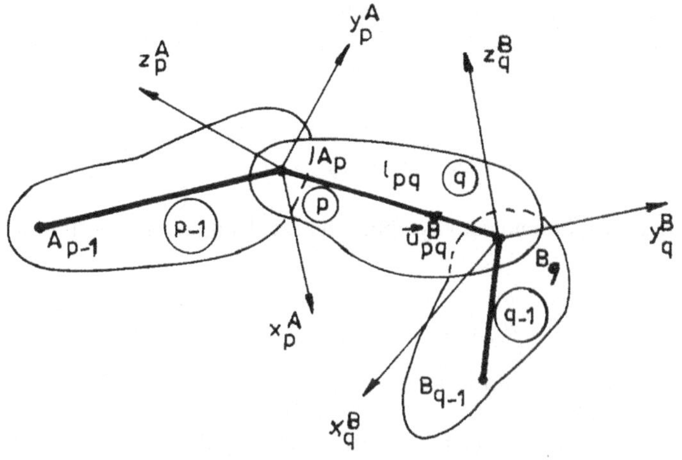

Fig. 7.3 Connection of adjoining links $T_{p-1}, T_p = T_q, T_{q-1}$

velocities $\vec{v}_{n0}^B$ and $\vec{\omega}_{n0}^B$ of the same platform, but computed into the direction of the second leg B (Fig. 7.3). The following system of equations is derived

$$\begin{aligned} &\vec{u}_i^T a_{n0}^T \vec{\omega}_{n0}^A = \vec{u}_i^T b_{n0}^T \vec{\omega}_{n0}^B = \vec{u}_i^T \vec{\omega}_n^0, \quad (i = 1, 2, 3) \\ &\vec{u}_i^T a_{n0}^T \{\vec{v}_{n0}^A + \tilde{\omega}_{n0}^A \vec{r}_n^{\,C_nA}\} = \vec{u}_i^T b_{n0}^T \{\vec{v}_{n0}^B + \tilde{\omega}_{n0}^B \vec{r}_n^{\,C_nB}\} = \vec{u}_i^T \dot{\vec{r}}_0^{\,C_n}, \end{aligned} \tag{7.10}$$

where $\vec{u}_1, \vec{u}_2$ and $\vec{u}_3$ are three orthogonal unit vectors pointing, respectively, along the three fixed axes x_0, y_0, z_0.

The characteristic relative velocities and the conventional *Jacobian* matrix are immediately obtained. This square invertible matrix is an essential element in the analysis of the singularities loci existing in the robot workspace. The differential kinematics equations, in terms of the geometric or the analytical Jacobian, establish a linear mapping between the joint space velocities and the task space velocities.

Performing the derivatives with respect to time of the relations (7.9), we obtain a recursive form (3.35) of the relative accelerations $\vec{\gamma}_{\sigma s}$ and $\vec{\varepsilon}_{\sigma s}$

$$\begin{aligned} \vec{\gamma}_{\sigma s} = {} & a_{\sigma,\sigma-1}\{\vec{\gamma}_{\sigma-1,s} + (\tilde{\omega}_{\sigma-1,s}^2 + \tilde{\varepsilon}_{\sigma-1,s})\vec{r}_{\sigma,\sigma-1}\} + \ddot{\lambda}_{\sigma,\sigma-1}\vec{u}_3 \\ & + 2\dot{\lambda}_{\sigma,\sigma-1} a_{\sigma,\sigma-1} \tilde{\omega}_{\sigma-1,s} a_{\sigma,\sigma-1}^T \vec{u}_3 \end{aligned} \tag{7.11}$$

$$\vec{\varepsilon}_{\sigma s} = a_{\sigma,\sigma-1}\vec{\varepsilon}_{\sigma-1,s} + \ddot{\varphi}_{\sigma,\sigma-1}\vec{u}_3 + \dot{\varphi}_{\sigma,\sigma-1} a_{\sigma,\sigma-1} \tilde{\omega}_{\sigma-1,s} a_{\sigma,\sigma-1}^T \vec{u}_3$$

and two useful square characteristic matrices [5]:

$$\begin{aligned}
\tilde{\omega}_{\sigma s} &= a_{\sigma,\sigma-1}\tilde{\omega}_{\sigma-1,s}a^{T}_{\sigma,\sigma-1} + \dot{\varphi}_{\sigma,\sigma-1}\tilde{u}_3 \\
\tilde{\omega}^2_{\sigma s} + \tilde{\varepsilon}_{\sigma s} &= a_{\sigma,\sigma-1}\{\tilde{\omega}^2_{\sigma-1,s} + \tilde{\varepsilon}_{\sigma-1,s}\}a^{T}_{\sigma,\sigma-1} + \dot{\varphi}^2_{\sigma,\sigma-1}\tilde{u}_3\tilde{u}_3 \\
&\quad + \ddot{\varphi}_{\sigma,\sigma-1}\tilde{u}_3 + 2\dot{\varphi}_{\sigma,\sigma-1}a_{\sigma,\sigma-1}\tilde{\omega}_{\sigma-1,s}a^{T}_{\sigma,\sigma-1}\tilde{u}_3.
\end{aligned} \tag{7.12}$$

The conditions of connectivity about the accelerations are the following

$$\begin{aligned}
&\vec{u}_i^T a^T_{n0}\vec{\varepsilon}^{\,A}_{n0} = \vec{u}_i^T b^T_{n0}\vec{\varepsilon}^{\,B}_{n0} = \vec{u}_i^T\vec{\varepsilon}^{\,0}_n \\
&\vec{u}_i^T a^T_{n0}\{\vec{\gamma}^{\,A}_{n0} + (\tilde{\omega}^A_{n0}\tilde{\omega}^A_{n0} + \tilde{\varepsilon}^A_{n0})\vec{r}^{\,C_nA}_n\} \\
&\quad = \vec{u}_i^T b^T_{n0}\{\vec{\gamma}^{\,B}_{n0} + (\tilde{\omega}^B_{n0}\tilde{\omega}^B_{n0} + \tilde{\varepsilon}^B_{n0})\vec{r}^{\,C_nB}_n\} = .\vec{u}_i^T\ddot{\vec{r}}^{\,C_n}_0.
\end{aligned} \tag{7.13}$$

These formulae give all the relative linear and angular accelerations of the robot. Equations (7.9)–(7.13) constitute the *matrix kinematics modelling* of the constrained robotic system.

7.3 Equations of Motion

The dynamics analysis of the parallel constrained robotic systems tends to be very complicated due to the existence of a spatial kinematical structure, which possesses a large number of passive degrees of freedom, the dominance of the inertial forces, frictional and gravitational components, and also the problem associated with real-time control in the inverse dynamics. The closed chains make the dynamics of parallel robots to be strongly coupled and highly nonlinear.

Three different methods produce the same results. The first one is the Newton-Euler approach, which consists of applying the free-body diagram (*FBD*) procedure for every rigid body where all internal joint forces and moments are unknown [6–9]. This formulation, takes into account the equations of motion for each rigid body of the mechanical system. Writing Newton-Euler equations for all bodies results in a large system of equations that must be solved simultaneously for all internal reaction joint forces between any two adjacent links.

The second method, as an alternative energy-based approach, is derived from Lagrange formalism, which introduces numerous scalar multipliers for each kinematical closure equation [10, 11]. This formulation describes the equations of motion of a mechanical system as a compact function of kinetic and potential energies, does not involve the unwanted reaction forces and moments and thus is more efficient than the Newton-Euler approach. Because of the analytical constraint relations imposed by all closed loops of the parallel robot, the deriving of the explicit equations of motion in terms of a set of independent generalized coordinates becomes a prohibitive task.

The third method for the dynamic analysis, which is the one that will be applied in the present study, is based on the general principle of virtual work [3, 12, 13]. Applying this principle to analyzing the inverse dynamics of a mechanism does not involve any unknown constraint forces or moments.

So, consider that a link T_τ of the mechanism is connected to other two neighbouring bodies by two types of joints: either a prismatic joint or a revolute joint. The displacement of T_τ with respect to the last link $T_{\tau-1}$ is described by two variables: the translation coordinate $\lambda_{\tau,\tau-1}$ and the joint angle of rotation $\varphi_{\tau,\tau-1}$. In the first case, the rigid body T_τ moves in a relative translation along the z_τ axis under the action of the internal force $\vec{f}_{\tau,\tau-1} = f_{\tau,\tau-1}\vec{u}_3$ which is applied by a hydraulic system located in the neighbouring frame $T_{\tau-1}$. In the second, the rigid body T_τ is rotated by an electric motor developing the internal torque of moment $\vec{m}_{\tau,\tau-1} = m_{\tau,\tau-1}\vec{u}_3$, pointed about the positive z_τ axis. We note that for a parallel robot, a single actuator that exerts either a force or a torque between two adjacent links can drive each leg of the mechanism.

Applying the free-body diagram procedure, Fig. 7.1 depicts the forces and the torques acting on a typical link $T_k = T_\tau$ that is connected to the link $T_{\tau-1}$ by the joint O_τ and to another link $T_{\tau+1}$ by the joint $O_{\tau+1}$. The forces acting on the link T_τ from $T_{\tau-1}$ can be reduced about O_τ to a force $\vec{f}'_{\tau,\tau-1}$ and a moment $\vec{m}'_{\tau,\tau-1}$. Similarly, the forces acting on the link T_τ from $T_{\tau+1}$ are reduced about $O_{\tau+1}$ to a resultant force $(-a^T_{\tau+1,\tau}\vec{f}'_{\tau+1,\tau})$ and a resultant moment $(-a^T_{\tau+1,\tau}\vec{m}'_{\tau+1,\tau})$, which are converted into the frame T_τ. Note that the vectors $\vec{f}_{n,n-1}$ and $\vec{m}_{n,n-1}$ represent the force and the moment exerted on the platform T_n by the link T_{n-1}, for $k = n$. When a robot's platform carries an object, the weight of the object becomes a known external load applied on the end-effector. Assuming that the frictional force at the joints is negligible, two vectors $\vec{f}^*_\tau$ and $\vec{m}^*_\tau$ composing the wrench about the point O_τ express precisely the action of external and internal forces, including the weight $m_\tau a_{\tau 0}\vec{g}$ of rigid body T_τ.

Now, we compute the inertia force and the resultant moment of inertia forces (6.15) exerted at the joint O_τ:

$$\begin{aligned} \vec{f}^{in}_\tau &= -m_\tau\{\vec{\gamma}_{\tau 0} + (\tilde{\omega}^2_{\tau 0} + \tilde{\varepsilon}_{\tau 0})\vec{r}^C_\tau\}, \\ \vec{m}^{in}_\tau &= -\{m_\tau\tilde{r}^C_\tau\vec{\gamma}_{\tau 0} + \hat{J}_\tau\vec{\varepsilon}_{\tau 0} + \tilde{\omega}_{\tau 0}\hat{J}_\tau\vec{\omega}_{\tau 0}\}, \end{aligned} \tag{7.14}$$

where m_τ and $\hat{J}_\tau = \oint_\tau \tilde{r}\tilde{r}^T dm$ are the *mass* and the symmetrical *tensor of inertia* (4.10) of the rigid body T_τ computed with respect O_τ, respectively.

Knowing the kinematics state of each link and the external forces acting on the robot, this analysis derives the final form of the explicit dynamics equations of a constrained robotic system, using the fundamental principle of virtual work. Finally, the active forces and the active torques, or some internal reaction joint forces, required in a given motion of the end-effector will be easily computed using a recursive procedure.

For parallel robots the classical procedure for developing the inverse dynamics is to open up every closed loop into individual open-loop structures and then to calculate the dynamics separately for each tree. The parallel robot can artificially be transformed into a set of open kinematical chains subject to the constraints. This is possible by cutting each joint for the moving platform and taking its effect into account by introducing the corresponding constraint conditions. The more complicated open tree system includes the acting link and ends with the moving platform present.

The virtual displacements should be consistent with the virtual motion imposed by all kinematic constraints and the joints at a given instant in time. Considering several independent virtual motions of the robot which are consistent with the constraints, the matrix connectivity equations (7.10) will generate the expressions of *virtual velocities*. By the intermediate of the complete Jacobian matrix, the absolute virtual velocities $\vec{v}^{\,v}_{\tau0}$, $\vec{\omega}^{\,v}_{\tau0}$ are related to the following relative virtual velocities $\vec{v}^{\,v}_{q,q-1} = v^{v}_{q,q-1}\vec{u}_3$, $\vec{\omega}^{\,v}_{q,q-1} = \omega^{v}_{q,q-1}\vec{u}_3$ associated with the actuated joints.

The fundamental principle of virtual work states that a robot is under dynamic equilibrium if and only if the total virtual work developed by all external, internal and inertia forces vanishes during any general virtual displacement, which is consistent with the kinematic constraints. Assuming that frictional forces at the joints are negligible, the virtual work produced by the forces of constraint at the joints is zero.

Replacing all virtual displacements $\delta^{v}_{k,k-1}$ with independent relative virtual velocities $\vec{v}^{\,v}_{k,k-1}$, $\omega^{v}_{k,k-1}$, the total virtual power contributed by the active forces $\vec{f}_{q,q-1}$ and the active torques $\vec{m}_{q,q-1}$, the known wrench $\vec{f}^{\,*}_{\tau}$, $\vec{m}^{\,*}_{\tau}$ of external and internal forces and, also, the wrench $\vec{f}^{\,in}_{\tau}$, $\vec{m}^{\,in}_{\tau}$ of inertia forces can be written in a general form (6.17) as follows:

$$\vec{v}^{\,vT}_{q,q-1}\vec{f}_{q,q-1} + \vec{\omega}^{\,vT}_{q,q-1}\vec{m}_{q,q-1} + \sum_{p=1}^{l}\sum_{\tau=1}^{n_p}\left\{\vec{v}^{\,vT}_{\tau0}\left(\vec{f}^{\,*}_{\tau} + \vec{f}^{\,in}_{\tau}\right) + \vec{\omega}^{\,vT}_{\tau0}\left(\vec{m}^{\,*}_{\tau} + \vec{m}^{\,in}_{\tau}\right)\right\} = 0, \tag{7.15}$$

where l denotes the number of serial independent trees fictitiously separated, starting from the fixed base and ending at the moving platform. The virtual power developed by the *active force* $\vec{f}_{q,q-1}$ and the *active torque* $\vec{m}_{q,q-1}$ we have isolated from those of the other known forces and the inertia forces applied to all legs of the robot.

We notice that the absolute virtual velocities $\vec{v}^{\,v}_{\tau0}$ and $\vec{\omega}^{\,v}_{\tau0}$ can easily be related to the virtual *relative* velocities $\vec{v}^{\,v}_{k,k-1}$, $\vec{\omega}^{\,v}_{k,k-1}$, as follows (3.22), (3.23):

$$\vec{v}^{v}_{\tau 0} = \sum_{k=1}^{\tau} v^{v}_{k,k-1} a_{\tau k} \vec{u}_3 + \sum_{k=1}^{\tau-1} \omega^{v}_{k,k-1} a_{\tau k} \tilde{\vec{r}}^{T}_{\tau k} \vec{u}_3$$
$$\vec{\omega}^{v}_{\tau 0} = \sum_{k=1}^{\tau} \omega^{v}_{k,k-1} a_{\tau k} \vec{u}_3, \quad \tilde{r}_{\sigma\tau} = \sum_{\upsilon=\tau}^{\sigma-1} a^{T}_{\upsilon\tau} \tilde{r}_{\upsilon+1,\upsilon} a_{\upsilon\tau} \tag{7.16}$$

Using two notations

$$\vec{F}_{\tau} = \sum_{\sigma=\tau}^{n} a^{T}_{\sigma\tau} \vec{F}_{\tau 0}, \quad \vec{M}_{\tau} = \sum_{\sigma=\tau}^{n} a^{T}_{\sigma\tau} \left\{ \vec{M}_{\tau 0} + \sum_{\upsilon=\tau}^{\sigma-1} a_{\sigma\upsilon} \tilde{r}_{\upsilon+1,\upsilon} a^{T}_{\sigma\upsilon} \vec{F}_{\tau 0} \right\}$$
$$\vec{F}_{\tau 0} = -\vec{f}^{*}_{\sigma} - \vec{f}^{in}_{\sigma}, \quad \vec{M}_{\tau 0} = -\vec{m}^{*}_{\sigma} - \vec{m}^{in}_{\sigma}, \tag{7.17}$$

we can remark that Eqs. (7.15)–(7.17) can be written in a recursive compact form, *only* based on the relative virtual velocities $\vec{v}^{v}_{k,k-1}$ and $\vec{\omega}^{v}_{k,k-1}$:

$$v^{v}_{q,q-1} f_{q,q-1} + \omega^{v}_{q,q-1} m_{q,q-1} = \vec{u}^{T}_3 \sum_{p=1}^{l} \sum_{\tau=1}^{n_p} \{ v^{v}_{\tau,\tau-1} \vec{F}_{\tau} + \omega^{v}_{\tau,\tau-1} \vec{M}_{\tau} \}, \tag{7.18}$$

where we notice the following *significant* recursive relations

$$\vec{F}_{\tau} = \vec{F}_{\tau 0} + a^{T}_{\tau+1,\tau} \vec{F}_{\tau+1}$$
$$\vec{M}_{\tau} = \vec{M}_{\tau 0} + a^{T}_{\tau+1,\tau} \vec{M}_{\tau+1} + \tilde{r}_{\tau+1,\tau} a^{T}_{\tau+1,\tau} \vec{F}_{\tau+1} \tag{7.19}$$

The *dynamics model* expressed by the recursive matrix equations (7.18) and (7.19) represents the *explicit dynamics equations of a constrained robotic system* [4, 14, 15]. These equations are valid for any virtual displacement, computing the active forces $f_{q,q-1}$ and the active torques $m_{q,q-1}$ in terms of output forces. Using the initial values $\vec{F}_{n_p+1} = \vec{0}$, $\vec{M}_{n_p+1} = \vec{0}$, the expressions and the time-history graphs plotted for the forces and the torques required by the actuators are immediately obtained. To obtain the force $f_{q,q-1}$, for example, we make $v^{v}_{q,q-1} = 1$ and $\omega^{v}_{q,q-1} = 0$, while to calculate the torque $m_{q,q-1}$ we consider $v^{v}_{q,q-1} = 0$ and $\omega^{v}_{q,q-1} = 1$.

Equation (7.18) are expressed in terms of the independent variables and their derivatives with respect to time, and the constraint forces are automatically eliminated. This new approach established here is highly useful for real-time control of the robot's evolution and computationally is more efficient than the Newton-Euler procedure or Lagrange formalism.

The recursive method presented can be applied to the dynamics analysis of closed-loop mechanical systems by making cuts at selected secondary joints to form spanning tree systems and analysing the equations of motion of the resulting open-loop branches.

In the following we can apply the Newton-Euler procedure to establish the set of analytical equations for each compounding rigid body of the mechanical system. These equations give all connecting forces in the external and internal joints. Several relations from the general system of equations could, if required, constitute verifications for the input forces or the active torques, obtained by the method based on the principle of virtual work. Now, we are able to calculate the friction forces and the friction torques in the joints, based on the friction coefficients and the maximum of the connecting forces in the joints. We apply again the explicit equations (7.18) and (7.19), where the contribution of the virtual work of friction forces in joints is added. The new active torques and input forces are compared to the values obtained in the first calculus.

The present method can also be rapidly applied in the *inverse dynamics* modelling of parallel robots and then the above recursive matrix equations become pure algebraic relations.

In the *direct problem* of dynamics, many velocities and accelerations expressed as functions of the input variables of the active joints and their first and second derivatives with respect to time of these independent variables are inserted in the same developed matrix relations. If the input forces or the active torques are all known, the set of differential equations established for the direct dynamics simulation is numerically integrated and ensures that the solutions represent the generalized coordinates of the mechanism, which determine the instantaneous position and the motion of the moving platform.

7.4 Advantage of Present Method

Compared with some other published methods [6, 16, 17], the novelties of the present approach, based on the principle of virtual work, are the followings:

- The theoretical modelling is accompanied by graphs of simulation through the MATLAB program, to warrant the accuracy, validity and effectiveness of the novel development.
- Geometrical constraint relations, under the matrix form, generate through successive recasting the connectivity conditions that will supply all the relative velocities and relative accelerations which characterize the independent kinematical chains.
- The accelerations of the mass centers, the angular accelerations and the wrench of the inertia forces are expressed through matrix formulae which contain the kinematic characteristics of the relative motion of the building elements of the manipulator.
- The explicit dynamics equation represents a *definitive formula*, obtained by the transformation of the general expression of the virtual work where the relative virtual velocities only appear, generated by recursive relations.

- All intermediate analytical calculations have been eliminated and the numerical computation is achieved through the numerical code, for each active force or active torque applied by the driving system.

Most dynamic models based on Lagrange formalism ignore the weight of intermediate bodies and take into consideration only the active forces or active moments and the wrench of external forces applied to the moving platform. The number of relations given by this approach is equal to the total number of the position variables and Lagrange multipliers inclusive. Also, the analytical calculi involved in these equations are very tedious, thus presenting an elevated risk of making errors.

The commonly-known Newton-Euler method, which takes into account the free-body diagrams of the mechanism, leads to a large number of equations with unknowns, including even the connecting forces in the joints. Finally, the actuating forces or torques can be obtained.

Within the inverse kinematics analysis, some exact conditions of connectivity given in Eqs. (7.10) and (7.13) which define in real-time the position, velocity and acceleration of each element of the parallel robot, have been established in this study.

The new approach based on the principle of virtual work takes into consideration the masses and the forces of inertia introduced by all component elements, can eliminate the forces of internal joints, and establishes a direct determination of the time-history evolution of the forces and the torques required by the actuators. The matrix modelling can easily be transformed into a model for automatic command of the parallel robots and can successfully be applied in forward and inverse mechanics of the parallel structures, their end-effectors having either a translation, a rotation or a mixed motion.

7.5 Application to Dynamics Analysis of Mechanisms

Knowing the position and kinematic state of each link as well as the external forces acting on a planar or spatial mechanism, the active forces or torques required in a given motion can easily be obtained first, based on the *principle of virtual work* [15].

7.5.1 Universal Hooke Mechanism

Leaving aside the friction forces, the key issue in the dynamics analysis for this spatial mechanism is to establish the inverse dynamic modelling when the required actuator torque m_{10}^{A} can be computed, if the time history $\varphi_{10}^{A} = \varphi_{10}^{A}(t)$ of rotation motion is given (Fig. 3.3).

Mounted in a Cardan suspension, the mechanism can artificially be transformed into a set of two open kinematic chains subject to constraints, by cutting the internal passive revolute joint B_2 and taking it into account by introducing the corresponding constraint conditions (3.39).

Applying *the equations of the parallel robots dynamics* (7.18), based on the fundamental principle of virtual work, the following matrix relation results for the active torque

$$m_{10}^{A} = \vec{u}_3^{T}\{\omega_{10}^{Av}\vec{M}_{10}^{A} + \omega_{10}^{Bv}\vec{M}_{1}^{B} + \omega_{21}^{Bv}\vec{M}_{20}^{B}\}, \tag{7.20}$$

where we denote

$$\begin{aligned} \vec{M}_{10}^{A} &= \hat{J}_{1}^{B}\vec{\varepsilon}_{10}^{A} + \tilde{\omega}_{10}^{A}\hat{J}_{1}^{A}\vec{\omega}_{10}^{A} \\ \vec{M}_{20}^{B} &= \hat{J}_{2}^{B}\vec{\varepsilon}_{20}^{B} + \tilde{\omega}_{20}^{B}\hat{J}_{2}^{B}\vec{\omega}_{20}^{B} \\ \vec{M}_{1}^{B} &= b_{21}^{T}\vec{M}_{20}^{B} + \hat{J}_{1}^{B}\vec{\varepsilon}_{10}^{B} + \tilde{\omega}_{10}^{B}\hat{J}_{1}^{B}\vec{\omega}_{10}^{B}. \end{aligned} \tag{7.21}$$

The characteristic *virtual velocities* ω_{10}^{Bv} and ω_{21}^{Bv} are expressed as functions of the pose of the spatial mechanism by kinematic equations (3.41), assuming $\omega_{10}^{Av} = 1$:

$$\begin{aligned} &\omega_{21}^{Av}\vec{u}_i^{T}a_{10}^{T}a_{21}^{T}\vec{u}_3 - \omega_{10}^{Bv}\vec{u}_i^{T}b_{10}^{T}\vec{u}_3 - \omega_{21}^{Bv}\vec{u}_i^{T}b_{10}^{T}b_{21}^{T}\vec{u}_3 \\ &= -\omega_{10}^{Av}\vec{u}_i^{T}a_{10}^{T}\vec{u}_3 \ \ (i = 1, 2, 3). \end{aligned} \tag{7.22}$$

A solution of the dynamics problem of this mechanism for one degree of freedom can be obtained based on the *theorem of kinetic energy* (5.17)

$$\dot{E} = m_{10}^{A}\omega_{10}^{A}, \tag{7.23}$$

with the total kinetic energy

$$E = \frac{1}{2}\vec{\omega}_{10}^{AT}\hat{J}_{1}^{A}\vec{\omega}_{10}^{A} + \frac{1}{2}\vec{\omega}_{10}^{BT}\hat{J}_{1}^{B}\vec{\omega}_{10}^{B} + \frac{1}{2}\vec{\omega}_{20}^{BT}\hat{J}_{2}^{B}\vec{\omega}_{20}^{B}. \tag{7.24}$$

Developing the calculus, we obtain the same expression (7.20) of the active torque, in analytical form

$$m_{10}^{A} = \omega_{10}^{Av}\vec{u}_3^{T}\hat{J}_{1}^{A}\vec{\varepsilon}_{10}^{A} + \omega_{10}^{Bv}\vec{u}_3^{T}\{\hat{J}_{1}^{B}\vec{\varepsilon}_{10}^{B} + b_{21}^{T}\hat{J}_{2}^{B}\vec{\varepsilon}_{20}^{B}\} + \omega_{21}^{Bv}\vec{u}_3^{T}\hat{J}_{2}^{B}\vec{\varepsilon}_{20}^{B}, \tag{7.25}$$

where ω_{10}^{Av}, ω_{10}^{Bv} and ω_{21}^{Bv} are virtual velocities

$$\omega_{10}^{Av} = 1, \quad \omega_{10}^{Bv} = \frac{\sin 2\varphi_{10}^{B}}{\sin 2\varphi_{10}^{A}}, \quad \omega_{21}^{Bv} = -\sin\alpha\cos\varphi_{10}^{B}. \tag{7.26}$$

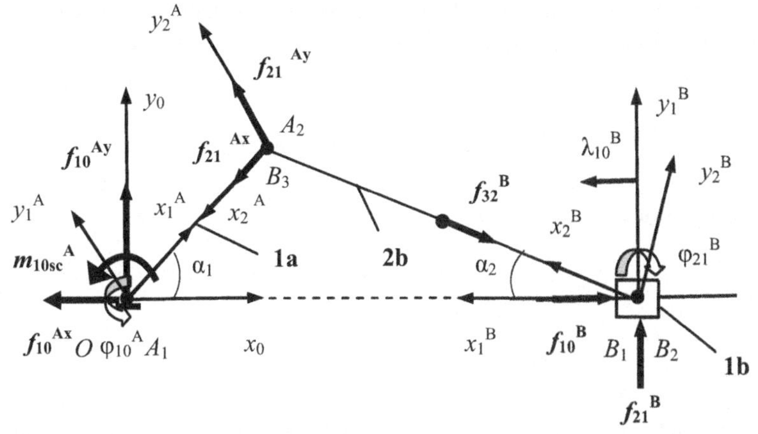

Fig. 7.4 Reaction forces in slider-crank mechanism

7.5.2 Slider-Crank Planar Mechanism

Under the active torque m_{10sc}^A, a specified input rotary motion of the crankshaft **1a** through the function $\varphi_{10}^A = \varphi_{10}^A(t)$ can control the motion of the piston **1b** and the intermediate connecting rod **2b** (Figs. 3.4 and 7.4).

The slider-crank planar mechanism is artificially transformed into two independent planar trees, by cutting the internal passive revolute joint $A_2 \equiv B_3$.

The force of inertia and the resulting torque of the inertia forces applied to the rigid bodies are computed with respect to the origins A_1, B_1 and B_2 of three frames, respectively, as follows

$$\begin{aligned}
\vec{f}_1^{inA} &= -m_1^A(\tilde{\omega}_{10}^A\tilde{\omega}_{10}^A + \tilde{\varepsilon}_{10}^A)\vec{r}_1^{CA}, \\
\vec{m}_1^{inA} &= -(\hat{J}_1^A\vec{\varepsilon}_{10}^A + \tilde{\omega}_{10}^A\hat{J}_1^A\vec{\omega}_{10}^A) \\
\vec{f}_1^{inB} &= -m_1^B\vec{\gamma}_{10}^B, \quad \vec{m}_1^{inB} = \vec{0} \\
\vec{f}_2^{inB} &= -m_2^B[\vec{\gamma}_{20}^B + (\tilde{\omega}_{20}^B\tilde{\omega}_{20}^B + \tilde{\varepsilon}_{20}^B)\vec{r}_2^{CB}], \\
\vec{m}_2^{inB} &= -(m_2^B\tilde{r}_2^{CB}\vec{\gamma}_{20}^B + \hat{J}_2^B\vec{\varepsilon}_{20}^B + \tilde{\omega}_{20}^B\hat{J}_2^B\vec{\omega}_{20}^B).
\end{aligned} \tag{7.27}$$

The influence of the action of the weights is evaluated by the wrench of following vectors:

$$\begin{aligned}
\vec{f}_1^{*A} &= -gm_1^A a_{10}\vec{u}_2, \quad \vec{m}_1^{*A} = \tilde{r}_1^{CA}\vec{f}_1^{*A}, \quad \vec{r}_1^{CA} = 0.5l_1\vec{u}_1, \quad \tilde{r}_1^{CA} = 0.5l_1\tilde{u}_1 \\
\vec{f}_1^{*B} &= -gm_1^B b_{10}\vec{u}_2 + f_{10}^B b_{10}\vec{u}_1, \quad \vec{m}_1^{*B} = \tilde{r}_1^{CB}\vec{f}_1^{*B}, \quad \vec{r}_1^{CB} = \vec{0}, \quad \tilde{r}_1^{CB} = 0 \\
\vec{f}_2^{*B} &= -gm_2^B b_{20}\vec{u}_2, \quad \vec{m}_2^{*B} = \tilde{r}_2^{CB}\vec{f}_2^{*B}, \quad \vec{r}_2^{CB} = 0.5l_2\vec{u}_1, \quad \tilde{r}_2^{CB} = 0.5l_2\tilde{u}_1.
\end{aligned} \tag{7.28}$$

Now, we can construct the set (7.19) of significant recursive relations for the *slider-crank planar mechanism*

$$\vec{F}_1^A = \vec{F}_{10}^A, \quad \vec{M}_1^A = \vec{M}_{10}^A, \quad \vec{F}_2^B = \vec{F}_{20}^B, \\ \vec{M}_2^B = \vec{M}_{20}^B, \quad \vec{F}_1^B = \vec{F}_{10}^B + b_{21}^T \vec{F}_2^B, \tag{7.29}$$

with the notations

$$\vec{F}_{10}^A = -\vec{f}_1^{in\ A} - \vec{f}_1^{*A}, \quad \vec{M}_{10}^A = -\vec{m}_1^{in\ A} - \vec{m}_1^{*A}, \\ \vec{F}_{\tau 0}^B = -\vec{f}_\tau^{in\ B} - \vec{f}_\tau^{*B}, \quad \vec{M}_{\tau 0}^B = -\vec{m}_\tau^{in\ B} - \vec{m}_\tau^{*B} \\ (\tau = 1, 2). \tag{7.30}$$

The characteristic *virtual velocities* v_{10m}^{Bv} and ω_{21m}^{Bv} are expressed as functions of the pose of the planar mechanism by kinematic equations (3.50), assuming $\omega_{10m}^{Av} = 1$:

$$v_{10m}^{Bv} \vec{u}_i^T \vec{u}_1 - l_2 \omega_{21m}^{Bv} \vec{u}_i^T b_{20}^T \vec{u}_2 = -l_1 \omega_{10m}^{Av} \vec{u}_i^T a_{10}^T \vec{u}_2 \quad (i = 1, 2). \tag{7.31}$$

The total virtual work or virtual power contributed by the input torque $\vec{m}_{10sc}^A$, the inertia forces, the torques of inertia forces and the wrench of weights can be written compactly. Because all frictional forces are negligible, the virtual work produced by the forces of constraint at the joints is zero. Applying the above (7.18) *equations of the parallel robots dynamics*, the following matrix relation is derived for the *torque* m_{10sc}^A of the *revolute actuator*, which defines the feed forward control law:

$$m_{10sc}^A = \vec{u}_3^T \omega_{10m}^{Av} \vec{M}_1^A + v_{10m}^{Bv} \vec{u}_1^T \vec{F}_1^B + \omega_{21m}^{Bv} \vec{u}_3^T \vec{M}_2^B, \tag{7.32}$$

where ω_{10m}^{Av}, v_{10m}^{Bv} and ω_{21m}^{Bv} are virtual velocities

$$\omega_{10m}^{Av} = 1, \quad v_{10m}^{Bv} = l_1 \frac{\sin(\phi_1 + \phi_2)}{\cos\phi_2}, \quad \omega_{21m}^{Bv} = \frac{l_1 \cos\phi_1}{l_2 \cos\phi_2}, \\ \phi_1 = \varphi_{10}^A + \alpha_1, \quad \phi_2 = \varphi_{21}^B + \alpha_2. \tag{7.33}$$

Based on the principle of virtual work, a recursive procedure can also be developed to obtain compact results concerning the time-history evolution for *external* and *internal forces in joints*. Adding the contributions of successive virtual translations during some fictitious displacements along the directions $A_2x_2^A, A_2y_2^A$ of the internal revolute joint A_2 (Fig. 7.4), for example, characteristic *virtual velocities* will be expressed as functions of the pose of the mechanism by kinematical equations

$$l_1 \omega_{10}^{Av} \vec{u}_i^T a_{10}^T \vec{u}_2 + v_{21}^{Axv} \vec{u}_i^T a_{10}^T \vec{u}_1 + v_{21}^{Ayv} \vec{u}_i^T a_{10}^T \vec{u}_2 \\ = -v_{10}^{Bv} \vec{u}_i^T \vec{u}_1 + l_2 \omega_{21}^{Bv} \vec{u}_i^T b_{20}^T \vec{u}_2 \quad (i = 1, 2). \tag{7.34}$$

Assuming that the planar mechanism has successively two virtual motions determined by the following sets of velocities:

$$\begin{aligned} &\omega_{10x}^{Av} = 0, \quad v_{21}^{Axv} = 1, \quad v_{21}^{Ayv} = 0, \\ &v_{10x}^{Bv} = \frac{\cos(\phi_1 + \phi_2)}{\cos\phi_2}, \quad \omega_{21x}^{Bv} = -\frac{\sin\phi_1}{l_2\cos\phi_2} \\ &\omega_{10y}^{Av} = 0, \quad v_{21}^{Axv} = 0, \quad v_{21}^{Ayv} = 1, \\ &v_{10y}^{Bv} = \frac{\sin(\phi_1 + \phi_2)}{\cos\phi_2}, \quad \omega_{21y}^{Bv} = \frac{\cos\phi_1}{l_2\cos\phi_2}, \end{aligned} \tag{7.35}$$

the orthogonal components f_{21}^{Ax} and f_{21}^{Ay} of *internal joint force* acting in the joint A_2 *on the intermediate connecting rod* **2b** have the following expressions

$$\begin{aligned} f_{21}^{Ax} &= v_{10x}^{Bv}\vec{u}_1^T\vec{F}_1^B + \omega_{21x}^{Bv}\vec{u}_3^T\vec{M}_2^B, \\ f_{21}^{Ay} &= v_{10y}^{Bv}\vec{u}_1^T\vec{F}_1^B + \omega_{21y}^{Bv}\vec{u}_3^T\vec{M}_2^B. \end{aligned} \tag{7.36}$$

New appropriate virtual velocities $v_{10h}^{Av} = 1$, $v_{10v}^{Av} = 1$, $v_{21n}^{Bv} = 1$, $v_{32a}^{Bv} = 1$ of the planar mechanism can successively be chosen for determining the components f_{10}^{Ax}, f_{10}^{Ay} of the joint force acting in the external joint A_1, the *external reaction force* f_{21}^B acting on the slider-block **1b** and the *internal axial force* f_{32}^B in the middle section of the connecting rod **2b**, respectively, as follows:

$$\begin{aligned} f_{10}^{Ax} &= v_{10h}^{Av}\vec{u}_1^T\theta\, a_{10}^T\vec{F}_1^A + v_{10h}^{Bv}\vec{u}_1^T\vec{F}_1^B, \quad \theta = rot(y,\pi) \\ f_{10}^{Ay} &= v_{10v}^{Av}\vec{u}_2^T a_{10}^T\vec{F}_1^A + v_{10v}^{Bv}\vec{u}_1^T\vec{F}_1^B + \omega_{21v}^{Bv}\vec{u}_3^T\vec{M}_2^B \\ f_{21}^B &= -\omega_{10n}^{Av} m_{10sc}^A + \omega_{10n}^{Av}\vec{u}_3^T\vec{M}_1^A + v_{21n}^{Bv}\vec{u}_2^T\vec{F}_1^B + \omega_{21n}^{Bv}\vec{u}_3^T\vec{M}_2^B \\ f_{32}^B &= v_{10a}^{Bv}\vec{u}_1^T\vec{F}_1^B + \omega_{21a}^{Bv}\vec{u}_3^T\vec{M}_2^B + v_{32a}^{Bv}\vec{u}_1^T\vec{F}_3^B, \end{aligned} \tag{7.37}$$

where

$$\begin{aligned} &\omega_{10h}^{Av} = 0, \quad v_{10h}^{Av} = 1, \quad v_{10h}^{Bv} = 1, \quad \omega_{21h}^{Bv} = 0 \\ &\omega_{10v}^{Av} = 0, \quad v_{10v}^{Av} = 1, \quad v_{10v}^{Bv} = \tan\phi_2, \quad \omega_{21v}^{Bv} = \frac{1}{l_2\cos\phi_2} \\ &\omega_{10n}^{Av} = \frac{\sin\phi_2}{l_1\sin(\phi_1+\phi_2)}, \quad v_{10n}^{Bv} = 0, \quad v_{21n}^{Bv} = 1, \quad \omega_{21n}^{Bv} = -\frac{\sin\phi_1}{l_2\sin(\phi_1+\phi_2)} \\ &\omega_{10a}^{Av} = 0, \quad v_{10a}^{Bv} = \frac{1}{\cos\phi_2}, \quad \omega_{21a}^{Bv} = \frac{\sin\phi_2}{l_2\cos\phi_2}, \quad v_{32a}^{Bv} = 1 \\ &\vec{F}_3^B = 0.5m_2^B[\theta\vec{\gamma}_{20}^B + 0.75l_2\theta(\tilde{\omega}_{20}^B\tilde{\omega}_{20}^B + \tilde{\varepsilon}_{20}^B)\vec{u}_1] + 0.5gm_2^B\theta b_{20}\vec{u}_2, \quad \theta = rot(y,\pi). \end{aligned} \tag{7.38}$$

Imagining a second set of virtual displacements with

$$\begin{aligned} \omega_{10b}^{Av} &= -\frac{1}{l_1 \sin(\phi_1 + \phi_2)}, \quad v_{10b}^{Bv} = 0, \\ \omega_{21b}^{Bv} &= -\frac{\cos(\phi_1 + \phi_2)}{l_2 \sin(\phi_1 + \phi_2)}, \quad v_{32b}^{Bv} = 1, \end{aligned} \tag{7.39}$$

we can obtain the same time-history evolution of the axial force f_{32}^B in the middle section of the connecting rod **2b**, as follows:

$$f_{32}^B = -\omega_{10b}^{Av} m_{10sc}^A + \omega_{10b}^{Av} \vec{u}_3^T \vec{M}_1^A + \omega_{21b}^{Bv} \vec{u}_3^T \vec{M}_2^B + v_{32b}^{Bv} \vec{u}_1^T \vec{F}_3^B. \tag{7.40}$$

A solution to the inverse dynamics problem of the slider-crank planar mechanism can be developed based on *Lagrange's equations* of second kind for a mechanical system with constraints. The generalized coordinates of the mechanism are represented by three variables

$$q_1 = \varphi_{10}^A, \quad q_2 = \lambda_{10}^B, \quad q_3 = \varphi_{21}^B. \tag{7.41}$$

Lagrange's equations with their *two multipliers* λ_1, λ_2 will be expressed by three differential relations [18]

$$\frac{d}{dt}\left\{\frac{\partial L}{\partial \dot{q}_k}\right\} - \frac{\partial L}{\partial q_k} = \Phi_k + \lambda_1 c_{1k} + \lambda_2 c_{2k} \quad (k = 1, 2, 3), \tag{7.42}$$

which contain following three generalized forces associated with the non-conservative applied forces

$$\Phi_1 = m_{10sc}^A, \quad \Phi_2 = -f_{10}^B, \quad \Phi_3 = 0. \tag{7.43}$$

It must be remark that the vector $\vec{\lambda} = \begin{bmatrix} \lambda_1 & \lambda_2 \end{bmatrix}^T$ of multipliers λ_1 and λ_2 represent precisely the vector that contains $\begin{bmatrix} X_2^A & Y_2^A \end{bmatrix}^T$ which contains the components of the internal reaction force in the virtually cut joint A_2. This vector $\vec{\lambda}$, whose dimension is equal to the number of constraint equations, is called the vector of *Lagrange multipliers*.

Two kinematical conditions of constraint $\sum_{k=1}^{3} c_{sk}\dot{q}_k = 0, (s = 1, 2)$, given by (3.50), can be written in a matrix form $C\dot{\vec{q}} = 0$ with the components

$$\begin{aligned} c_{11} &= l_1 \vec{u}_1^T a_{10}^T \vec{u}_2, \quad c_{12} = 1, \quad c_{13} = -l_2 \vec{u}_1^T b_{20}^T \vec{u}_2 \\ c_{21} &= l_1 \vec{u}_2^T a_{10}^T \vec{u}_2, \quad c_{22} = 0, \quad c_{23} = -l_2 \vec{u}_2^T b_{20}^T \vec{u}_2 \\ \dot{q}_1 &= \dot{\varphi}_{10}^A, \quad \dot{q}_2 = \dot{\lambda}_{10}^B, \quad \dot{q}_3 = \dot{\varphi}_{21}^B. \end{aligned} \tag{7.44}$$

The general Lagrange function

$$L = L_1^A + L_1^B + L_2^B = \frac{1}{2}[J_{1z}^A(\dot{\varphi}_{10}^A)^2 + J_{2z}^B(\dot{\varphi}_{21}^B)^2 \\ + (m_1^B + m_2^B)(\dot{\lambda}_{10}^B)^2 - l_2 m_2^B \dot{\lambda}_{10}^B \dot{\varphi}_{21}^B \sin(\varphi_{21}^B + \alpha_2) \\ - l_1 m_1^A g \sin(\varphi_{10}^A + \alpha_1) - l_2 m_2^B g \sin(\varphi_{21}^B + \alpha_2)] \tag{7.45}$$

can be expressed in a compact form as an analytical function of the generalized coordinates and their first derivatives

$$L_1^A = \frac{1}{2}\vec{\omega}_{10}^{AT} \hat{J}_1^A \vec{\omega}_{10}^A - m_1^A g \vec{u}_2^T a_{10}^T \vec{r}_1^{CA}, \quad L_1^B = \frac{1}{2} m_1^B \vec{v}_{10}^{BT} \vec{v}_{10}^B \\ L_2^B = \frac{1}{2} m_2^B \vec{v}_{20}^{BT} \vec{v}_{20}^B + \frac{1}{2}\vec{\omega}_{20}^{BT} \hat{J}_2^B \vec{\omega}_{20}^B + m_2^B \vec{v}_{20}^{BT} \tilde{\omega}_{20}^B \vec{r}_2^{CB} - m_2^B g \vec{u}_2^T b_{20}^T \vec{r}_2^{CB}. \tag{7.46}$$

The first derivatives of orthogonal matrices a_{10} and b_{21} are computed as follows:

$$\dot{a}_{10} = \dot{\varphi}_{10}^A \tilde{u}_3^T a_{10}, \quad \dot{a}_{10}^T = \dot{\varphi}_{10}^A a_{10}^T \tilde{u}_3, \quad \frac{\partial a_{10}}{\partial \varphi_{10}^A} = \tilde{u}_3^T a_{10}, \quad \frac{\partial a_{10}^T}{\partial \varphi_{10}^A} = a_{10}^T \tilde{u}_3 \\ \dot{b}_{21} = \dot{\varphi}_{21}^B \tilde{u}_3^T b_{21}, \quad \dot{b}_{21}^T = \dot{\varphi}_{21}^B b_{21}^T \tilde{u}_3, \quad \frac{\partial b_{21}}{\partial \varphi_{21}^B} = \tilde{u}_3^T b_{21}, \quad \frac{\partial b_{21}^T}{\partial \varphi_{21}^B} = b_{21}^T \tilde{u}_3. \tag{7.47}$$

Based on the relations (3.53), we can calculate the partial derivatives and the derivatives with respect to time of the Lagrange functions, using the above first derivatives (7.47) of orthogonal transformation matrices

$$\frac{\partial L_1^A}{\partial \varphi_{10}^A} = -m_1^A g \vec{u}_2^T a_{10}^T \tilde{u}_3 \vec{r}_1^{CA}, \quad \frac{\partial L_1^A}{\partial \dot{\varphi}_{10}^A} = \dot{\varphi}_{10}^A \vec{u}_3^T \hat{J}_1^A \vec{u}_3 \\ \frac{\partial L_1^B}{\partial \lambda_{10}^B} = 0, \quad \frac{\partial L_1^B}{\partial \dot{\lambda}_{10}^B} = m_1^B \dot{\lambda}_{10}^B \\ \frac{\partial L_2^B}{\partial \lambda_{10}^B} = 0, \quad \frac{\partial L_2^B}{\partial \dot{\lambda}_{10}^B} = m_2^B \dot{\lambda}_{10}^B + m_2^B \dot{\varphi}_{21}^B \vec{u}_1^T b_{21}^T \tilde{u}_3 \vec{r}_2^{CB} \\ \frac{\partial L_2^B}{\partial \varphi_{21}^B} = m_2^B \dot{\lambda}_{10}^B \dot{\varphi}_{21}^B \vec{u}_1^T b_{21}^T \tilde{u}_3 \tilde{u}_3 \vec{r}_2^{CB} - m_2^B g \vec{u}_2^T b_{20}^T \tilde{u}_3 \vec{r}_2^{CB} \\ \frac{\partial L_2^B}{\partial \dot{\varphi}_{21}^B} = \dot{\varphi}_{21}^B \vec{u}_3^T \hat{J}_2^B \vec{u}_3 + m_2^B \dot{\lambda}_{10}^B \vec{u}_1^T b_{21}^T \tilde{u}_3 \vec{r}_2^{CB}$$

$$\frac{d}{dt}\left\{\frac{\partial L}{\partial \dot{\varphi}_{10}^{A}}\right\} = \ddot{\varphi}_{10}^{A}\vec{u}_3^T\hat{J}_1^A\vec{u}_3$$
$$\frac{d}{dt}\left\{\frac{\partial L}{\partial \dot{\lambda}_{10}^{B}}\right\} = (m_1^B + m_2^B)\ddot{\lambda}_{10}^B + m_2^B\ddot{\varphi}_{21}^B\vec{u}_1^T b_{21}^T\tilde{u}_3\vec{r}_2^{CB} + m_2^B\dot{\varphi}_{21}^B\dot{\varphi}_{21}^B\vec{u}_1^T b_{21}^T\tilde{u}_3\tilde{u}_3\vec{r}_2^{CB}$$
$$\frac{d}{dt}\left\{\frac{\partial L}{\partial \dot{\varphi}_{21}^{B}}\right\} = \ddot{\varphi}_{21}^{B}\vec{u}_3^T\hat{J}_2^B\vec{u}_3 + m_2^B\ddot{\lambda}_{10}^B\vec{u}_1^T b_{21}^T\tilde{u}_3\vec{r}_2^{CB} + m_2^B\dot{\lambda}_{10}^B\dot{\varphi}_{21}^B\vec{u}_1^T b_{21}^T\tilde{u}_3\tilde{u}_3\vec{r}_2^{CB}. \tag{7.48}$$

In the direct or inverse dynamics problem, after elimination of two multipliers λ_1, λ_2 from the relations (7.42), finally we obtain the same compact expression (7.32) for the input torque m_{10sc}^A.

In particular, if the planar mechanism has the following geometrical and architectural characteristics:

$$\begin{gathered} m_1^A = m_2^B = m = 0.5\,\text{kg}, \quad m_1^B = M = 2\,\text{kg}, \; f_{10}^B = 10\,\text{N}, \quad g = 9.81\,\text{ms}^{-2} \\ l_1 = l_2 = l, \quad \alpha_1 = \alpha_2 = \alpha, \; \varphi_{10}^A = \varphi_{21}^B = \varphi, \; \phi_1 = \phi_2 = \phi = \varphi + \alpha, \end{gathered} \tag{7.49}$$

we obtains the following expression for the active torque

$$\begin{aligned} m_{10sc}^A = & \frac{2}{3}ml^2\ddot{\varphi} + l^2(m+2M)(2\ddot{\varphi}\sin^2\phi + \dot{\varphi}^2\sin 2\phi) \\ & + mgl\cos\phi + 2f_{10}^B l\sin\phi. \end{aligned} \tag{7.50}$$

Using the characteristics (3.55), (3.56) and (7.49) of the slider-crank planar mechanism (Figs. 3.4 and 7.4), we sketch the graphs for the input torque m_{10sc}^A (Fig. 7.5), the components f_{21}^{Ax}, f_{21}^{Ay} (Fig. 7.6) of the reaction force developed in the internal joint A_2, the components f_{10}^{Ax} (Fig. 7.7), f_{10}^{Ay} (Fig. 7.8) of the joint force acting in the external joint A_1, the external reaction force f_{21}^B (Fig. 7.9) acting on the slider-block and the *internal axial force* f_{32}^B in the middle section of the connecting rod (Fig. 7.10).

The simple one-degree-of-freedom slider-crank mechanism example, analyzed in detail here, demonstrates the use of the theoretical formulation and of general-purpose multi-body computer programs to solve the inverse dynamics problem. A similar procedure can be used if the mechanical system has multiple degrees of freedom, where one can obtain the driving joint forces associated with the independent joint coordinates.

Remark

In order to validate the efficiency of the above theoretical modelling, the time-history of all internal and external reaction forces can quickly be *verified* using appropriate equations delivered by the Newton-Euler procedure, as follows:

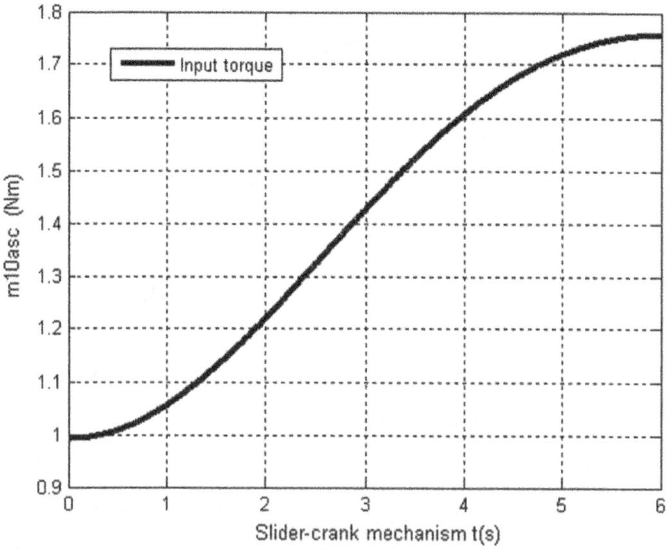

Fig. 7.5 Input torque m^A_{10sc}

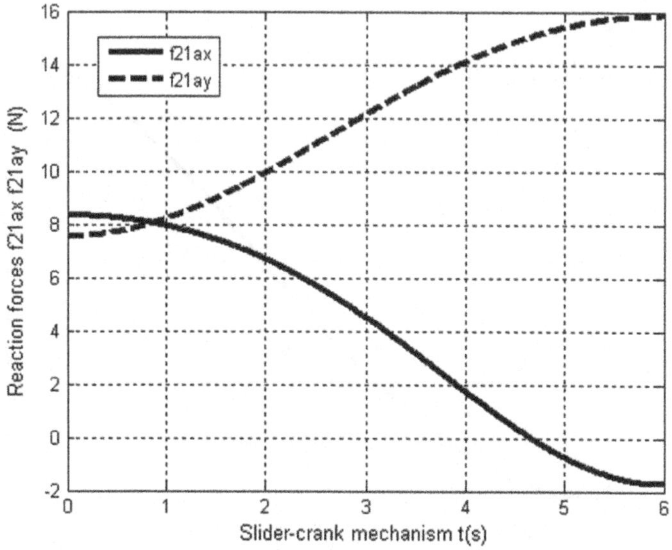

Fig. 7.6 Internal reaction forces f^{Ax}_{21} and f^{Ay}_{21}

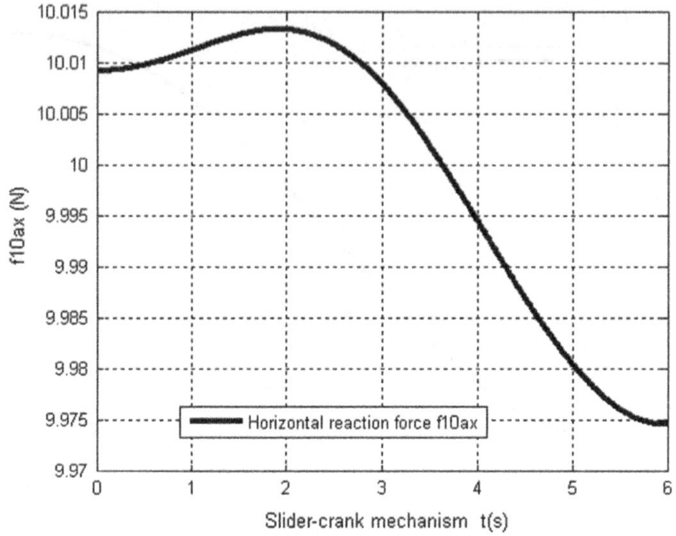

Fig. 7.7 Horizontal reaction force f_{10}^{Ax}

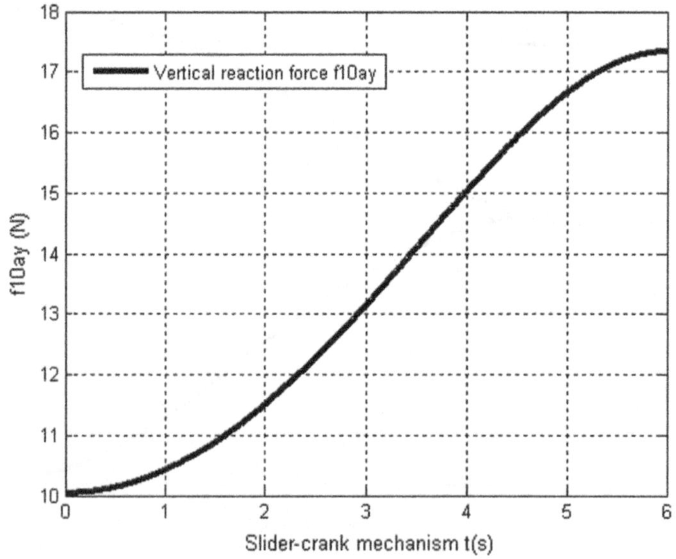

Fig. 7.8 Vertical reaction force f_{10}^{Ay}

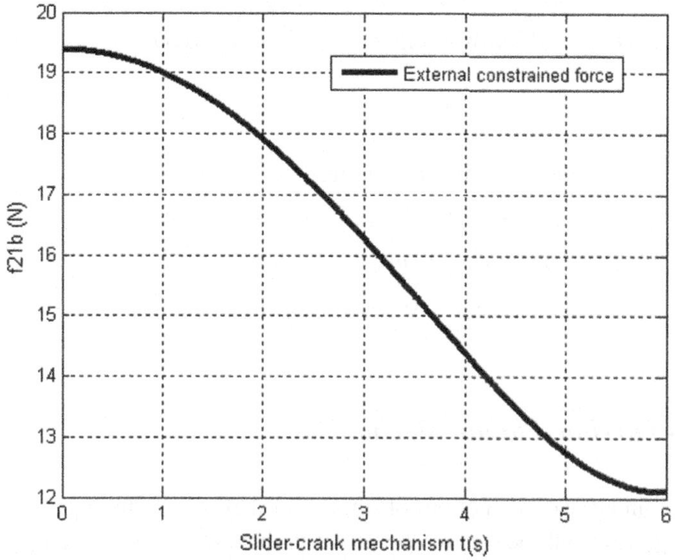

Fig. 7.9 External constrained force f_{21}^{B}

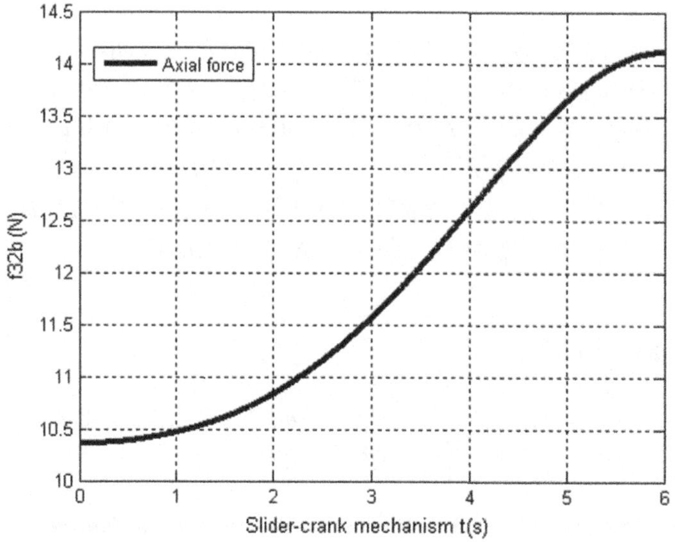

Fig. 7.10 Internal axial force f_{32}^{B}

$$
\begin{aligned}
f_{21}^{B} &= (M+m)g - 0.5\left(ml\ddot{\varphi} - \frac{m_{10sc}^{A}}{l}\right) / \cos\phi \\
f_{32}^{B} &= \left[f_{10}^{B} + l\left(2M + \frac{7}{8}m\right)(\ddot{\varphi}\sin\phi + \dot{\varphi}^2\cos\phi)\right] / \cos\phi \\
f_{10}^{Ax} &= f_{10}^{B} + 2l(M+m)(\ddot{\varphi}\sin\phi + \dot{\varphi}^2\cos\phi) \\
f_{10}^{Ax} &= (2M+m)g - f_{21}^{B} + ml(\ddot{\varphi}\cos\phi - \dot{\varphi}^2\sin\phi) \\
f_{21}^{Ax} &= mg\sin\phi - 0.5ml\dot{\varphi}^2 + f_{10}^{Ax}\cos\phi - f_{10}^{Ay}\sin\phi \\
f_{21}^{Ay} &= -mg\cos\phi - 0.5ml\ddot{\varphi} + f_{10}^{Ax}\sin\phi + f_{10}^{Ay}\cos\phi.
\end{aligned}
\tag{7.51}
$$

7.5.3 *Four-Bar Planar Mechanism*

Given the input rotary motion of the crankshaft A_1A_2 through the function $\varphi_{10}^{A} = \varphi_{10}^{A}(t)$, we will determine the active torque m_{10fb}^{A} of this planar mechanism (Fig. 3.11). To apply the principle of virtual work, the mechanical system is artificially transformed into two independent planar trees, by separating the passive revolute joint $A_3 \equiv B_2$, for example.

We will compute with respect to the origins A_1, A_2 and B_1 of three frames the torques of the weights and the forces of inertia acting on the rigid bodies

$$
\begin{aligned}
\vec{f}_{1}^{*A} &= -gm_{1}^{A}a_{10}\vec{u}_2, \quad \vec{m}_{1}^{*A} = \tilde{r}_{1}^{CA}\vec{f}_{1}^{*A}, \quad \vec{r}_{1}^{CA} = 0.5l_1\vec{u}_1, \quad \tilde{r}_{1}^{CA} = 0.5l_1\tilde{u}_1 \\
\vec{f}_{2}^{*A} &= -gm_{2}^{A}a_{20}\vec{u}_2, \quad \vec{m}_{2}^{*A} = \tilde{r}_{2}^{CA}\vec{f}_{2}^{*A}, \quad \vec{r}_{2}^{CA} = -0.5l_2\vec{u}_1, \quad \tilde{r}_{2}^{CA} = -0.5l_2\tilde{u}_1 \\
\vec{f}_{1}^{*B} &= -gm_{1}^{B}b_{10}\vec{u}_2, \quad \vec{m}_{1}^{*B} = \tilde{r}_{1}^{CB}\vec{f}_{1}^{*B}, \quad \vec{r}_{1}^{CB} = -0.5l_3\vec{u}_1, \quad \tilde{r}_{1}^{CB} = -0.5l_3\tilde{u}_1, \\
\vec{f}_{1}^{\,inA} &= -m_{1}^{A}(\tilde{\omega}_{10}^{A}\tilde{\omega}_{10}^{A} + \tilde{\varepsilon}_{10}^{A})\vec{r}_{1}^{CA}, \quad \vec{m}_{1}^{\,in\,A} = -(\hat{J}_{1}^{A}\vec{\varepsilon}_{10}^{A} + \tilde{\omega}_{10}^{A}\hat{J}_{1}^{A}\vec{\omega}_{10}^{A}) \\
\vec{f}_{2}^{\,in\,A} &= -m_{2}^{A}[\vec{\gamma}_{20}^{A} + (\tilde{\omega}_{20}^{A}\tilde{\omega}_{20}^{A} + \tilde{\varepsilon}_{20}^{A})\vec{r}_{2}^{CA}], \quad \vec{m}_{2}^{\,in\,A} = -(m_{2}^{A}\tilde{r}_{2}^{CA}\vec{\gamma}_{20}^{A} + \hat{J}_{2}^{A}\vec{\varepsilon}_{20}^{A} + \tilde{\omega}_{20}^{B}\hat{J}_{2}^{A}\vec{\omega}_{20}^{A}) \\
\vec{f}_{1}^{\,in\,B} &= -m_{1}^{B}(\tilde{\omega}_{10}^{B}\tilde{\omega}_{10}^{B} + \tilde{\varepsilon}_{10}^{B})\vec{r}_{1}^{CB}, \quad \vec{m}_{1}^{\,in\,B} = -(\hat{J}_{1}^{B}\vec{\varepsilon}_{10}^{B} + \tilde{\omega}_{10}^{B}\hat{J}_{1}^{B}\vec{\omega}_{10}^{B}).
\end{aligned}
\tag{7.52}
$$

Using the notations

$$
\begin{aligned}
\vec{F}_{\tau 0}^{A} &= -\vec{f}_{\tau}^{\,in\,A} - \vec{f}_{\tau}^{*A}, \quad \vec{M}_{\tau 0}^{A} = -\vec{m}_{\tau}^{\,in\,A} - \vec{m}_{\tau}^{*A}, \quad (\tau = 1, 2) \\
\vec{F}_{10}^{B} &= -\vec{f}_{1}^{\,in\,B} - \vec{f}_{1}^{*B}, \quad \vec{M}_{10}^{B} = -\vec{m}_{1}^{\,in\,b} - \vec{m}_{1}^{*B},
\end{aligned}
\tag{7.53}
$$

we can obtain recursive relations for the *four-bar planar mechanism*

$$
\vec{F}_{2}^{A} = \vec{F}_{20}^{A}, \quad \vec{M}_{2}^{A} = \vec{M}_{20}^{A}, \quad \vec{M}_{1}^{A} = \vec{M}_{10}^{A} + a_{21}^{T}\vec{M}_{2}^{A} + \tilde{r}_{21}^{A}a_{21}^{T}\vec{F}_{2}^{A}, \quad \vec{M}_{1}^{B} = \vec{M}_{10}^{B}. \tag{7.54}
$$

Starting from general *equations of the parallel robots dynamics* (7.18), the following relation is derived for the *active torque*:

$$m^A_{10fb} = \vec{u}^T_3 \vec{M}^A_1 + \omega^{Av}_{21} \vec{u}^T_3 \vec{M}^A_2 + \omega^{Bv}_{10} \vec{u}^T_3 \vec{M}^B_1, \tag{7.55}$$

where the virtual velocities ω^{Av}_{21} and ω^{Bv}_{10} are given by the equations

$$\begin{aligned} &\omega^{Av}_{21} \vec{u}^T_i a^T_{10} a^T_{21} \tilde{u}_3 \vec{r}^A_{32} - \omega^{Bv}_{10} \vec{u}^T_i b^T_{10} \tilde{u}_3 \vec{r}^B_{21} \\ &\quad = -\vec{u}^T_i a^T_{10} \tilde{u}_3 (\vec{r}^A_{21} + a^T_{21} \vec{r}^A_{32}) \quad (i = 1, 2). \end{aligned} \tag{7.56}$$

7.5.4 Quick-Return Planar Mechanism

Knowing the kinematics state of this planar mechanism (Fig. 3.12), we can determine the torque m^A_{10qr} of the revolute actuator A_1, based on the principle of virtual work. The mechanism system is first artificially transformed into two independent planar trees, by cutting the passive revolute joint $A_2 \equiv B_2$.

The torques of the weights and the forces of inertia forces are computed with respect to the origins A_1, B_1 and B_2 of three frames, respectively, as follows:

$$\begin{aligned} &\vec{f}^{*A}_1 = -g m^A_1 a_{10} \vec{u}_2, \quad \vec{m}^{*A}_1 = \tilde{r}^{CA}_1 \vec{f}^{*A}_1, \quad \vec{r}^{CA}_1 = 0.5 l_1 \vec{u}_1, \quad \tilde{r}^{CA}_1 = 0.5 l_1 \tilde{u}_1 \\ &\vec{f}^{*B}_1 = -g m^B_1 b_{10} \vec{u}_2, \quad \vec{m}^{*B}_1 = \tilde{r}^{CB}_1 \vec{f}^{*B}_1, \quad \vec{r}^{CB}_1 = 0.5 l_2 \vec{u}_1, \quad \tilde{r}^{CB}_1 = 0.5 l_2 \tilde{u}_1 \\ &\vec{f}^{*B}_2 = -g m^B_2 b_{20} \vec{u}_2, \quad \vec{m}^{*B}_2 = \vec{0} \\ &\vec{f}^{inA}_1 = -m^A_1 (\tilde{\omega}^A_{10} \tilde{\omega}^A_{10} + \tilde{\varepsilon}^A_{10}) \vec{r}^{CA}_1, \quad \vec{m}^{in\,A}_1 = -(\hat{J}^A_1 \vec{\varepsilon}^A_{10} + \tilde{\omega}^A_{10} \hat{J}^A_1 \vec{\omega}^A_{10}) \\ &\vec{f}^{in\,B}_1 = -m^B_1 (\tilde{\omega}^B_{10} \tilde{\omega}^B_{10} + \tilde{\varepsilon}^B_{10}) \vec{r}^{CB}_1, \quad \vec{m}^{in\,B}_1 = -(\hat{J}^B_1 \vec{\varepsilon}^B_{10} + \tilde{\omega}^B_{10} \hat{J}^B_1 \vec{\omega}^B_{10}) \\ &\vec{f}^{in\,B}_2 = -m^B_2 \vec{\gamma}^B_{20}, \quad \vec{m}^{in\,B}_2 = \vec{0} \end{aligned} \tag{7.57}$$

Starting from the notations (7.30), one expresses the following vectors for the *quick-return planar mechanism*

$$\vec{M}^A_1 = \vec{M}^A_{10}, \quad \vec{F}^B_2 = \vec{F}^B_{20}, \quad \vec{M}^B_1 = \vec{M}^B_{10} + \tilde{r}^B_{21} b^T_{21} \vec{F}^B_2. \tag{7.58}$$

Pursuing a similar procedure, we can obtain the *active torque*

$$m^A_{10qr} = \vec{u}^T_3 \vec{M}^A_1 + \omega^{Bv}_{10} \vec{u}^T_3 \vec{M}^B_1 + v^{Bv}_{21} \vec{u}^T_1 \vec{F}^B_2, \tag{7.59}$$

where the virtual velocities

$$\omega^{Bv}_{10} = l_1 \frac{\cos(\phi_1 + \phi_2)}{l_0 + \lambda^B_{21}}, \quad v^{Bv}_{21} = l_1 \sin(\phi_1 + \phi_2) \tag{7.60}$$

are delivered from the constraint conditions

$$(l_0 + \lambda_{21}^{B})\omega_{10}^{Bv}\vec{u}_i^T b_{10}^T \vec{u}_2 + v_{21}^{Bv}\vec{u}_i^T b_{10}^T \vec{u}_1 = l_1 \vec{u}_i^T a_{10}^T \vec{u}_2 \quad (i = 1, 2). \tag{7.61}$$

References

1. Tsai, L.-W.: Robot Analysis: The Mechanics of Serial and Parallel Manipulators. Wiley, New York (1999)
2. McCarthy, J.M.: Dual orthogonal matrix in manipulator kinematics. Int. J. Robot. Res. **5**(2), 45–51 (1986)
3. Staicu, S.: Mecanica Teoretica. Edit. Didactica & Pedagogica, Bucharest (1998)
4. Staicu, S., Zhang, D.: A novel dynamic modelling approach for parallel mechanisms analysis. Robot. Comput.-Integr. Manuf. **24**(1), 167–172 (2008)
5. Staicu, S.: Dynamics analysis of the star parallel manipulator. Robot. Auton. Syst. **57**(11), 1057–1064 (2009)
6. Dasgupta, B., Mruthyunjaya, T.S.: A Newton-Euler formulation for the inverse dynamics of the Stewart platform manipulator. Mech. Mach. Theory **33**(8), 1135–1152 (1998)
7. Guegan, S., Khalil, W., Chablat, D., Wenger, P.: Modélisation dynamique d'un robot parallèle à 3-DDL: l'Orthoglide. In: Proceedings of Conférence Internationale Francophone d'Automatique, Nantes (2002)
8. Li, Y., Xu, Q.: Kinematics and inverse dynamics analysis for a general 3-PRS spatial parallel mechanism. Robotica **23**(2), 219–229 (2005)
9. Zanganeh, R., Sinatra, R., Angeles, J.: Kinematics and dynamics of a six-degrees-of-freedom parallel manipulator with revolute legs. Robotica **15**(4), 385–394 (1997)
10. Miller, K., Clavel, R.: The lagrange-based model of Delta-4 robot dynamics. Robotersysteme **8**, 49–54 (1992)
11. Xin, G., Deng, H., Zhong, G.: Closed-form dynamics of a 3-DOF spatial parallel manipulator by combining the lagrangian formulation with the virtual work principle. Nonlinear Dyn. **86** (2), 1329–1347 (2016)
12. Tsai, L.-W.: Solving the inverse dynamics of Stewart-Gough manipulator by the principle of virtual work. J. Mech. Des. **122**(1), 3–9 (2000)
13. Zhang, C.-D., Song, S.-M.: An efficient method for inverse dynamics of manipulators based on virtual work principle. J. Robot. Syst. **10**(5), 605–627 (1993)
14. Li, Y., Staicu, S.: Inverse dynamics of a 3-$\underline{P}$RC parallel kinematic machine. Nonlinear Dyn. **67**(2), 1031–1041 (2012)
15. Staicu, S., Liu, X.-J., Wang, J.: Inverse dynamics of the HALF parallel manipulator with revolute actuators. Nonlinear Dyn. **50**(1–2), 1–12 (2007)
16. Khalil, W., Ibrahim, O.: General solution for the dynamic modeling of parallel robots. J. Intell. Robot. Syst. **49**(1), 19–37 (2007)
17. Tsai, L-W., Stamper, R.: A parallel manipulator with only translational degrees of freedom. In: Proceedings of ASME Design Engineering Technical Conferences, Irvine, CA (1996)
18. Voinea, R., Stroe, I., Predoi, M.: Technical Mechanics. Edit. Politehnica Press, Bucharest (2010)

Chapter 8
Planar Parallel Robots

Symmetric parallel mechanisms for which the number of chains is strictly equal to the number of degrees of freedom of the end-effectors are called *fully parallel manipulators*. Planar parallel robots are useful for manipulating an object on a plane. A mechanism is said to be a *planar robot* if all the moving links in the mechanism perform planar motions. For a planar mechanism, the loci of all points in all links can be drawn conveniently on a plane. In a planar linkage, the axes of all revolute joints must be normal to the plane of motion, while the direction of translation of a prismatic joint must be parallel to the plane of motion.

Three planar independent kinematical chains, actuated by three actuators, are attached at three points to the moving platform, which has *three degrees offreedom*: two orthogonal translations and one rotation, which are computed from the *Grübler-Kutzbach* mobility criterion [1]. Assuming that each chain of the mechanism is constituted of two rigid bodies linked together by three joints, each pair must be a 1-DOF joint. Introducing revolute joints and prismatic pairs between the elements that form the assembly, each of these rigid bodies is linked by a joint either with the fixed base or with the end-effector. As each of these chains must be linked to the ground and to the moving platform at the same time, there will be three attachment points on the ground and another three on the platform.

A planar parallel manipulator is capable of performing a multitude of tasks by actuating each of three prismatic pairs or, alternatively, each of three revolute joints. Using revolute and prismatic joints as kinematical pairs, we obtain *seven* possible arrangements for 3-DOF fully planar parallel manipulators with three identical limb structures: *RRR*, *RRP*, *RPR*, *RPP*, *PRR*, *PRP* and *PPR*, where the revolute joint axes are perpendicular to the plane of motion, while the prismatic joint axes lies on the plane of motion. The simplest form of these manipulators, given by Merlet [2], consists of a traveling platform, three revolute or prismatic actuators, and three closed kinematical chains with two links in each chain, where *R* denotes a revolute joint while *P* is a prismatic joint.

In their paper, Aradyfio and Qiao [3] examine the inverse kinematics solution for the three different 3-DOF planar parallel robots. Gosselin and Angeles [4] and

S. Staicu, *Dynamics of Parallel Robots*, Parallel Robots: Theory and Applications,
https://doi.org/10.1007/978-3-319-99522-9_8

Pennock and Kassner [5] each present a kinematical study of a planar parallel robot where a moving platform is connected to a fixed base by three links, every leg consisting of two binary links and three parallel revolute joints. Sefrioui and Gosselin [6] give an interesting numerical solution in the inverse and direct kinematics of this kind of planar robot. Daniali et al. [7] present a study of velocity relationships and singular conditions for general planar parallel robots. Merlet [2] solved the forward pose kinematics problem for a broad class of planar parallel manipulators. Williams et al. [8] analysed the dynamics and the control of a planar three-degrees-of-freedom parallel manipulator at Ohio University, while Yang et al. [9] concentrate on the singularity analysis of a class of 3-$\underline{R}RR$ planar parallel robots developed in their laboratory. Bonev et al. [10] describe several types of singular configurations by studying the direct kinematics model of a 3-$\underline{R}PR$ planar parallel robot with actuated base joints.

Due to the possibility of interference between the legs which are connected to same moving platform, we find some disadvantages of planar parallel robots: limited workspace, many singular configurations within its workspace and reduced robot dexterity.

In the following, elaborative case studies are provided for three significant planar parallel manipulators.

8.1 Power Requirement Comparison in Dynamics of the 3-*PRR* Planar Parallel Robot

A recursive method is developed to reduce significantly the number of equations and computation operations by using a set of matrices for kinematics and inverse dynamics modelling of a 3-*PRR* planar parallel manipulator. Several equations offer expressions and graphs for the power requirement comparison of each of three actuators in two different actuation schemes: prismatic actuators and revolute actuators.

8.1.1 Kinematics Analysis

Provided with a closed-loop structure, the planar parallel 3-*PRR* manipulator is a special symmetrical mechanism composed of three planar kinematical chains with identical topology, all connecting the fixed base to the moving platform. The points A_0, B_0, C_0 represent the summits of a fixed triangular base and three moving revolute joints A_3, B_3, C_3 define the geometry of the moving platform. Every leg consists of two links, with one prismatic joint and two revolute joints. The parallel mechanism with seven links $(T_k, k = 1, 2, \ldots, 7)$ consists of three prismatic and six revolute joints (Fig. 8.1).

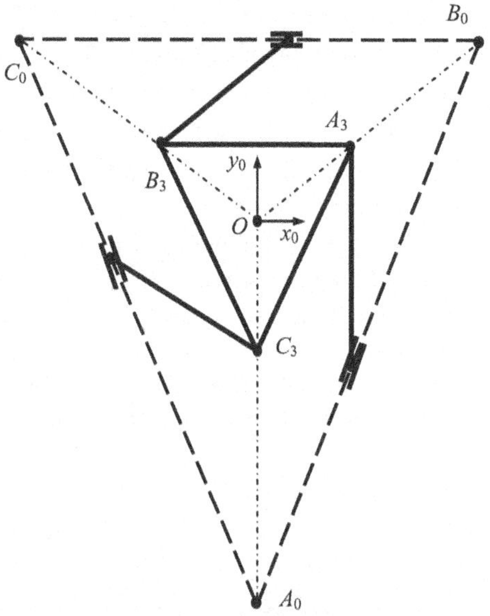

Fig. 8.1 The 3-*PRR* planar parallel manipulator

Whereas the kinematics degrees of freedom, or mobility, of a serial chain robot can be obtained as the sum of the degrees of freedom of each of the joints, the situation for parallel robots or closed chains in general is somewhat more complex, since only a subset of joints can be independently actuated. The Grübler-Kutzbach equation predicts that the device has certainly three degrees of freedom [1].

In the actuation scheme $\underline{P}RR$ each prismatic joint is an actively controlled prismatic cylinder. Thus, all prismatic actuators can be installed on the fixed base. In the second configuration $P\underline{R}R$ we consider the moving platform as the output link and A_2A_3, B_2B_3, C_2C_3 as the input links of three mobile revolute actuators.

For the purpose of analysis, we attach a Cartesian frame $Ox_0y_0z_0(T_0)$ to the fixed base with its origin located at triangle centre O, the z_0 axis perpendicular to the base and the x_0 axis pointing along the direction C_0B_0. Another mobile reference frame $Gx_Gy_Gz_G$ is attached to the moving platform. The origin of this central coordinate system is located just at the centre G of the moving triangle.

In the following, we will represent the intermediate reference frames by only two axes as is the practice in many robotics papers [1, 11]. It is noted that the relative translation $\lambda_{k,k-1}$ and the rotation angle $\varphi_{k,k-1}$ always point along or about the direction of z_k axis.

We consider that the moving platform is initially located at a *central configuration*, where the platform is not rotated with respect to the fixed base and the mass centre G coincides with the origin O of fixed frame.

One of three active legs (for example leg A) consists of a prismatic joint, which is also a piston **1** of mass m_1 linked at the $A_1x_1^Ay_1^Az_1^A$ frame, having a rectilinear motion of displacement λ_{10}^A, velocity $v_{10}^A = \dot{\lambda}_{10}^A$ and acceleration $\gamma_{10}^A = \ddot{\lambda}_{10}^A$. The second element of the leg is a rigid rod **2** linked at the $A_2x_2^Ay_2^Az_2^A$ frame, having a relative rotation about z_2^A axis with the angle φ_{21}^A, velocity $\omega_{21}^A = \dot{\varphi}_{21}^A$ and acceleration $\varepsilon_{21}^A = \ddot{\varphi}_{21}^A$. It has the length l_2, mass m_2 and tensor of inertia $\hat{J}_2$ with respect to T_2^A frame. Finally, a revolute joint is introduced at a planar moving platform, which is schematised as an equilateral triangle with edge $l = r\sqrt{3}$, mass m_3 and inertia tensor $\hat{J}_3$ with respect to A_3, which rotates with the angle φ_{32}^A and the angular velocity $\omega_{32}^A = \dot{\varphi}_{32}^A$ about z_3^A (Fig. 8.2).

The affixing the origin of any body coordinate system to a lower joint centre and attaching it to the body leads to a significant simplification in the resulting dynamics equations.

At the central configuration, we also consider that all legs are symmetrically extended and that the angles of orientation of three edges of the fixed platform are given by

$$\alpha_A = \frac{\pi}{3}, \quad \alpha_B = \pi, \quad \alpha_C = -\frac{\pi}{3}. \tag{8.1.1}$$

The determining of the location and orientation of the bodies in the mechanical system draws upon *position analysis*.

In the following, we apply the method of successive displacements to the geometric analysis of planar closed-loop chains and we note that a joint variable is the displacement required to move a link from the initial location to the actual position. If every link is connected to least two other links, the chain forms one or more independent closed loops. The variable angle $\varphi_{\tau,\tau-1}^j$ of rotation about the joint axis

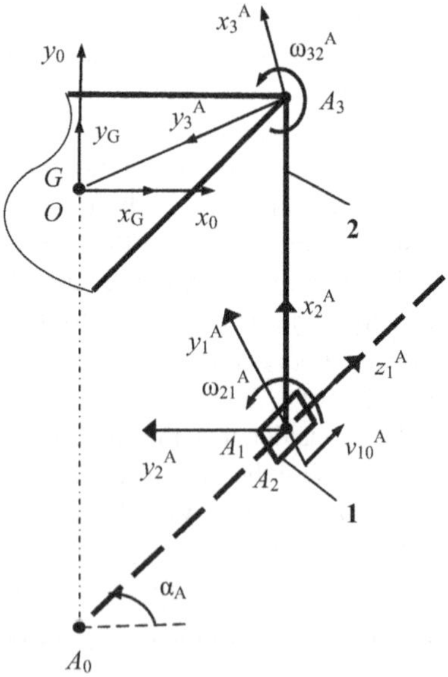

Fig. 8.2 Kinematical scheme of first leg *A* of the mechanism

z_τ^j is the parameter needed to bring the next link from a reference configuration to the next configuration. We call the matrix $q_{\tau,\tau-1}^{\varphi}$, for example, the orthogonal transformation 3×3 matrix of relative rotation with the angle $\varphi_{\tau,\tau-1}^j$ of link T_τ^j around z_τ^j.

In the study of the kinematics of planar robot manipulators, we are interested in deriving a matrix equation relating the location of an arbitrary T_τ^j body to the joint variables. When the change of coordinates is successively considered, the corresponding matrices are pre-multiplied. So, starting from the reference origin O and pursuing the three legs $OA_0A_1A_2A_3$, $OB_0B_1B_2B_3$, $OC_0C_1C_2C_3$, we obtain the following transformation matrices [12]:

$$q_{10} = \theta_1 a_\alpha^j, \quad q_{21} = q_{21}^{\varphi}\theta_2\theta_1^T, \quad q_{32} = q_{32}^{\varphi}\theta_2, \quad q_{20} = q_{21}q_{10}, \quad q_{30} = q_{32}q_{20}$$
$$(q = a, b, c) \quad (j = A, B, C) \tag{8.1.2}$$

where

$$q_{\tau,\tau-1}^{\varphi} = rot(z, \varphi_{\tau,\tau-1}^j) = \begin{bmatrix} \cos\varphi_{\tau,\tau-1}^j & \sin\varphi_{\tau,\tau-1}^j & 0 \\ -\sin\varphi_{\tau,\tau-1}^j & \cos\varphi_{\tau,\tau-1}^j & 0 \\ 0 & 0 & 1 \end{bmatrix} (\tau = 2,3), \quad a_\alpha^j = rot(z, \alpha_j)$$

$$\theta_1 = rot(y, \pi/2) = \begin{bmatrix} 0 & 0 & -1 \\ 0 & 1 & 0 \\ 1 & 0 & 0 \end{bmatrix}, \quad \theta_2 = rot(z, \pi/6) = \frac{1}{2}\begin{bmatrix} \sqrt{3} & 1 & 0 \\ -1 & \sqrt{3} & 0 \\ 0 & 0 & 2 \end{bmatrix}. \tag{8.1.3}$$

Three displacements $\lambda_{10}^A, \lambda_{10}^B, \lambda_{10}^C$ of the active links are the joint variables that give the input vector $\vec{\lambda}_{10} = [\lambda_{10}^A \quad \lambda_{10}^B \quad \lambda_{10}^C]^T$ of the instantaneous position of the mechanism in an *initial configuration*. But in the inverse geometric problem, we can consider that the position of the mechanism is completely given by the coordinates x_0^G, y_0^G of the mass centre G of the moving platform and the orientation angle ϕ of the movable frame $Gx_Gy_Gz_G$. The known orthogonal rotation matrix $R = rot(z, \phi)$ of the platform from $Ox_0y_0z_0$ to $Gx_Gy_Gz_G$ reference system is used.

Further, we suppose that the position vector $\vec{r}_0^G = [x_0^G \; y_0^G \; 0]^T$ of G centre and the orientation angle ϕ, which are expressed by the following analytical functions

$$\frac{x_0^G}{x_{0G}^*} = \frac{y_0^G}{y_{0G}^*} = \frac{\phi}{\phi^*} = 1 - \cos\frac{\pi}{3}t, \tag{8.1.4}$$

are sufficient to describe the general absolute motion of the moving platform in its *vertical plane*. The values $2x_{0G}^*, 2y_{0G}^*, 2\phi^*$ denote the final pose of the moving platform.

The conditions concerning the absolute orientation of the moving platform are expressed by three identities

$$q_{30}^{\circ T} q_{30} = R, \quad (q = a, b, c) \tag{8.1.5}$$

where $q_{30}^{\circ} = q_{30}(t = 0) = \theta_2 \theta_2 a_{\alpha}^{j}$; we obtain

$$\varphi_{21}^{A} + \varphi_{32}^{A} = \varphi_{21}^{B} + \varphi_{32}^{B} = \varphi_{21}^{C} + \varphi_{32}^{C} = \phi. \tag{8.1.6}$$

Six independent variables λ_{10}^{A}, φ_{21}^{A}, λ_{10}^{B}, φ_{21}^{B}, λ_{10}^{C}, φ_{21}^{C} will be determined by several vector-loop relations

$$\vec{r}_{10}^{\,j} + q_{10}^{T} \vec{r}_{21}^{\,j} + q_{20}^{T} \vec{r}_{32}^{\,j} + q_{30}^{T} \vec{r}_{3j}^{\,G} = \vec{r}_{0}^{\,G} \quad (q = a, b, c) \ (j = A, B, C), \tag{8.1.7}$$

where

$$\begin{aligned} &\vec{r}_{10}^{\,j} = \vec{r}_{00}^{\,j} + (l_1 + \lambda_{10}^{j}) q_{10}^{T} \vec{u}_3, \quad \vec{r}_{21}^{\,j} = \vec{0}, \quad \vec{r}_{32}^{\,j} = l_2 \vec{u}_1, \quad \vec{r}_{3j}^{\,G} = r \vec{u}_2 \\ &\vec{r}_{00}^{\,A} = l_0 [0 \quad -1 \quad 0]^{T}, \quad \vec{r}_{00}^{\,B} = 0.5 l_0 \left[\sqrt{3} \quad 1 \quad 0\right]^{T}, \quad \vec{r}_{00}^{\,C} = 0.5 l_0 \left[-\sqrt{3} \quad 1 \quad 0\right]^{T} \\ &\vec{u}_1 = [1 \quad 0 \quad 0]^{T}, \quad \vec{u}_2 = [0 \quad 1 \quad 0]^{T}, \quad \vec{u}_3 = [0 \quad 0 \quad 1]^{T}. \end{aligned} \tag{8.1.8}$$

We obtain the following analytical equations

$$\begin{aligned} &(l_1 + \lambda_{10}^{j}) \cos \alpha_i + l_2 \cos\left(\varphi_{21}^{j} + \frac{\pi}{6} + \alpha_j\right) = x_0^{G} - x_{00}^{j} + r \sin\left(\phi + \frac{\pi}{3} + \alpha_j\right) \\ &(l_1 + \lambda_{10}^{j}) \sin \alpha_i + l_2 \sin\left(\varphi_{21}^{j} + \frac{\pi}{6} + \alpha_j\right) = y_0^{G} - y_{00}^{j} - r \cos\left(\phi + \frac{\pi}{3} + \alpha_j\right). \end{aligned} \tag{8.1.9}$$

The second step in the kinematics study is the velocity analysis, which is used to determine the respective velocities of the bodies in the system as function of the time rate of the degrees of freedom. This can be achieved by differentiating the relationships obtained from the position analysis.

Known the planar motion of the moving platform through its angular velocity $\vec{\omega}_0^{G} = \dot{\phi}\, \vec{u}_3$ and the centre's velocity $\vec{v}_0^{G} = \dot{\vec{r}}_0^{G}$, we compute first the linear and angular velocities of every leg.

The rotations of the compounding elements of an arbitrary leg j are characterized by recursive relations of skew-symmetric matrices

$$\tilde{\omega}_{k0}^{j} = q_{k,k-1} \tilde{\omega}_{k-1,0}^{j} q_{k,k-1}^{T} + \omega_{k,k-1}^{j} \tilde{u}_3, \quad \omega_{k,k-1}^{j} = \dot{\varphi}_{k,k-1}^{j} \ (k = 1, 2, 3), \tag{8.1.10}$$

which are *associated* with the absolute angular velocities

$$\begin{aligned} &\vec{\omega}^j_{10} = \vec{0}, \quad \vec{\omega}^j_{\tau,\tau-1} = \dot{\varphi}^j_{k,k-1}\vec{u}_3, \quad \vec{\omega}^j_{20} = q_{21}\vec{\omega}^j_{10} + \vec{\omega}^j_{21} = \dot{\varphi}^j_{21}\vec{u}_3 \\ &\vec{\omega}^j_{30} = q_{32}\vec{\omega}^j_{20} + \vec{\omega}^j_{32} = (\dot{\varphi}^j_{21} + \dot{\varphi}^j_{32})\vec{u}_3 \quad (\tau = 2, 3). \end{aligned} \tag{8.1.11}$$

Following relations give the velocities $\vec{v}^j_{k0}$ of the joints

$$\begin{aligned} &\vec{v}^j_{k0} = q_{k,k-1}\vec{v}^j_{k-1,0} + q_{k,k-1}\tilde{\omega}^j_{k-1,0}\vec{r}^j_{k,k-1} + v^j_{k,k-1}\vec{u}_3, \quad v^j_{10} = \dot{\lambda}^j_{10}, \quad v^j_{\tau,\tau-1} = 0 \\ &(\tau = 2, 3). \end{aligned} \tag{8.1.12}$$

Equations of geometrical constraints (8.1.6) and (8.1.7) when differentiated with respect to time lead to the following *matrix conditions of connectivity* [13]:

$$\begin{aligned} &\omega^j_{21} + \omega^j_{32} = \dot{\phi}, v^j_{10}\vec{u}^T_i q^T_{10}\vec{u}_3 + \omega^j_{21}\vec{u}^T_i q^T_{20}\tilde{u}_3\left\{\vec{r}^j_{32} + q^T_{32}\vec{r}^G_{3j}\right\} + \omega^j_{32}\vec{u}^T_i q^T_{30}\tilde{u}_3\vec{r}^G_{3j} = \vec{u}^T_i\dot{\vec{r}}^G_0 \\ &(i = 1, 2) \;\; (q = a, b, c) \;\; (j = A, B, C), \end{aligned} \tag{8.1.13}$$

where $\widetilde{u}_3$ is a skew-symmetric matrix associated with the unit vector $\vec{u}_3$ pointing in the positive direction of the z_k axis. From these equations, we obtain the relative velocities $v^j_{10}, \omega^j_{21}, \omega^j_{32}$ as functions of the angular velocity of the platform $\dot{\phi}$ and the velocity of mass centre G. However, these conditions give also the *complete* Jacobian matrix, as a fundamental prerequisite for the analysis of the robot workspace and the particular configurations of singularities where the planar manipulator becomes uncontrollable.

Jacobian matrices are interesting because they allow us to determine the minimum necessary accuracy of the robot internal sensors in order to obtain a given error in the generalized coordinates. The expression of these matrices remains relatively complex as soon as the number of degrees of freedom increases, obtaining the procedure through a symbolic inversion is, generally, a difficult task. For planar robots, this inversion is possible and the Jacobian matrix could even be obtained directly by differentiating of the closure equations.

By rearranging, the six constraint equations in (8.1.9) can immediately be written as

$$\begin{aligned} &\left[x^G_0 - x^j_{00} + r\sin\left(\phi + \frac{\pi}{3} + \alpha_i\right) - (l_1 + \lambda^j_{10})\cos\alpha_i\right]^2 \\ &\quad + \left[y^G_0 - y^j_{00} - r\cos\left(\phi + \frac{\pi}{3} + \alpha_i\right) - (l_1 + \lambda^j_{10})\sin\alpha_i\right]^2 = l^2_2 \;\; (j = A, B, C) \end{aligned} \tag{8.1.14}$$

where the "zero" position $x_{00}^G = 0, y_{00}^G = 0, \phi^0 = 0$ corresponds to the joints variables $\vec{\lambda}_{10}^0 = [0 \quad 0 \quad 0]^T$. The derivative with respect to time of conditions (8.1.14) leads to the matrix equation

$$J_{1p}\dot{\vec{\lambda}}_{10} = J_{2p}\begin{bmatrix}\dot{x}_0^G & \dot{y}_0^G & \dot{\phi}\end{bmatrix}^T \tag{8.1.15}$$

for the planar robot with *fixed prismatic actuators*. Matrices J_{1p} and J_{2p} are, respectively, the inverse and forward Jacobian of the manipulator and can be expressed as

$$J_{1p} = diag\{\delta_{Ap} \quad \delta_{Bp} \quad \delta_{Cp}\}, \quad J_{2p} = \begin{bmatrix} \beta_{1p}^A & \beta_{2p}^A & \beta_{3p}^A \\ \beta_{1p}^B & \beta_{2p}^B & \beta_{3p}^B \\ \beta_{1p}^C & \beta_{2p}^C & \beta_{3p}^C \end{bmatrix}, \tag{8.1.16}$$

with

$$\begin{aligned}
\delta_{jp} &= (x_0^G - x_{00}^j)\cos\alpha_i + (y_0^G - y_{00}^j)\sin\alpha_j + r\sin\left(\phi + \frac{\pi}{3}\right) - l_1 - \lambda_{10}^j \\
\beta_{1p}^j &= x_0^G - x_{00}^j + r\sin\left(\phi + \frac{\pi}{3} + \alpha_j\right) - (l_1 + \lambda_{10}^j)\cos\alpha_j \\
\beta_{2p}^j &= y_0^G - y_{00}^j - r\cos\left(\phi + \frac{\pi}{3} + \alpha_j\right) - (l_1 + \lambda_{10}^j)\sin\alpha_j \\
\beta_{3p}^j &= r\left[(x_0^G - x_{00}^j)\cos\left(\phi + \frac{\pi}{3} + \alpha_j\right) + (y_0^G - y_{00}^j)\sin\left(\phi + \frac{\pi}{3} + \alpha_j\right) - (l_1 + \lambda_{10}^j)\cos\left(\phi + \frac{\pi}{3}\right)\right].
\end{aligned} \tag{8.1.17}$$

The three kinds of singularities of the three closed-loop kinematical chains can be determined through the analysis of the two Jacobian matrices J_{1p} and J_{2p} [14].

Since λ_{10}^j is a passive variable in a *second kind* of the planar robot with *mobile revolute actuators*, it should be eliminated from Eq. (8.1.9) as follows:

$$l_2\sin\left(\varphi_{21}^j + \frac{\pi}{6}\right) = (x_{00}^j - x_0^G)\sin\alpha_j - (y_{00}^j - y_0^G)\cos\alpha_j - r\cos\left(\phi + \frac{\pi}{3}\right). \tag{8.1.18}$$

A new matrix relation is obtained, taking the derivative of this equation with respect to time

$$J_{1r}\dot{\vec{\varphi}}_{21} = J_{2r}\begin{bmatrix}\dot{x}_0^G & \dot{y}_0^G & \dot{\phi}\end{bmatrix}^T, \tag{8.1.19}$$

where

$$
\begin{aligned}
&\vec{\varphi}_{21} = \begin{bmatrix} \varphi_{21}^{A} & \varphi_{21}^{B} & \varphi_{21}^{C} \end{bmatrix}^{T} \\
&J_{1r} = diag\{\, \delta_{Ar} \quad \delta_{Br} \quad \delta_{Cr} \,\}, \quad J_{2r} = \begin{bmatrix} \beta_{1r}^{A} & \beta_{2r}^{A} & \beta_{3r}^{A} \\ \beta_{1r}^{B} & \beta_{2r}^{B} & \beta_{3r}^{B} \\ \beta_{1r}^{C} & \beta_{2r}^{C} & \beta_{3r}^{C} \end{bmatrix} \\
&\delta_{jr} = l_2 \cos\left(\varphi_{21}^{j} + \frac{\pi}{6}\right) \\
&\beta_{1r}^{j} = -\sin\alpha_j, \quad \beta_{2r}^{j} = -\cos\alpha_j, \quad \beta_{3r}^{j} = r\sin\left(\phi + \frac{\pi}{3}\right)
\end{aligned}
\tag{8.1.20}
$$

The matrix kinematical relations (8.1.10)–(8.1.13) will be further required in the computing of virtual velocity distribution of the elements of the manipulator.

Let us assume, for example, that the robot has successively two kinds of *virtual motions* determined by the linear velocities $v_{10a}^{Av} = 1$, $v_{10a}^{jv} = 0\ (j \neq A)$; $v_{10b}^{Bv} = 1$, $v_{10b}^{jv} = 0\,(j \neq B)$; $v_{10c}^{Cv} = 1$, $v_{10c}^{jv} = 0\,(j \neq C)$ or the angular velocities $\omega_{21a}^{Av} = 1$, $\omega_{21a}^{jv} = 0\,(j \neq A)$; $\omega_{21b}^{Bv} = 1$, $\omega_{21b}^{jv} = 0\,(j \neq B)$; $\omega_{21c}^{Cv} = 1$, $\omega_{21c}^{jv} = 0\,(j \neq C)$. The characteristic virtual velocities are expressed as functions of the position of the mechanism by the general kinematical constraints Eq. (8.1.13). These virtual velocities are required into the expressing of the virtual work of all forces applied to the elements of the robot.

Once the displacements and velocities are determined, one can proceed to the third step in the kinematics study, which is referred to as the acceleration analysis. The velocity relationships are differentiated with respect to time in order to obtain the respective accelerations of the bodies in the system.

As for the relative accelerations $\gamma_{10}^{j}, \varepsilon_{21}^{j}, \varepsilon_{32}^{j}$ of the robot, new connectivity conditions are obtained by the time derivative of equations in (8.1.13), which are

$$
\begin{aligned}
\varepsilon_{21}^{j} + \varepsilon_{32}^{j} = \dot{\phi}, \gamma_{10}^{j}\vec{u}_i^{T} q_{10}^{T}\vec{u}_3 + \varepsilon_{21}^{j}\vec{u}_i^{T} q_{20}^{T}\tilde{u}_3\{\vec{r}_{32}^{\,j} + q_{32}^{T}\vec{r}_{3j}^{\,G}\} \\
+ \varepsilon_{32}^{j}\vec{u}_i^{T} q_{30}^{T}\tilde{u}_3\vec{r}_{3j}^{\,G} = \vec{u}_i^{T}\ddot{\vec{r}}_0^{\,G} \\
- \omega_{21}^{j}\omega_{21}^{j}\vec{u}_i^{T} q_{20}^{T}\tilde{u}_3\tilde{u}_3\{\vec{r}_{32}^{\,j} + q_{32}^{T}\vec{r}_{3j}^{\,G}\} - \omega_{32}^{j}\omega_{32}^{j}\vec{u}_i^{T} q_{30}^{T}\tilde{u}_3\tilde{u}_3\vec{r}_{3j}^{\,G} \\
- 2\omega_{21}^{j}\omega_{32}^{j}\vec{u}_i^{T} q_{20}^{T}\tilde{u}_3 q_{32}^{T}\tilde{u}_3\vec{r}_{3j}^{\,G} \\
(q = a, b, c)\ \ (j = A, B, C)\ \ (i = 1, 2).
\end{aligned}
\tag{8.1.21}
$$

Finally, recursive general relations (7.11) give the square matrix $\tilde{\omega}_{k0}^{j}\tilde{\omega}_{k0}^{j} + \tilde{\varepsilon}_{k0}^{j}$, the angular accelerations $\vec{\varepsilon}_{k0}^{\,j}$ and the linear accelerations $\vec{\gamma}_{k0}^{\,j}$ of the joints

$$
\begin{aligned}
\tilde{\omega}^j_{k0}\tilde{\omega}^j_{k0} + \tilde{\varepsilon}^j_{k0} &= q_{k,k-1}\{\tilde{\omega}^j_{k-1,0}\tilde{\omega}^j_{k-1,0} + \tilde{\varepsilon}^j_{k-1,0}\}q^T_{k,k-1} \\
&\quad + \omega^j_{k,k-1}\omega^j_{k,k-1}\tilde{u}_3\tilde{u}_3 + \varepsilon^j_{k,k-1}\tilde{u}_3 + 2\omega^j_{k,k-1}q_{k,k-1}\tilde{\omega}^j_{k-1,0}q^T_{k,k-1}\tilde{u}_3 \\
\vec{\varepsilon}^j_{k0} &= q_{k,k-1}\vec{\varepsilon}^j_{k-1,0} + \varepsilon^j_{k,k-1}\vec{u}_3 + \omega^j_{k,k-1}q_{k,k-1}\tilde{\omega}^j_{k-1,0}q^T_{k,k-1}\vec{u}_3, \;\; \varepsilon^j_{10} = 0, \;\; \varepsilon^j_{\tau,\tau-1} = \ddot{\varphi}^j_{\tau,\tau-1} \\
\vec{\gamma}^j_{k0} &= q_{k,k-1}\vec{\gamma}^j_{k-1,0} + q_{k,k-1}\{\tilde{\omega}^j_{k-1,0}\tilde{\omega}^j_{k-1,0} + \tilde{\varepsilon}^j_{k-1,0}\}\vec{r}^j_{k,k-1} \\
&\quad + 2v^j_{k,k-1}q_{k,k-1}\tilde{\omega}^j_{k-1,0}q^T_{k,k-1}\vec{u}_3 + \gamma^j_{k,k-1}\vec{u}_3, \;\; \gamma^j_{10} = \ddot{\lambda}^j_{10}, \;\; \gamma^j_{\tau,\tau-1} = 0, \;\; (\tau = 2,3)\;(k = 1,2,3)
\end{aligned}
\tag{8.1.22}
$$

Relations (8.1.13) and (8.1.21) constitute *the inverse kinematics model* of the 3-*PRR* planar parallel robot.

8.1.2 Dynamics Modelling

Knowing the position and kinematics state of every link as well as the external forces acting on the planar parallel robot, the torques of actuators or the active forces required in a given motion of the moving platform can be recursively formulated, based on the *principle of virtual work*.

Three independent mechanical systems acting along the planar directions z_1^A, z_1^B, z_1^C, with the forces $\vec{f}^A_{10} = f^A_{10}\vec{u}_3, \vec{f}^B_{10} = f^B_{10}\vec{u}_3, \vec{f}^C_{10} = f^C_{10}\vec{u}_3$—or the other three electric motors A_2, B_2, C_2 that generate three couples of moments $\vec{m}^A_{21} = m^A_{21}\vec{u}_3$, $\vec{m}^B_{21} = m^B_{21}\vec{u}_3$, $\vec{m}^C_{21} = m^C_{21}\vec{u}_3$ oriented about parallel axes—can control the platform's motion.

Using an *equivalent graph* representation, the parallel robot can artificially be transformed into a set of three open planar kinematical chains subject to the constraints, by cutting two passive joints B_3 and C_3 for moving the platform and taking its effect into account by introducing the corresponding constraint conditions. The third and more complicated open tree system $1a - 2a - p$ includes the leg A and the moving platform (Fig. 8.3).

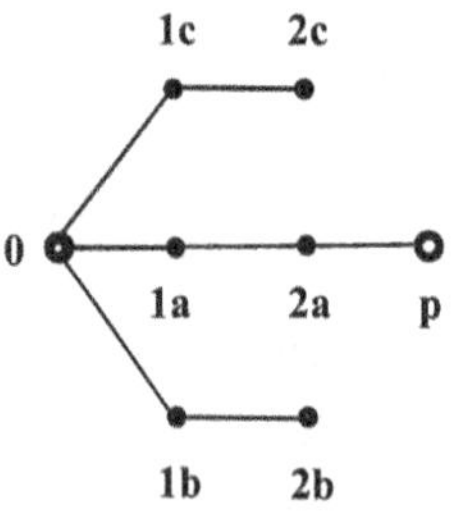

Fig. 8.3 Three open trees of the planar parallel robot

The force of inertia and the resulting torque of the inertia forces applied on an arbitrary rigid body T_k^j

$$\begin{aligned} \vec{f}_k^{inj} &= -m_k^j[\vec{\gamma}_{k0}^j + (\tilde{\omega}_{k0}^j\tilde{\omega}_{k0}^j + \tilde{\varepsilon}_{k0}^j)\vec{r}_{kC}^j], \\ \vec{m}_k^{inj} &= -[m_k^j\tilde{r}_{kC}^j\vec{\gamma}_{k0}^j + \hat{J}_k^j\vec{\varepsilon}_{k0}^j + \tilde{\omega}_{k0}^j\hat{J}_k^j\vec{\omega}_{k0}^j] \end{aligned} \tag{8.1.23}$$

are computed with respect to the centre of joint. On the other hand, the wrench of two vectors $\vec{f}_k^{*j}$ and $\vec{m}_k^{*j}$ evaluates the influence of the action of the weight $m_k^j\vec{g}$ and of other external and internal forces applied to the same element T_k^j of the manipulator, for example:

$$\vec{f}_{k*}^j = -gm_k^j q_{k0}\vec{u}_2 \quad \vec{m}_{k*}^j = -gm_k^j\tilde{r}_{kC}^j q_{k0}\vec{u}_2 \ \ (j = A, B, C). \tag{8.1.24}$$

Two significant recursive relations (7.19) generate the vectors

$$\vec{F}_k^j = \vec{F}_{k0}^j + q_{k+1,k}^T\vec{F}_{k+1}^j, \quad \vec{M}_k^j = \vec{M}_{k0}^j + q_{k+1,k}^T\vec{M}_{k+1}^j + \tilde{r}_{k+1,k}^j q_{k+1,k}^T\vec{F}_{k+1}^j, \tag{8.1.25}$$

with the notations

$$\vec{F}_{k0}^j = -\vec{f}_k^{inj} - \vec{f}_k^{*j}, \quad \vec{M}_{k0}^j = -\vec{m}_k^{inj} - \vec{m}_k^{*j}. \tag{8.1.26}$$

The principle of virtual work states that the planar mechanism is under dynamic equilibrium if and only if the virtual work developed by all external, internal and inertia forces vanish during any general virtual displacement, which is consistent with the constraints imposed on the mechanism. During three successive independent virtual motions of the robot, the displacements and the velocities should be compatible to all kinematical constraints at a given instant in time. By the intermediate of the complete Jacobian matrix derived from above conditions of connectivity (8.1.13), the absolute virtual velocities associated with all moving links are related to a set of independent *relative virtual velocities* $v_{k,k-1}^{vj}$ and $\omega_{k,k-1}^{vj}$. The total virtual work contributed by the input forces or active torques, the inertia forces, the torques of inertia forces and the wrench of known external forces can be written in a compact form, based on the relative virtual angular velocities *only*.

Assuming that frictional forces at the joints are negligible, the virtual work produced by all forces of constraint in the joints is zero. Pursuing all trees of the equivalent graph (Fig. 8.4), following compact matrix relations results, from *the equations of the parallel robots dynamics* (7.18) and (7.19), the following compact matrix relations results

$$f_{10}^A = \vec{u}_3^T\{\vec{F}_1^A + \omega_{21a}^{Av}\vec{M}_2^A + \omega_{32a}^{Av}\vec{M}_3^A + \omega_{21a}^{Bv}\vec{M}_2^B + \omega_{21a}^{Cv}\vec{M}_2^C\} \tag{8.1.27}$$

for the force of the first fixed *prismatic actuator* and

$$m_{21}^{A} = \vec{u}_3^T\{v_{10a}^{Av}\vec{F}_1^A + \vec{M}_2^A + \omega_{32a}^{Av}\vec{M}_3^A + v_{10a}^{Bv}\vec{F}_1^B + v_{10a}^{Cv}\vec{F}_1^C\}. \quad (8.1.28)$$

for the torque of the first mobile *revolute actuator*. The relations (8.1.25), (8.1.27) and (8.1.28) represent the *inverse dynamics model* of the 3-*PRR* planar parallel manipulator.

The computer program is developed in MATLAB to solve the inverse kinematics and the dynamics of the planar parallel robot, assuming that in a period of three second the platform starts from the central configuration, evolving successively a rotation around a horizontal axis and a rectilinear planar translation.

For simulation purposes, let us consider a planar robot which has the following characteristics:

$$x_{0G}^* = -0.025\,\text{m},\ \ y_{0G}^* = 0.025\,\text{m},\ \ \phi^* = \frac{\pi}{12},\ \ r = 0.1\,\text{m},\ \ l_1 = l = r\sqrt{3}$$
$$l_0 = 0.3\,\text{m},\ \ l_2 = 0.2\,\text{m},\ \ \Delta t = 3\,\text{s},\ \ m_1 = 1\,\text{kg},\ \ m_2 = 1.5\,\text{kg},\ \ m_3 = 3\,\text{kg}$$

Numerical computations are first started by the determination of the displacements λ_{10}^j (Figs. 8.4 and 8.5) and the angles of rotation φ_{21}^j (Figs. 8.6 and 8.7) required by each actuator during the platform's evolutions.

Considering that there are no external forces and moments acting on the moving platform, a comparative study of the robot dynamics in both configurations (with either prismatic or revolute actuators) is carried out by computation of the powers $P_{10}^j = v_{10}^j f_{10}^j$, $P_{21}^j = \omega_{21}^j m_{21}^j$ required by each actuator during the platform's evolution.

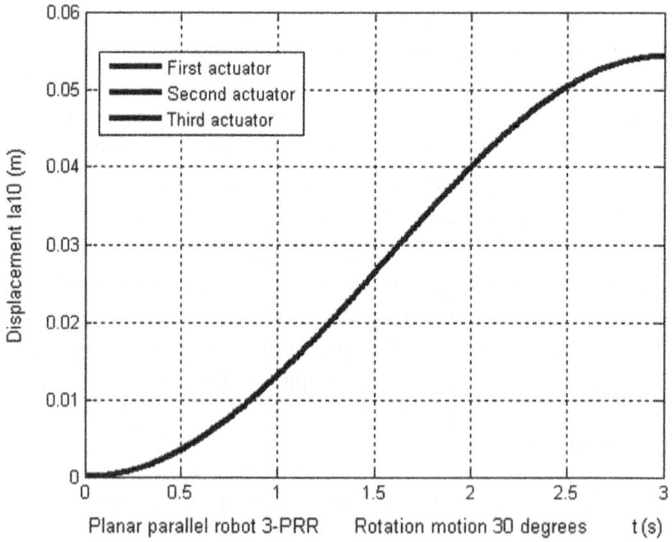

Fig. 8.4 Displacements $\lambda_{10}^A, \lambda_{10}^B, \lambda_{10}^C$ (rotation)

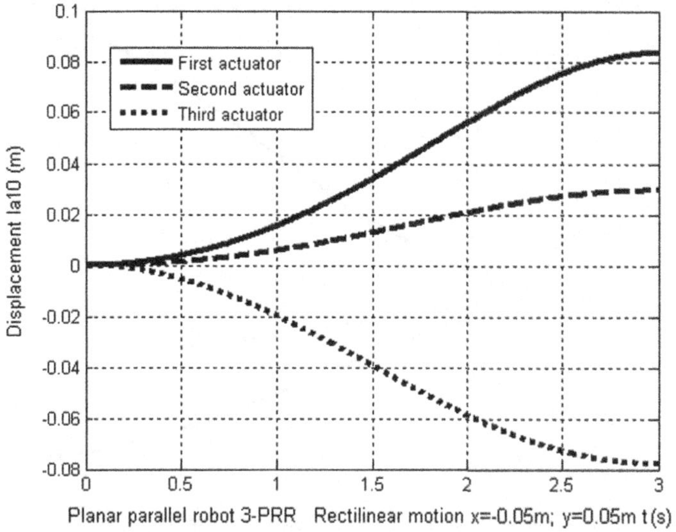

Fig. 8.5 Displacements $\lambda_{10}^{A}, \lambda_{10}^{B}, \lambda_{10}^{C}$ (translation)

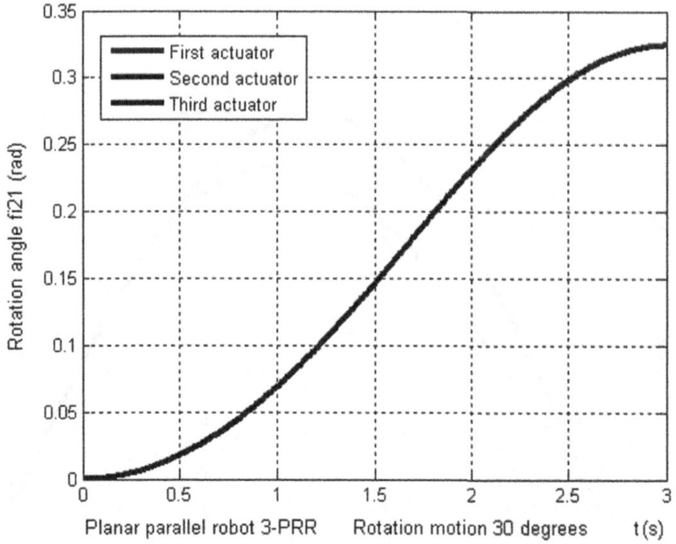

Fig. 8.6 Rotation angles $\varphi_{21}^{A}, \varphi_{21}^{B}, \varphi_{21}^{C}$ (rotation)

To illustrate the simulation, in the first example we consider the *rotation motion* of the moving platform about the z_0 axis with variable angular acceleration while all the other positional parameters are held to be equal to zero. As can be seen from P_{10}^{A}, P_{21}^{A} (Fig. 8.8), P_{10}^{B}, P_{21}^{B} (Fig. 8.9) and P_{10}^{C}, P_{21}^{C} (Fig. 8.10), for the third leg C only, both actuating powers are permanently of opposite signs.

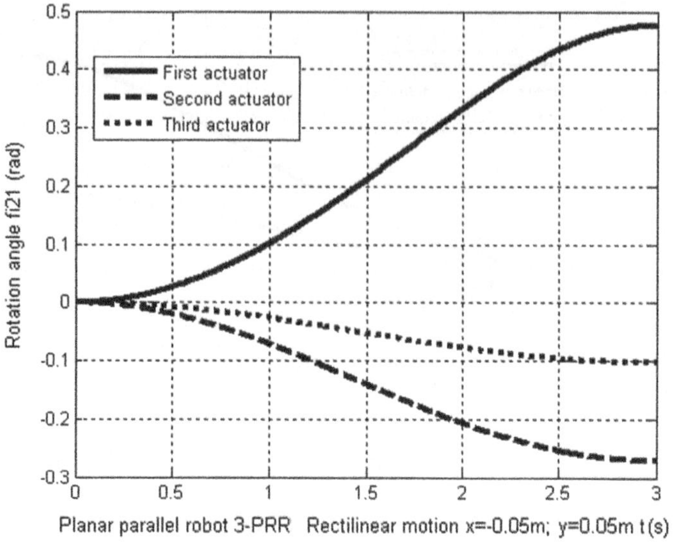

Fig. 8.7 Rotation angles $\varphi_{21}^{A}, \varphi_{21}^{B}, \varphi_{21}^{C}$ (translation)

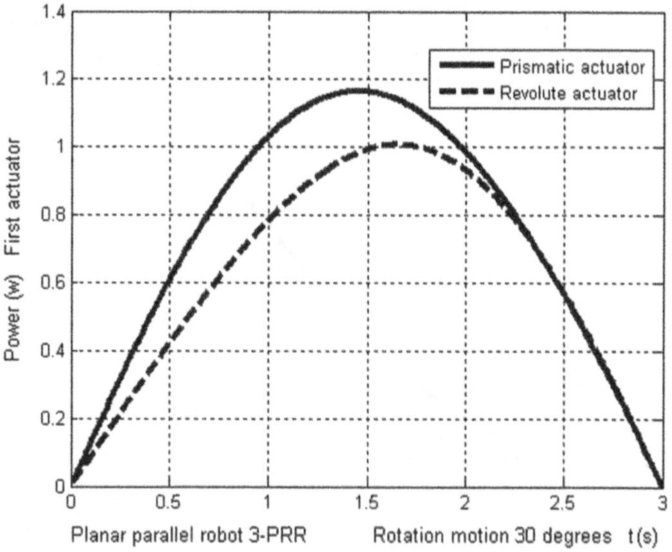

Fig. 8.8 Powers P_{10}^{A}, P_{21}^{A} of first actuator (rotation)

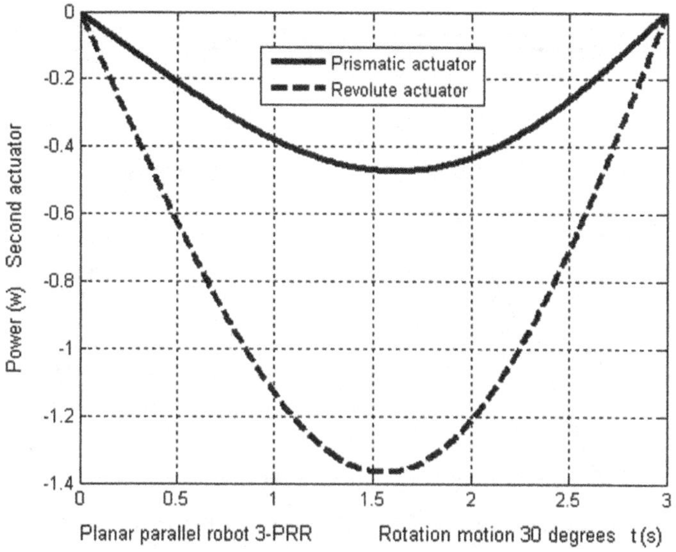

Fig. 8.9 Powers P_{10}^{B}, P_{21}^{B} of second actuator (rotation)

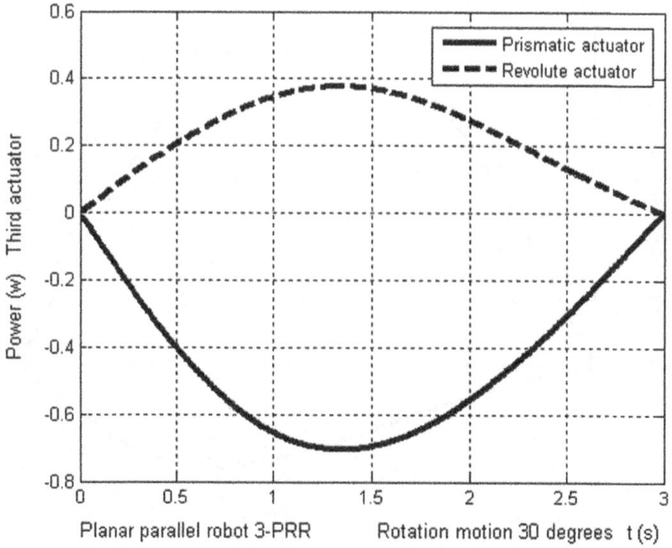

Fig. 8.10 Powers P_{10}^{C}, P_{21}^{C} of third actuator (rotation)

For the case where the platform's centre G moves along a *rectilinear planar trajectory* without any rotation of the platform, the powers required by the actuators are calculated and depicted in Figs. 8.11, 8.12 and 8.13.

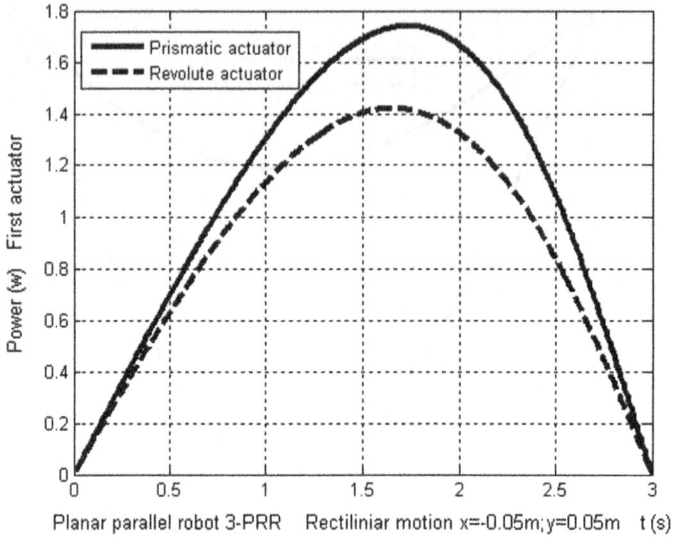

Fig. 8.11 Powers P_{10}^A, P_{21}^A of first actuator (translation)

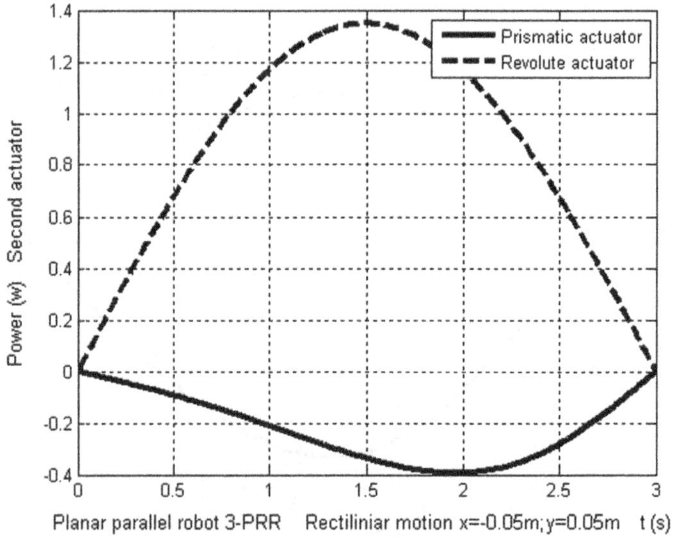

Fig. 8.12 Powers P_{10}^B, p_{21}^B of second actuator (translation)

The total powers developed by the complex of three actuators in the case of the two driving configurations, during the period of a given displacement of the platform, are completely computed and plotted for the rotation motion (Fig. 8.14) and the rectilinear translation (Fig. 8.15) of the moving platform.

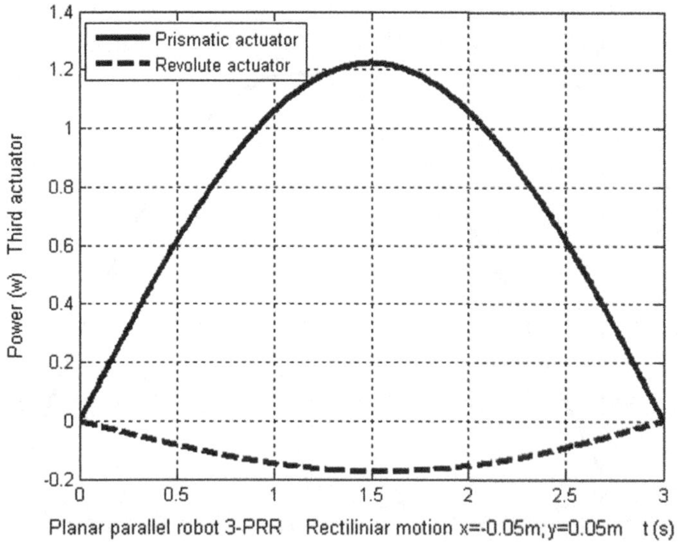

Fig. 8.13 Powers P^C_{10}, P^C_{21} of third actuator (translation)

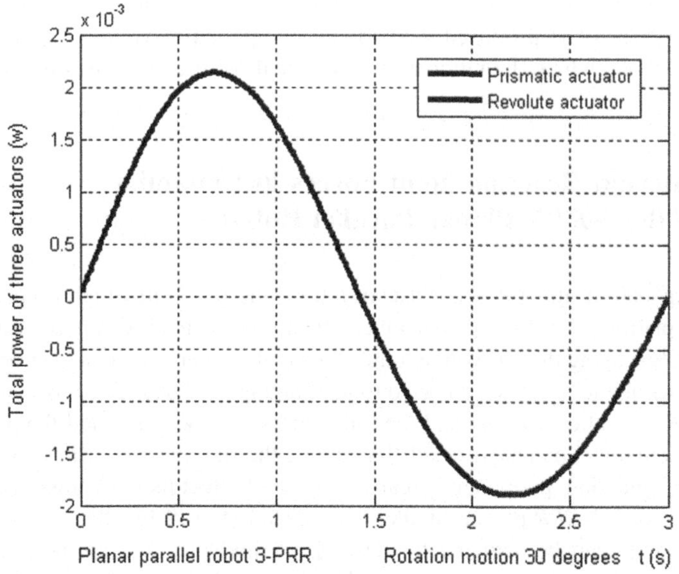

Fig. 8.14 Total powers $P^{tot}_{10}, P^{tot}_{21}$ of actuators (rotation)

Concerning the power requirement comparison, we remark that for the same evolution of the moving platform into its vertical plane, the power distribution upon the three actuators depends on the actuating configuration, but the total power P^{tot}_{10}

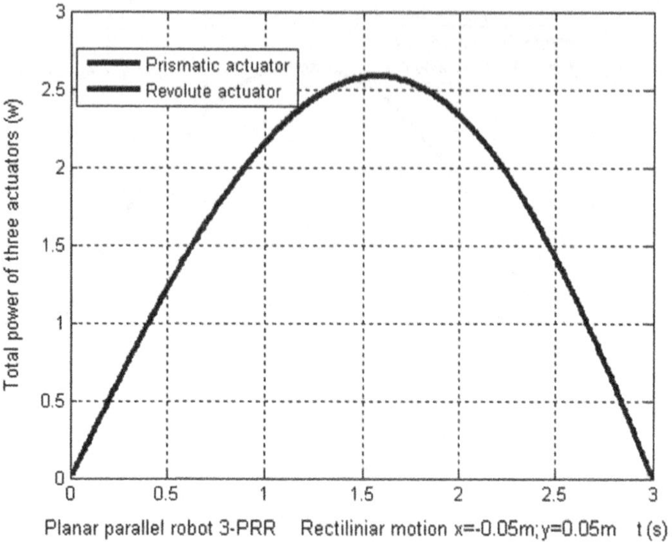

Fig. 8.15 Total powers $P_{10}^{tot}, P_{21}^{tot}$ of actuators (translation)

or P_{21}^{tot} absorbed by the set of three actuators is the same, at any instant, for both driving systems. The proposed method can be applied to various types of spatial complex robots, where the number of components of the mechanisms is increased.

8.2 Internal Reaction Joint Forces in Dynamics of the 3-*RRR* Planar Parallel Robot

The mechanism of this robot consists of a traveling platform which is connected to the base platform by three planar kinematical chains. Each chain consists of two links attached using three revolute joints. One end of every chain is connected to the actuator, while the other is connected to the moving platform. The link connected to the actuator is called a *crank* and the link connecting the crank and the platform is called a *coupler*. Connected to all the chains, the moving platform is as an equilateral triangle and permanently carries the end-effectors. Pursuing the general schematics of a 3-*RRR* planar parallel manipulator (Fig. 8.16), though any three of the nine joints can be actuated for desired manipulation and the actuators under consideration, they have wisely been placed at the base of each link, in order to lighten the weight of the moving components.

Based on the principle of virtual work, a recursive procedure is developed to obtain compact results concerning the time-history evolution for the input torques of the three actuators, but also for external and internal forces in the joints.

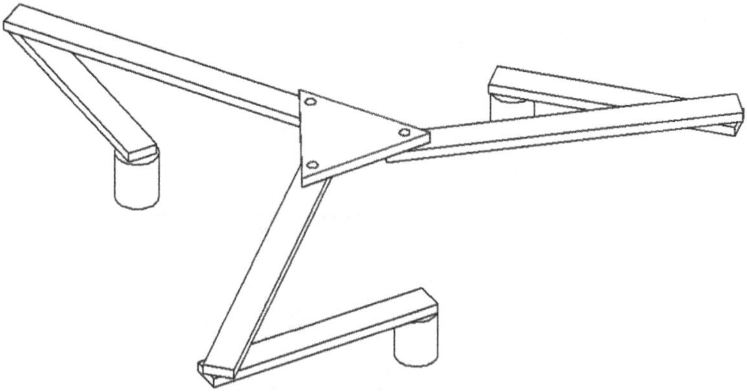

Fig. 8.16 General layout of a 3-*RRR* planar parallel robot

8.2.1 Kinematics Modelling

Having a closed-loop structure, the planar parallel robot 3-*RRR* is a symmetrical mechanism composed of three planar chains with identical topology, all connecting the fixed base to the moving platform. Centres A_1, B_1, C_1 of three fixed pivots determine the position of a fixed base, while another three moving revolute joints A_3, B_3, C_3 define the orientation of the planar moving platform. Each leg of the mechanism is represented by two binary links with three parallel revolute joints. Together, the three-degrees-of-freedom manipulator consists of seven moving links, nine revolute joints and three revolute actuators installed on the fixed base.

We attach a Cartesian frame $Ox_0y_0z_0(T_0)$ to the fixed base with its origin located at the triangle centre, the z_0 axis perpendicular to the base and the x_0 axis pointing along the direction C_1O. Another mobile reference frame $Gx_Gy_Gz_G$ is attached to the moving platform. The origin of this central coordinate system is located just at the centre G of the moving triangle (Fig. 8.17).

In what follows, we consider that the moving platform is initially located at a *central configuration*, where the platform is not rotated with respect to the fixed base and the mass centre G is at the origin O of the fixed frame. Relative rotation of any body T_k with angle $\varphi_{k,k-1}$ must be always pointing about the direction of z_k axis.

First active leg A, for example, consists of a fixed revolute joint A_1, a moving crank **1** of length l_1, mass m_1 and tensor of inertia $\hat{J}_1$, which has a rotation about z_1^A axis with the angle φ_{10}^A, the angular velocity $\omega_{10}^A = \dot{\varphi}_{10}^A$ and the angular acceleration $\varepsilon_{10}^A = \ddot{\varphi}_{10}^A$. A new element of the leg is a rigid rod **2** linked at the $A_2x_2^Ay_2^Az_2^A$ frame, having a relative rotation with the angle φ_{21}^A, velocity $\omega_{21}^A = \dot{\varphi}_{21}^A$ and acceleration $\varepsilon_{21}^A = \ddot{\varphi}_{21}^A$. It has the length l_2, mass m_2 and tensor of inertia $\hat{J}_2$. Finally, a revolute joint is introduced at the moving platform, which is schematised as an equilateral triangle with the edge $l = r\sqrt{3}$, mass m_3 and a symmetrical tensor of inertia $\hat{J}_3$ with respect to central frame $Gx_Gy_Gz_G$.

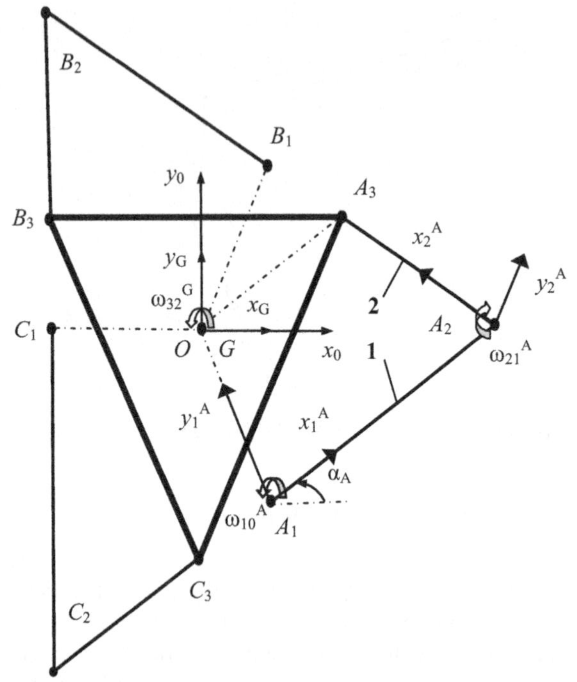

Fig. 8.17 Planar 3-*RRR* parallel robot

At the central configuration, we also consider that all legs are symmetrically extended with three angles $\alpha_A = \pi/6, \alpha_B = 5\alpha_A, \alpha_C = -3\alpha_A$, giving the initial pose of the mechanism.

In the study of the kinematics of this manipulator, we are interested in deriving a matrix equation relating the location of an arbitrary body to the joint variables. We call the matrix $q^{\varphi}_{\tau,\tau-1}$ for example, the orthogonal transformation 3×3 matrix of relative rotation with the angle $\varphi^j_{\tau,\tau-1}$ of link T^j_τ around the z^j_τ axis.

Pursuing three independent legs $OA_1A_2A_3$, $OB_1B_2B_3B_3$ and $OC_1C_2C_3$, we obtain the following transformation matrices

$$q_{10} = q^{\varphi}_{10}a^j_\alpha, \quad q_{21} = q^{\varphi}_{21}\theta_2\theta_1, \quad q_{20} = q_{21}q_{10} \;\; (q = a, b, c) \;\; (j = A, B, C), \tag{8.2.1}$$

where we denote [15]:

$$q^{\varphi}_{\tau,\tau-1} = rot(z, \varphi^j_{\tau,\tau-1}) = \begin{bmatrix} \cos\varphi^j_{\tau,\tau-1} & \sin\varphi^j_{\tau,\tau-1} & 0 \\ -\sin\varphi^j_{\tau,\tau-1} & \cos\varphi^j_{\tau,\tau-1} & 0 \\ 0 & 0 & 1 \end{bmatrix}, \quad (\tau = 1, 2), \;\; a^j_\alpha = rot(z, \alpha_j)$$

$$\theta_1 = rot(y, \pi) = \begin{bmatrix} -1 & 0 & 0 \\ 0 & 1 & 0 \\ 0 & 0 & -1 \end{bmatrix}, \quad \theta_2 = rot(z, \pi/3) = \frac{1}{2}\begin{bmatrix} 1 & \sqrt{3} & 0 \\ -\sqrt{3} & 1 & 0 \\ 0 & 0 & 2 \end{bmatrix}. \tag{8.2.2}$$

In the *inverse geometric problem*, it can be considered that the position of the planar mechanism is completely given through the coordinates $\lambda_{10}^G = x_0^G$, $\lambda_{21}^G = y_0^G$ of the mass centre G of the moving platform and the orientation angle $\varphi_{32}^G = \phi$ of the central frame $Gx_G y_G z_G$. An orthogonal rotation matrix of the platform from $Ox_0 y_0 z_0$ to $Gx_G y_G z_G$ is used:

$$R = rot(z, \phi) = \begin{bmatrix} \cos\phi & \sin\phi & 0 \\ -\sin\phi & \cos\phi & 0 \\ 0 & 0 & 1 \end{bmatrix}. \tag{8.2.3}$$

Furthermore, we suppose that the position vector of G centre $\vec{r}_0^G = \begin{bmatrix} x_0^G & y_0^G & 0 \end{bmatrix}^T$ and the orientation angle ϕ, which are expressed by following the analytical functions

$$\frac{x_0^G}{x_0^{G*}} = \frac{y_0^G}{y_0^{G*}} = \frac{\phi}{\phi^*} = \left(1 - \cos\frac{\pi}{3}t\right), \tag{8.2.4}$$

are capable of describing the general absolute motion of the moving platform in its *vertical plane.*

Six variables $\varphi_{10}^A, \varphi_{21}^A$, $\varphi_{10}^B, \varphi_{21}^B$, $\varphi_{10}^C, \varphi_{21}^C$ will be determined by several vector-loop equations, as follows

$$\vec{r}_{10}^{\,j} + q_{10}^T \vec{r}_{21}^{\,j} + q_{10}^T q_{21}^T \vec{r}_{32}^{\,j} = \vec{r}_0^{\,G} + R^T \vec{r}_G^{\,j_3} \ (j = A, B, C) \ (q = a, b, c), \tag{8.2.5}$$

where we denote

$$\vec{u}_1 = \begin{bmatrix} 1 \\ 0 \\ 0 \end{bmatrix}, \quad \vec{u}_2 = \begin{bmatrix} 0 \\ 1 \\ 0 \end{bmatrix}, \quad \vec{u}_3 = \begin{bmatrix} 0 \\ 0 \\ 1 \end{bmatrix}, \quad \tilde{u}_3 = \begin{bmatrix} 0 & -1 & 0 \\ 1 & 0 & 0 \\ 0 & 0 & 0 \end{bmatrix}$$
$$\vec{r}_{10}^{\,A} = 0.25r\begin{bmatrix} \sqrt{3} & -3 & 0 \end{bmatrix}^T, \quad \vec{r}_{10}^{\,B} = 0.25r\begin{bmatrix} \sqrt{3} & 3 & 0 \end{bmatrix}^T, \quad \vec{r}_{10}^{\,C} = \begin{bmatrix} -0.5r\sqrt{3} & 0 & 0 \end{bmatrix}^T$$
$$\vec{r}_G^{\,A_3} = 0.5r\begin{bmatrix} \sqrt{3} & 1 & 0 \end{bmatrix}^T, \quad \vec{r}_G^{\,B_3} = 0.5r\begin{bmatrix} -\sqrt{3} & 1 & 0 \end{bmatrix}^T, \quad \vec{r}_G^{\,C_3} = [0 \quad -r \quad 0]^T$$
$$\vec{r}_{21}^{\,j} = 1.5r\vec{u}_1, \quad \vec{r}_{32}^{\,j} = r\vec{u}_1, \quad \vec{r}_{10}^{\,G} = x_0^G \vec{u}_1, \quad \vec{r}_{21}^{\,G} = y_0^G \vec{u}_2, \quad \vec{r}_{32}^{\,G} = \vec{0}. \tag{8.2.6}$$

There is only one inverse geometric solution for the position of the manipulator, and the following relationships between the relative and absolute coordinates hold:

$$\begin{aligned} l_1 \cos(\varphi_{10}^j + \alpha_j) - l_2 \cos(\varphi_{10}^j + \alpha_j - \varphi_{21}^j - \pi/3) &= x_0^G - x_{10}^j + \vec{u}_1^T R^T \vec{r}_G^{\,j_3} \\ l_1 \sin(\varphi_{10}^j + \alpha_j) - l_2 \sin(\varphi_{10}^j + \alpha_j - \varphi_{21}^j - \pi/3) &= y_0^G - y_{10}^j + \vec{u}_2^T R^T \vec{r}_G^{\,j_3}. \end{aligned} \tag{8.2.7}$$

The *inverse kinematics problem* determines the values of the actuated joint variables, given the position and orientation of the traveling plate. Evaluation of the inverse kinematics is very important for position control of any parallel robot. Once the dependent coordinates are determined, the dependent velocities and accelerations can be computed. We determine the velocities and accelerations, supposing that the planar motion of the moving platform is known through the relations (8.2.4). First, we compute the linear and angular velocities of each leg in terms of the angular velocity $\vec{\omega}_{30}^{G} = \dot{\varphi}_{32}^{G}\,\vec{u}_3$ and the centre's velocity $\dot{\vec{r}}_0^{G} = [\dot{\lambda}_{10}^{G} \;\; \dot{\lambda}_{21}^{G} \;\; 0]^T$ of the moving platform.

The rotations of compounding elements of each leg j are characterized by the recursive relations of following skew-symmetric matrices

$$\tilde{\omega}_{10}^{j} = \dot{\varphi}_{10}^{j}\tilde{u}_3, \quad \tilde{\omega}_{20}^{j} = q_{21}\tilde{\omega}_{10}^{j}q_{21}^{T} + \tilde{\omega}_{21}^{j} = (\dot{\varphi}_{21}^{j} - \dot{\varphi}_{10}^{j})\tilde{u}_3, \tag{8.2.8}$$

which are *associated* with the absolute angular velocities

$$\vec{\omega}_{10}^{j} = \dot{\varphi}_{10}^{j}\vec{u}_3, \quad \vec{\omega}_{20}^{j} = q_{21}\vec{\omega}_{10}^{j} + \vec{\omega}_{21}^{j} = (\dot{\varphi}_{21}^{j} - \dot{\varphi}_{10}^{j})\vec{u}_3. \tag{8.2.9}$$

The following expressions give the absolute velocities of the joint's centres.

$$\vec{v}_{10}^{j} = \vec{0}, \quad \vec{v}_{20}^{j} = q_{21}\tilde{\omega}_{10}^{j}\vec{r}_{21}^{j}. \tag{8.2.10}$$

Equations of geometrical constraints (8.2.5) can be derivate with respect to time to obtain the following *matrix conditions of connectivity* [16]

$$\omega_{10}^{j}c_{i1}^{j} + \omega_{21}^{j}c_{i2}^{j} + \dot{x}_0^{G}c_{i1}^{G} + \dot{y}_0^{G}c_{i2}^{G} + \dot{\phi}\,c_{i3}^{G} = 0, \tag{8.2.11}$$

with the notations

$$\begin{aligned} c_{i1}^{j} &= \vec{u}_i^{T}q_{10}^{T}\tilde{u}_3\{\vec{r}_{21}^{j} + q_{21}^{T}\vec{r}_{32}^{j}\} \\ c_{i2}^{j} &= \vec{u}_i^{T}q_{20}^{T}\tilde{u}_3\vec{r}_{32}^{j} \\ c_{i1}^{G} &= -\vec{u}_i^{T}\vec{u}_1 \\ c_{i2}^{G} &= -\vec{u}_i^{T}\vec{u}_2 \\ c_{i3}^{G} &= -\vec{u}_i^{T}\tilde{u}_3R^{T}\vec{r}_G^{j3} \quad (i = 1, 2) \;\; (j = A, B, C). \end{aligned} \tag{8.2.12}$$

From these equations, we immediately obtain the *complete* Jacobian matrix of the manipulator and the relative angular velocities $\omega_{10}^{j}, \omega_{21}^{j}$ as functions of angular velocity of the platform and velocity of its mass centre G.

As for the relative angular accelerations $\varepsilon_{10}^{j}, \varepsilon_{21}^{j}$, the derivatives with respect to time of Eq. (8.2.11) give other conditions of connectivity. Once the coordinates and velocities are determined, the equations for the accelerations can be conducted as

$$
\begin{aligned}
&\varepsilon_{10}^{j}\vec{u}_i^{T}q_{10}^{T}\tilde{u}_3\{\vec{r}_{21}^{j}+q_{21}^{T}\vec{r}_{32}^{j}\}+\varepsilon_{21}^{j}\vec{u}_i^{T}q_{20}^{T}\tilde{u}_3\vec{r}_{32}^{j}=\vec{u}_i^{T}\ddot{\vec{r}}_0^{G}+\vec{u}_i^{T}\{\dot{\varphi}^2\tilde{u}_3\tilde{u}_3+\ddot{\varphi}\,\tilde{u}_3\}R^{T}\vec{r}_G^{j_3}\\
&\quad-\omega_{10}^{j}\omega_{10}^{j}\vec{u}_i^{T}q_{10}^{T}\tilde{u}_3\tilde{u}_3\{\vec{r}_{21}^{j}+q_{21}^{T}\vec{r}_{32}^{j}\}-\omega_{21}^{j}\omega_{21j}^{i}\vec{u}_i^{T}q_{20}^{T}\tilde{u}_3\tilde{u}_3\vec{r}_{32}^{j}-2\omega_{10}^{j}\omega_{21}^{j}\vec{u}_i^{T}q_{10}^{T}\tilde{u}_3q_{21}^{T}\tilde{u}_3\vec{r}_{32}^{j}\\
&\quad(i=1,2)\;(j=A,B,C).
\end{aligned}
\tag{8.2.13}
$$

Starting from (8.1.22), the following recursive relations give the angular accelerations $\vec{\varepsilon}_{k0}^{j}$ and the accelerations $\vec{\gamma}_{k0}^{j}$ of the joints:

$$
\begin{aligned}
\vec{\varepsilon}_{k0}^{j}&=q_{k,k-1}\vec{\varepsilon}_{k-1,0}^{j}+\varepsilon_{k,k-1}^{j}\vec{u}_3+\omega_{k,k-1}^{j}q_{k,k-1}\tilde{\omega}_{k-1,0}^{j}q_{k,k-1}^{T}\vec{u}_3,\;\varepsilon_{k,k-1}^{j}\\
&=\ddot{\varphi}_{k,k-1}^{j}\tilde{\omega}_{k0}^{j}\tilde{\omega}_{k0}^{j}+\tilde{\varepsilon}_{k0}^{j}=q_{k,k-1}\left\{\tilde{\omega}_{k-1,0}^{j}\tilde{\omega}_{k-1,0}^{j}+\tilde{\varepsilon}_{k-1,0}^{j}\right\}q_{k,k-1}^{T}\\
&\quad+\omega_{k,k-1}^{j}\omega_{k,k-1}^{j}\tilde{u}_3\tilde{u}_3+\varepsilon_{k,k-1}^{j}\tilde{u}_3+2\omega_{k,k-1}^{j}q_{k,k-1}\tilde{\omega}_{k-1,0}^{j}q_{k,k-1}^{T}\tilde{u}_3\\
\vec{\gamma}_{k0}^{j}&=q_{k,k-1}\vec{\gamma}_{k-1,0}^{j}+q_{k,k-1}\left\{\tilde{\omega}_{k-1,0}^{j}\tilde{\omega}_{k-1,0}^{j}+\tilde{\varepsilon}_{k-1,0}^{j}\right\}\vec{r}_{k,k-1}^{j}\quad(k=1,2).
\end{aligned}
\tag{8.2.14}
$$

Matrix relations (8.2.13) and (8.2.14) will be further used for the computation of wrench of the inertia forces for each link of the robot's legs.

8.2.2 *Dynamics Equations*

Dynamics of the 3-*RRR* planar parallel manipulator is a generalized approach in which the *inverse dynamics* model is presented in terms of the dynamics models of the legs. The relevant objective of a complete dynamics is first to determine the input torques, which must be exerted by the actuators in order to produce a given trajectory of the end-effector, but also to calculate some internal joint forces or torques.

8.2.2.1 Principle of Virtual Work

Three electric motors A_1, B_1, C_1 that generate the moments $\vec{m}_{10}^{A}=m_{10}^{A}\vec{u}_3$, $\vec{m}_{10}^{B}=m_{10}^{B}\vec{u}_3$, $\vec{m}_{10}^{C}=m_{10}^{C}\vec{u}_3$ oriented about fixed parallel axes, control the motion of the platform. To apply the *principle of virtual work* in order to establish definitive recursive matrix relations for the calculus of input torques, the robot can artificially be transformed into a set of *four open kinematical chains*. In this method first we form the minimal open-loop structure, by separating the legs from the traveling platform. Thus we have a moving platform whose dynamics can be easily modeled in Cartesian coordinates and three legs, each of which can be modeled as a planar RR serial manipulator. In general, it is easier to model the traveling platform as a function of the end-effector's position, velocity and acceleration; whereas the dynamics of the legs are studied as a function of the joint coordinates. Using a *graph* representation, this can be achieved by cutting each passive joint of the moving platform and taking its effect into account by introducing the applicable constraint conditions. First tree $0-1g-2g-3g$ represent three successive mobile frames used to decompose the planar motion of the moving platform (Fig. 8.18).

The wrench of two vectors $\vec{f}^{j}_{k*}=-gm^{j}_{k}q_{k0}\vec{u}_{2}$, $\vec{m}^{j}_{k*}=-gm^{j}_{k}\tilde{r}^{C}_{kj}q_{k0}\vec{u}_{2}$ $j=(A,B,C)\,q=(a,b,c)$ evaluates here the influence of the action of the weight $m^{j}_{k}\vec{g}$ of the element $T^{j}_{k}(k=1,2)$ of the robot. The force of inertia $\vec{f}^{inj}_{k}$ and the resulting moment of inertia forces $\vec{m}^{inj}_{k}$ of the same arbitrary rigid body T^{j}_{k} of mass m^{j}_{k} and the tensor of inertia $\hat{J}^{j}_{k}$ are computed with respect to the centre of its first joint:

$$\begin{aligned}\vec{f}^{inj}_{k}&=-m^{j}_{k}\left\{\vec{\gamma}^{j}_{k0}+(\tilde{\omega}^{j}_{k0}\tilde{\omega}^{j}_{k0}+\tilde{\varepsilon}^{j}_{k0})\vec{r}^{C}_{kj}\right\},\\ \vec{m}^{inj}_{k}&=-\left\{m^{j}_{k}\tilde{r}^{C}_{kj}\vec{\gamma}^{j}_{k0}+\hat{J}^{j}_{k}\vec{\varepsilon}^{j}_{k0}+\tilde{\omega}^{j}_{k0}\hat{J}^{j}_{k}\vec{\omega}^{j}_{k0}\right\}\\ &j=(A,B,C)\;\;q=(a,b,c).\end{aligned}\tag{8.2.15}$$

Using the notations $\vec{F}^{j}_{k0}=-\vec{f}^{inj}_{k}-\vec{f}^{*j}_{k}$, $\vec{M}^{j}_{k0}=-\vec{m}^{inj}_{k}-\vec{m}^{*j}_{k}$, the above matrix relations generate a set of interesting recursive relations for the leg j:

$$\begin{aligned}&\vec{F}^{j}_{2}=\vec{F}^{j}_{20},\quad \vec{F}^{j}_{1}=\vec{F}^{j}_{10}+q^{T}_{21}\vec{F}^{j}_{2}\\ &\vec{M}^{j}_{2}=\vec{M}^{j}_{20},\quad \vec{M}^{j}_{1}=\vec{M}^{j}_{10}+q^{T}_{21}\vec{M}^{j}_{2}+\tilde{r}^{j}_{21}q^{T}_{21}\vec{F}^{j}_{2}.\end{aligned}\tag{8.2.16}$$

Especially, for the tree of the moving platform we have

$$\begin{aligned}&\vec{F}^{G}_{3}=m_{3}R(g\vec{u}_{2}+\ddot{x}^{G}_{0}\vec{u}_{1}+\ddot{y}^{G}_{0}\vec{u}_{2}),\quad \vec{F}^{G}_{2}=\vec{F}^{G}_{1}=R^{T}\vec{F}^{G}_{3}\\ &\vec{M}^{G}_{3}=J^{z}_{3}\ddot{\varphi}\,\vec{u}_{3},\quad \vec{M}^{G}_{2}=R^{T}\vec{M}^{G}_{3},\quad \vec{M}^{G}_{1}=\vec{M}^{G}_{2}+\tilde{r}^{G}_{21}\vec{F}^{G}_{2}.\end{aligned}\tag{8.2.17}$$

The characteristic *virtual velocities* are expressed as functions of the pose of the mechanism by the general kinematical equation (8.2.12), where we add the contributions of successive virtual translations during some fictitious displacements along the directions $A_1x^{A}_{1},A_1y^{A}_{1}$, $B_1x^{B}_{1},B_1y^{B}_{1}$, $C_1x^{C}_{1},C_1y^{C}_{1}$ of the middle revolute joints A_2, B_2,C_2:

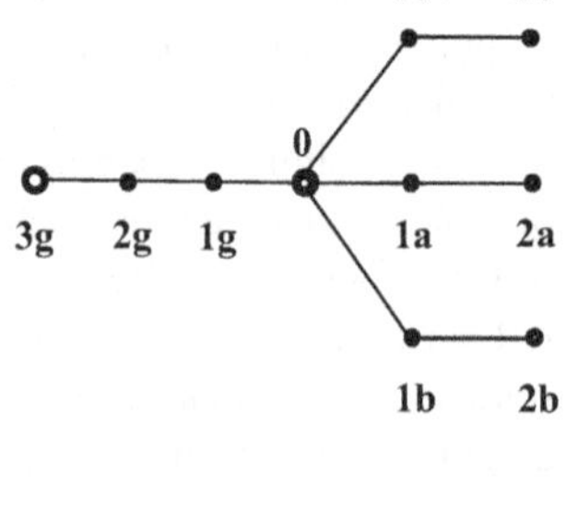

Fig. 8.18 Four open trees of the planar parallel robot

$$\begin{aligned}&\omega_{10j}^{v}\vec{u}_i^T q_{10}^T\tilde{u}_3\{\vec{r}_{21}^{\,j}+q_{21}^T\vec{r}_{32}^{\,j}\}+v_{21}^{jxv}\vec{u}_i^T q_{10}^T\vec{u}_1+v_{21}^{jyv}\vec{u}_i^T q_{10}^T\vec{u}_2+\omega_{21j}^{v}\vec{u}_i^T q_{20}^T\tilde{u}_3\vec{r}_{32}^{\,j}\\&=v_{10G}^{v}\vec{u}_j^T\vec{u}_1+v_{21G}^{v}\vec{u}_i^T\vec{u}_2+\omega_{32G}^{v}\vec{u}_i^T\tilde{u}_3R^T\vec{r}_G^{\,j3}\ \ (i=1,2).\end{aligned}\tag{8.2.18}$$

Considering some independent virtual motions of the planar mechanism, virtual displacements and velocities should be compatible with the virtual motions imposed by all kinematical constraints and joints at any instant in time. Let us assume that the robot has successively nine virtual motions determined by the following sets of velocities:

$$\begin{aligned}&\omega_{10a}^{Av}=1,\ \ \omega_{10a}^{jv}=0\ (j\neq A),\ \ v_{21a}^{jxv}=0,\ \ v_{21a}^{jyv}=0\\&\omega_{10b}^{Bv}=1,\ \ \omega_{10b}^{jv}=0\ (j\neq B),\ \ v_{21b}^{jxv}=0,\ \ v_{21b}^{jyv}=0\\&\omega_{10c}^{Bv}=1,\ \ \omega_{10c}^{jv}=0\ (j\neq C),\ \ v_{21c}^{jxv}=0,\ \ v_{21c}^{jyv}=0\\&\omega_{10a}^{jv}=0,\ \ v_{21a}^{Axv}=1,\ \ v_{21a}^{jxv}=0\ (j\neq A),\ \ v_{21a}^{jyv}=0\\&\omega_{10b}^{jv}=0,\ \ v_{21b}^{Axv}=1,\ \ v_{21b}^{jxv}=0\ (j\neq B),\ \ v_{21b}^{jyv}=0\\&\omega_{10c}^{jv}=0,\ \ v_{21c}^{Axv}=1,\ \ v_{21c}^{jxv}=0\ (j\neq C),\ \ v_{21c}^{jyv}=0\\&\omega_{10a}^{jv}=0,\ \ v_{21a}^{jxv}=0,\ \ v_{21a}^{Ayv}=1,\ \ v_{21a}^{jyv}=0\ (j\neq A)\\&\omega_{10b}^{jv}=0,\ \ v_{21b}^{jxv}=0,\ \ v_{21b}^{Ayv}=1,\ \ v_{21b}^{jyv}=0\ (j\neq B)\\&\omega_{10c}^{jv}=0,\ \ v_{21c}^{jxv}=0,\ \ v_{21c}^{Ayv}=1,\ \ v_{21c}^{jyv}=0\ (j\neq C)\ (j=A,B,C).\end{aligned}\tag{8.2.19}$$

The total virtual work contributed by the inertia forces, external forces and the first actuator torque $\vec{m}_{10}^A$ or any internal joint forces which may be present, for example, can be written in a compact matrix form, using the set of equations of parallel robots dynamics (7.18) and (7.19). Finally, we obtain the *torque* m_{10}^A of the first revolute actuator A_1

$$\begin{aligned}m_{10}^A&=\vec{u}_3^T\{\vec{M}_1^A+\omega_{21a}^{Av}\vec{M}_2^A+\omega_{21a}^{Bv}\vec{M}_2^B+\omega_{21a}^{Cv}\vec{M}_2^C\}\\&\quad+v_{10a}^{Gv}\vec{u}_1^T\vec{F}_1^G+v_{21a}^{Gv}\vec{u}_2^T\vec{F}_2^G+\omega_{32a}^{Gv}\vec{u}_3^T\vec{M}_3^G.\end{aligned}\tag{8.2.20}$$

and the two orthogonal components of the *internal joint force* acting in the intermediary joint A_2

$$\begin{aligned}f_{21A}^x&=\vec{u}_1^T a_{21}^T\vec{F}_2^A+\vec{u}_3^T\{\omega_{21a}^{Av}\vec{M}_2^A+\omega_{21a}^{Bv}\vec{M}_2^B+\omega_{21a}^{Cv}\vec{M}_2^C\}\\&\quad+v_{10a}^{Gv}\vec{u}_1^T\vec{F}_1^G+v_{21a}^{Gv}\vec{u}_2^T\vec{F}_2^G+\omega_{32a}^{Gv}\vec{u}_3^T\vec{M}_3^G\\f_{21A}^y&=\vec{u}_2^T a_{21}^T\vec{F}_2^A+\vec{u}_3^T\{\omega_{21a}^{Av}\vec{M}_2^A+\omega_{21a}^{Bv}\vec{M}_2^B+\omega_{21a}^{Cv}\vec{M}_2^C\}\\&\quad+v_{10a}^{Gv}\vec{u}_1^T\vec{F}_1^G+v_{21a}^{Gv}\vec{u}_2^T\vec{F}_2^G+\omega_{32a}^{Gv}\vec{u}_3^T\vec{M}_3^G.\end{aligned}\tag{8.2.21}$$

The relations (8.2.15)–(8.2.21) represent the *complete inverse dynamics model* of the 3-*RRR* planar parallel robot and take into account the contribution of each link in the above final expressions. The various dynamic effects, including the Coriolis and centrifugal forces coupling and the gravitational actions are considered in the equations.

8.2.2.2 Lagrange Equations

A solution of the dynamics problem of the planar 3-*RRR* parallel robot can be developed, using *Lagrange's equations* of second kind for a mechanical system with constraints. The generalized coordinates of the robot are represented by nine parameters $s_1^A = \varphi_{10}^A$, $s_2^A = \varphi_{21}^A$, $s_1^B = \varphi_{10}^B$, $s_2^B = \varphi_{21}^B$, $s_1^C = \varphi_{10}^C$, $s_2^C = \varphi_{21}^C$, $s_1^G = x_0^G$, $s_2^G = y_0^G$, $s_3^G = \varphi$.

Lagrange's equations with their six multipliers $\lambda_1^j, \lambda_2^j\,(j = A, B, C)$ will be expressed by nine differential relations

$$\begin{aligned}
&\frac{d}{dt}\left\{\frac{\partial L}{\partial \dot{\varphi}_{10}^j}\right\} - \frac{\partial L}{\partial \varphi_{10}^j} = \Phi_1^j + \lambda_1^j c_{11}^j + \lambda_2^j c_{21}^j \\
&\frac{d}{dt}\left\{\frac{\partial L}{\partial \dot{\varphi}_{21}^j}\right\} - \frac{\partial L}{\partial \varphi_{21}^j} = \Phi_2^j + \lambda_1^j c_{12}^j + \lambda_2^j c_{22}^j \; (j = A, B, C) \\
&\frac{d}{dt}\left\{\frac{\partial L}{\partial \dot{x}_0^G}\right\} - \frac{\partial L}{\partial x_0^G} = \Phi_1^G - \lambda_1^A - \lambda_1^B - \lambda_1^C \\
&\frac{d}{dt}\left\{\frac{\partial L}{\partial \dot{y}_0^G}\right\} - \frac{\partial L}{\partial y_0^G} = \Phi_2^G - \lambda_2^A - \lambda_2^B - \lambda_2^C \\
&\frac{d}{dt}\left\{\frac{\partial L}{\partial \dot{\phi}}\right\} - \frac{\partial L}{\partial \phi} = \Phi_3^G + (\lambda_1^A + \lambda_1^B + \lambda_1^C)c_{13}^G + (\lambda_2^A + \lambda_2^B + \lambda_2^C)c_{32}^G,
\end{aligned} \tag{8.2.22}$$

which contain the following nine generalized forces

$$\begin{aligned}
&\Phi_1^A = m_{10}^A, \quad \Phi_2^A = 0, \quad \Phi_1^B = m_{10}^B, \quad \Phi_2^B = 0, \quad \Phi_1^C = m_{10}^C, \\
&\Phi_2^C = 0, \quad \Phi_1^G = 0, \quad \Phi_2^G = 0, \quad \Phi_3^G = 0.
\end{aligned} \tag{8.2.23}$$

Starting from the relations (8.2.11) and (8.2.12), a number of six kinematical conditions of constraint are composed in a matrix form

$$\begin{bmatrix} c_{11}^{A} & c_{12}^{A} & 0 & 0 & 0 & 0 & -1 & 0 & c_{13}^{G} \\ c_{21}^{A} & c_{22}^{A} & 0 & 0 & 0 & 0 & 0 & -1 & c_{23}^{G} \\ 0 & 0 & c_{11}^{B} & c_{12}^{B} & 0 & 0 & -1 & 0 & c_{13}^{G} \\ 0 & 0 & c_{21}^{B} & c_{22}^{B} & 0 & 0 & 0 & -1 & c_{23}^{G} \\ 0 & 0 & 0 & 0 & c_{11}^{C} & c_{12}^{C} & -1 & 0 & c_{13}^{G} \\ 0 & 0 & 0 & 0 & c_{21}^{C} & c_{22}^{C} & 0 & -1 & c_{23}^{G} \end{bmatrix} \begin{bmatrix} \dot{s}_1^A \\ \dot{s}_2^A \\ \dot{s}_1^B \\ \dot{s}_2^B \\ \dot{s}_1^C \\ \dot{s}_2^C \\ \dot{s}_1^G \\ \dot{s}_2^G \\ \dot{s}_3^G \end{bmatrix} = 0. \qquad (8.2.24)$$

The components of the general expression of the Lagrange function $L = L_G + \sum_{\mu=1}^{2}\left(L_\mu^A + L_\mu^B + L_\mu^C\right)$ are expressed as analytical functions of the generalized coordinates and their first derivatives with respect to time:

$$\begin{aligned} L_G &= \frac{1}{2}m_3(\dot{x}_{0G}^2 + \dot{y}_{0G}^2) + \frac{1}{2}J_3^z\dot{\phi}^2 - m_3 g y_0^G \\ L_1^j &= \frac{1}{2}\vec{\omega}_{10j}^T\hat{J}_1\vec{\omega}_{10}^j - m_1 g\vec{u}_2^T(\vec{r}_{10}^j + q_{10}^T\vec{r}_1^{Cj}) \\ L_2^j &= \frac{1}{2}m_2\vec{v}_{20j}^T\vec{v}_{20}^j + \frac{1}{2}\vec{\omega}_{20j}^T\hat{J}_2\vec{\omega}_{20}^j + m_2\vec{v}_{20j}^T\tilde{\omega}_{20}^j\vec{r}_2^{Cj} - m_2 g\vec{u}_2^T \\ &\quad (\vec{r}_{10}^j + q_{10}^T\vec{r}_{21}^j + q_{20}^T\vec{r}_2^{Cj}) \quad q = (a,b,c), \;\; j = (A,B,C). \end{aligned} \qquad (8.2.25)$$

A long calculus of the derivatives with respect to time $\frac{d}{dt}\left\{\frac{\partial L}{\partial \dot{s}_l}\right\}$ $(l = 1,2,\ldots,9)$ of the above functions leads to a system of nine relations, starting from the relations (8.2.9), (8.2.10) and using the first derivatives of orthogonal transformation matrices $q_{k,k-1}$:

$$\begin{aligned} &\dot{q}_{k,k-1} = \dot{\varphi}_{k,k-1}^j\tilde{u}_3^T q_{k,k-1}, \quad \dot{q}_{k,k-1}^T = \dot{\varphi}_{k,k-1}^j q_{k,k-1}^T\tilde{u}_3, \\ &\frac{\partial q_{k,k-1}}{\partial \varphi_{k,k-1}^j} = \tilde{u}_3^T q_{k,k-1}, \quad \frac{\partial q_{k,k-1}^T}{\partial \varphi_{k,k-1}^j} = q_{k,k-1}^T\tilde{u}_3. \end{aligned} \qquad (8.2.26)$$

In the inverse dynamics problem, after *elimination of six multipliers*, finally we obtain the same expressions (8.2.20) for the six input torques $m_{10}^A, m_{10}^B, m_{10}^C$ required by the revolute actuators.

As an application we consider a 3-*RRR* robot, having the following geometrical and architectural characteristics:

$$x_{0G}^{*} = 0.025\,\text{m},\ \ y_{0G}^{*} = 0.025\,\text{m},\ \ \phi^{*} = \frac{\pi}{12},\ \ \Delta t = 3\,\text{s}$$
$$r = 0.3\,\text{m},\ \ l_1 = 1.5r,\ \ l_2 = r,\ \ l = r\sqrt{3},$$
$$m_1 = 2.25\,\text{kg},\ \ m_2 = 1.5\,\text{kg},\ \ m_3 = 5\,\text{kg}.$$

Using the MATLAB software, a dynamic simulation is based on the computation of three input torques m_{10}^{j} required by every actuator during the platform's evolution and the components f_{21j}^{x}, f_{21j}^{y} of internal joint forces acting in the middle joints. A computer program was developed to solve the inverse dynamics of the planar parallel robot, assuming that for a period of three seconds the platform starts at rest from a central configuration and rotates or translates along rectilinear directions.

For the first example, we consider the *translation motion* of the moving platform in its vertical plane, along the *horizontal* axis x_0, with variable acceleration while all the other positional parameters are held to be equal to zero. The input torques of three actuators and the internal joint forces are calculated by the program and plotted against time as follows: Figs. 8.19, 8.20 and 8.21.

If the platform's centre G moves along the *vertical* axis y_0 with no rotation of the platform, the graphs are sketched in Figs. 8.22, 8.23 and 8.24.

For the third example of simulation, we consider the *rotation motion* of the moving platform around the *horizontal axis* z_0 with variable angular acceleration: Figs. 8.25, 8.26, 8.27.

Based on the principle of virtual work, the dynamics model takes into consideration the mass, the tensor of inertia and the action of inertia force introduced by all compounding elements of the parallel mechanism.

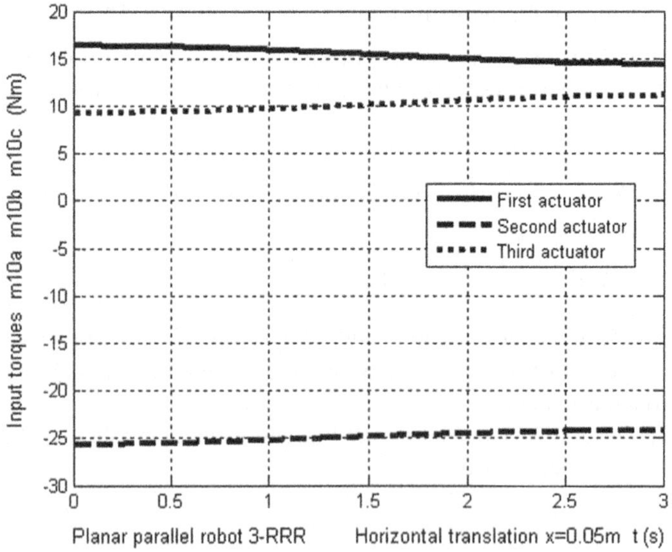

Fig. 8.19 Input torques of actuators (horizontal translation)

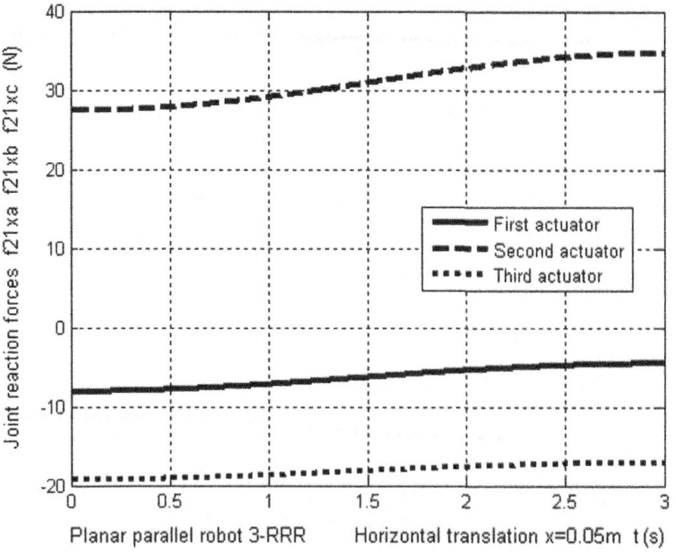

Fig. 8.20 Joint forces $f_{21x}^{A}, f_{21x}^{B}, f_{21x}^{C}$ (horizontal translation)

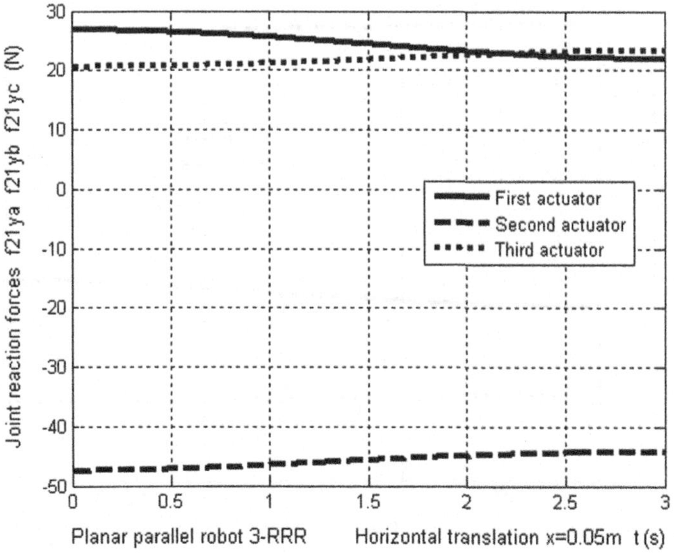

Fig. 8.21 Joint forces $f_{21y}^{A}, f_{21y}^{B}, f_{21y}^{C}$ (horizontal translation)

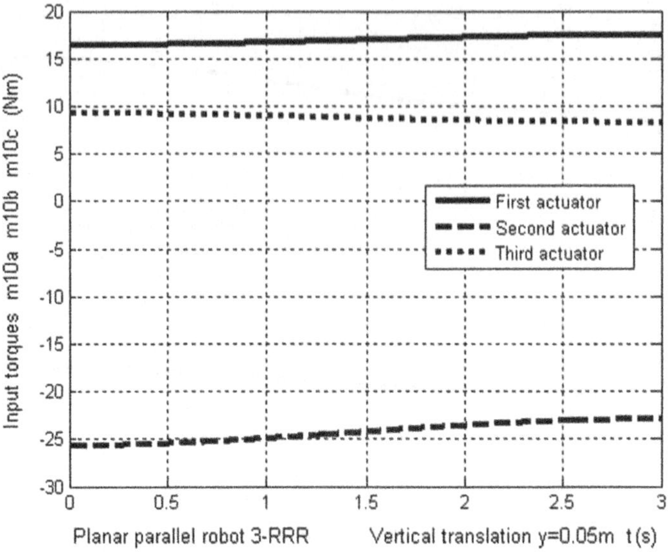

Fig. 8.22 Input torques of actuators (vertical translation)

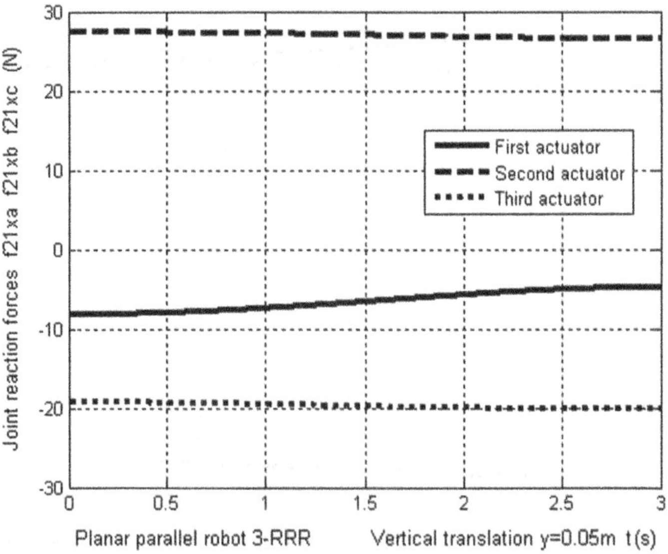

Fig. 8.23 Joint forces $f^A_{21x}, f^B_{21x}, f^C_{21x}$ (vertical translation)

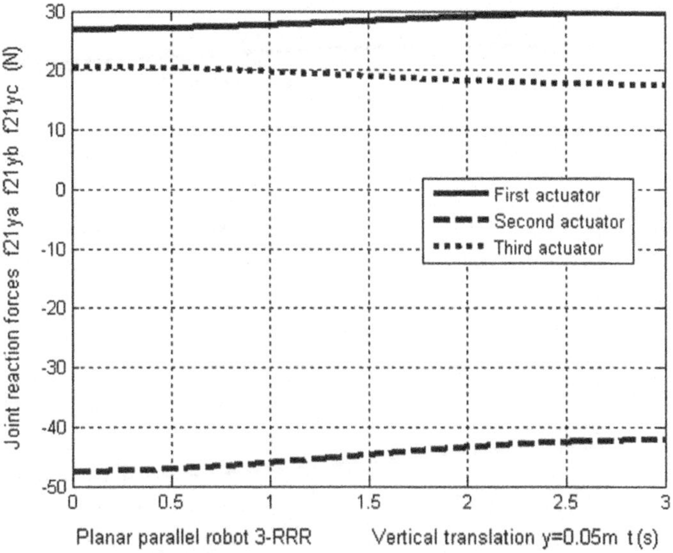

Fig. 8.24 Joint forces $f^A_{21y}, f^B_{21y}, f^C_{21y}$ (vertical translation)

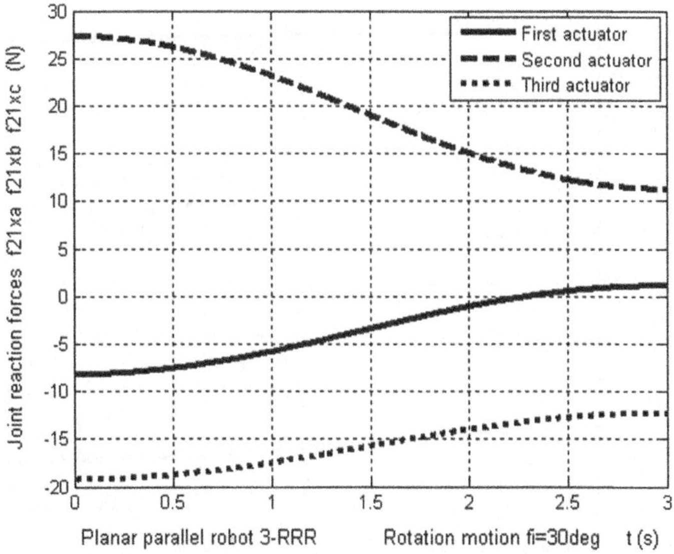

Fig. 8.25 Input torques of three actuators (rotation)

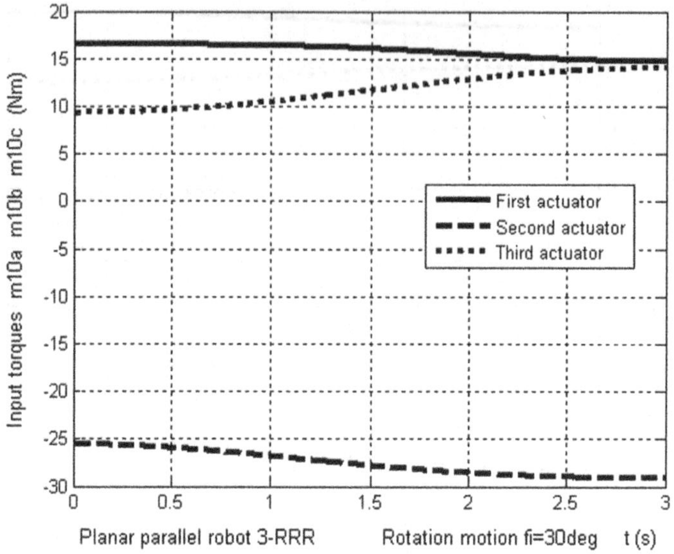

Fig. 8.26 Joint forces $f_{21x}^{A}, f_{21x}^{B}, f_{21x}^{C}$ (rotation)

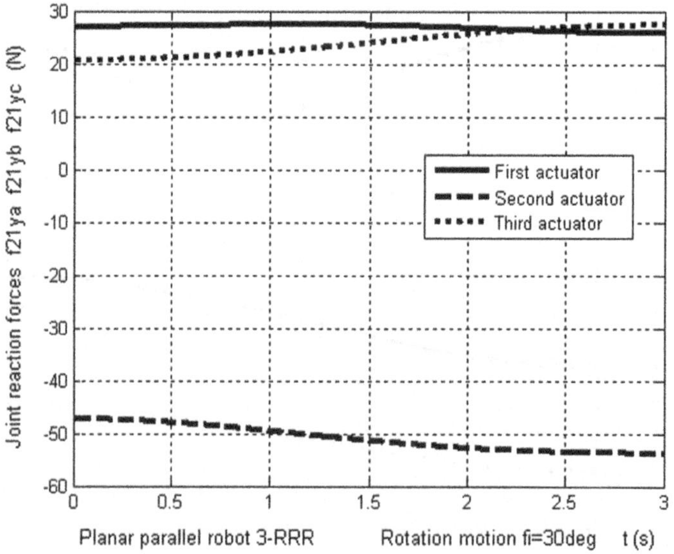

Fig. 8.27 Joint forces $f_{21y}^{A}, f_{21y}^{B}, f_{21y}^{C}$ (rotation)

8.3 Inverse Kinematics and Dynamics of a 3-*PRP* Planar Parallel Robot

Daniali and Zsombor-Murray [7] discuss a manipulator of the 3-*PRP* type having a triangular fixed base and a triangular mobile platform, where three intermediary revolute joints allow their platform to turn, sliding simultaneously along the sides of the fixed platform.

Recursive modelling for the kinematics and complete dynamics of the *PRP* planar parallel robot will be found to be developed in a numerically efficient manner.

8.3.1 Kinematics Analysis

A planar 3-*PRP* parallel robot is a symmetrical closed-loop mechanism composed of three planar kinematical chains with identical topology, each consisting of two links with one revolute and two prismatic joints, which connect the base to the moving platform. With seven links $(T_k, k = 1, 2, \ldots, 7)$, three revolute joints and six prismatic joints, the mechanism has three degrees of freedom. Three controlled prismatic cylinders can be located on the fixed vertical base $A_0B_0C_0$ (Fig. 8.28).

We attach a Cartesian frame $Ox_0y_0z_0(T_0)$ to the fixed base with its origin located at triangle centre O, the z_0 axis perpendicular to the base and the x_0 axis pointing along the C_0B_0 direction. A mobile reference frame $Gx_Gy_Gz_G$ is attached to the moving platform. The origin of this central frame is located at the centre of the moving triangle, which is considered to be located initially at a *central*

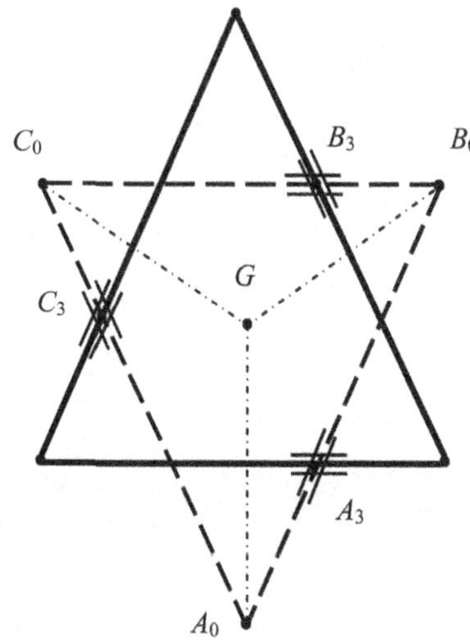

Fig. 8.28 The 3-*P*RP planar parallel robot

configuration, where the platform is not rotated with respect to the fixed base and its mass centre is at the origin O of fixed frame.

One of three active legs (for example leg A) consists of a prismatic joint, which is like as a linear drive **1** linked at the $A_1x_1^Ay_1^Az_1^A$ frame, having a rectilinear motion of displacement λ_{10}^A, velocity $v_{10}^A = \dot{\lambda}_{10}^A$ and acceleration $\gamma_{10}^A = \ddot{\lambda}_{10}^A$. The following link of the leg is a rigid body **2** linked at the $A_2x_2^Ay_2^Az_2^A$ frame, which has a relative rotation about z_2^A axis with angle φ_{21}^A, velocity $\omega_{21}^A = \dot{\varphi}_{21}^A$ and acceleration $\varepsilon_{21}^A = \ddot{\varphi}_{21}^A$. A prismatic passive joint is introduced at the edge of a moving platform as an equilateral triangle of side $l = l_0\sqrt{3}$, which translates relatively at the displacement λ_{32}^A and the velocity $v_{32}^A = \dot{\lambda}_{32}^A$ along z_3^A axis (Fig. 8.29). All legs are symmetrically extended at the central configuration, so that the three angles of orientation of the edges of fixed platform are given by

$$\alpha_A = \frac{\pi}{3}, \quad \alpha_B = \pi, \quad \alpha_C = -\frac{\pi}{3}. \tag{8.3.1}$$

The displacement of body **2** is only along a fixed line, and its velocity and acceleration are equal to those of the actuated joint associated. We call the matrix $q_{k,k-1}^{\varphi}$ the orthogonal transformation 3×3 matrix of relative rotation with the angle $\varphi_{k,k-1}^i$ of link T_k^i around the z_k^i axis.

Pursuing three independent legs $OA_0A_1A_2A_3$, $OB_0B_1B_2B_3$, $OC_0C_1C_2C_3$, we obtain following transformation matrices [17]

$$q_{10} = \theta_1\theta_{\alpha}^j, \quad q_{21} = q_{21}^{\varphi}\theta_1^T, \quad q_{32} = \theta_1\theta_2^T \quad (q = a, b, c) \quad (j = A, B, C), \tag{8.3.2}$$

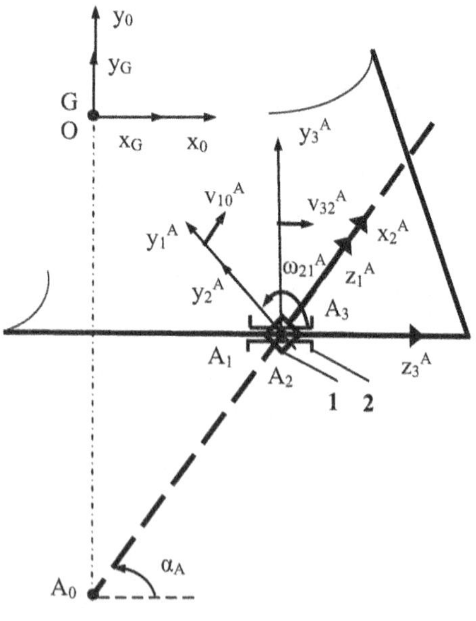

Fig. 8.29 Kinematical scheme of first leg A of the mechanism

where

$$q_{k,k-1}^{\varphi} = rot(z, \varphi_{k,k-1}^{i}), \quad q_{k0} = \prod_{s=1}^{k} q_{k-s+1,k-s} \ (k = 1, 2, 3)$$
$$\theta_1 = rot(y, \pi/2), \quad \theta_2 = rot(z, \pi/3), \quad \theta_{\alpha}^{j} = rot(z, \alpha_j). \tag{8.3.3}$$

The displacements $\lambda_{10}^{A}, \lambda_{10}^{B}, \lambda_{10}^{C}$ of the active links are the joint variables that give the input vector $\vec{\lambda}_{10} = \begin{bmatrix} \lambda_{10}^{A} & \lambda_{10}^{B} & \lambda_{10}^{C} \end{bmatrix}^{T}$ of the instantaneous position of the mechanism. In the inverse geometric problem, the position of the mechanism is completely given by the coordinates x_0^G, y_0^G of the mass centre G of the moving platform and the orientation angle ϕ of the movable frame $Gx_Gy_Gz_G$. We suppose that the position vector $\vec{r}_0^{\,G} = \begin{bmatrix} x_0^G & y_0^G & 0 \end{bmatrix}^{T}$ of G and the angle ϕ, which are expressed by the following analytical functions

$$\frac{x_0^G}{x_{0G}^{*}} = \frac{y_0^G}{y_{0G}^{*}} = \frac{\phi}{\phi^{*}} = 1 - \cos\frac{\pi}{3}t, \tag{8.3.4}$$

can describe the general absolute motion of the moving platform in its *vertical plane*.

Using the orthogonal rotation matrix $R = rot(z, \varphi)$, the conditions on the absolute orientation of the moving platform are expressed by the three identities

$$q_{30}^{\circ T} q_{30} = R, \quad q_{30}^{\circ} = q_{30}(t = 0) = \theta_1 \theta_2 \theta_{\alpha}^{j} \ (q = a, b, c) \ (j = A, B, C) \tag{8.3.5}$$

from which we obtain the initial relations between the angles of rotation $\varphi_{21}^{A} = \varphi_{21}^{B} = \varphi_{21}^{C} = \phi$.

Six independent variables $\lambda_{10}^{A}, \lambda_{32}^{A}, \lambda_{10}^{B}, \lambda_{32}^{B}, \lambda_{10}^{C}, \lambda_{32}^{C}$ will be determined by vector-loop equations as follows:

$$\vec{r}_{10}^{\,j} + \sum_{k=1}^{2} q_{k0}^{T} \vec{r}_{k+1,k}^{\,j} + q_{30}^{T} \vec{r}_{3}^{\,jG} = \vec{r}_{0}^{\,G} \quad (q = a, b, c) \ (j = A, B, C) \tag{8.3.6}$$

where

$$\vec{r}_{10}^{\,j} = \vec{r}_{00}^{\,j} + (l_0/\sqrt{3} + \lambda_{10}^{j}) q_{10}^{T} \vec{u}_3, \quad \vec{r}_{00}^{\,A} = l_0 \begin{bmatrix} 0 & -1 & 0 \end{bmatrix}^{T}$$
$$\vec{r}_{00}^{\,B} = 0.5 l_0 \begin{bmatrix} \sqrt{3} & 1 & 0 \end{bmatrix}^{T}, \quad \vec{r}_{00}^{\,C} = 0.5 l_0 \begin{bmatrix} -\sqrt{3} & 1 & 0 \end{bmatrix}^{T} \tag{8.3.7}$$
$$\vec{r}_{21}^{\,j} = \vec{0}, \quad \vec{r}_{32}^{\,j} = \lambda_{32}^{j} q_{32}^{T} \vec{u}_3, \quad \vec{r}_{3j}^{\,G} = 0.5 l_0 \begin{bmatrix} 0 & 1 & -1/\sqrt{3} \end{bmatrix}^{T}.$$

Interesting analytical relations are derived:

$$\begin{aligned}
&\left(\frac{l_0}{\sqrt{3}} + \lambda_{10}^j\right)\cos\alpha_i + \lambda_{32}^j \cos\left(\phi - \frac{\pi}{3} + \alpha_j\right) \\
&\quad = x_0^G - x_{00}^j - \frac{l_0}{2\sqrt{3}}\cos(\phi + \alpha_j) + \frac{l_0}{2}\sin(\phi + \alpha_j) \\
&\left(\frac{l_0}{\sqrt{3}} + \lambda_{10}^j\right)\cos\alpha_i + \lambda_{32}^j \cos\left(\phi - \frac{\pi}{3} + \alpha_j\right) \\
&\quad = x_0^G - x_{00}^j - \frac{l_0}{2\sqrt{3}}\cos(\phi + \alpha_j) + \frac{l_0}{2}\sin(\phi + \alpha_j).
\end{aligned} \tag{8.3.8}$$

The rotations of elements of each leg j are characterized by the following skew-symmetric matrices

$$\begin{aligned}
&\tilde{\omega}_{k0}^j = q_{k,k-1}\tilde{\omega}_{k-1,0}^j q_{k,k-1}^T + \omega_{k,k-1}^j \tilde{u}_3, \quad \omega_{21}^j = \dot{\varphi}, \quad \omega_{\tau,\tau-1}^j = 0 \\
&(\tau = 1, 3) \ \ (k = 1, 2, 3),
\end{aligned} \tag{8.3.9}$$

which are *associated* with the angular velocities

$$\begin{aligned}
&\vec{\omega}_{10}^j = \vec{0}, \quad \vec{\omega}_{21}^j = \dot{\phi}\,\vec{u}_3, \\
&\vec{\omega}_{32}^j = \vec{0}, \quad \vec{\omega}_{20}^j = q_{21}\vec{\omega}_{10}^j + \vec{\omega}_{21}^j = \dot{\varphi}\vec{u}_3, \\
&\vec{\omega}_{30}^j = q_{32}\vec{\omega}_{20}^j + \vec{\omega}_{32}^j = \dot{\phi}\vec{u}_3.
\end{aligned} \tag{8.3.10}$$

Following expressions give the velocities $\vec{v}_{k0}^j$ of the centres of the joints

$$\vec{v}_{k0}^j = q_{k,k-1}\vec{v}_{k-1,0}^j + q_{k,k-1}\tilde{\omega}_{k-1,0}^j \vec{r}_{k,k-1}^j + v_{k,k-1}^j \vec{u}_3, \quad v_{21}^j = 0, \quad v_{\tau,\tau-1}^j = \dot{\lambda}_{\tau,\tau-1}^j. \tag{8.3.11}$$

When differentiated with respect to time, equations of geometrical constraints (8.3.6) lead to the following *matrix conditions of connectivity* [12]:

$$v_{10}^j \vec{u}_i^T q_{10}^T \vec{u}_3 + v_{32}^j \vec{u}_i^T q_{30}^T \vec{u}_3 = \vec{u}_i^T \dot{\vec{r}}_0^G - \dot{\phi}\,\vec{u}_i^T q_{20}^T \tilde{u}_3 \{\vec{r}_{32}^j + q_{32}^T \vec{r}_{3G}^j\} \quad (i = 1, 2), \tag{8.3.12}$$

where $\tilde{u}_3$ is the skew-symmetric matrix associated with unit vector $\vec{u}_3$. The right-hand side of (8.3.12) is assumed to be known since the terms depend on time and the coordinates that are assumed to be known from the position analysis. Since the independent velocities $\dot{x}_0^G, \dot{y}_0^G$ and $\dot{\phi}$ are assumed to be known, the equations delivered by (8.3.12) can be considered as a linear system of algebraic equations in the dependent velocities. Within an inverse kinematics, from these equations, we obtain the relative velocities v_{10}^j and v_{32}^j as functions of the angular velocity $\dot{\phi}$ of the platform and the velocity of mass centre G.

The derivatives with respect to time of the six constraint equation (8.3.8) lead to the matrix equation

$$J_1 \dot{\vec{\lambda}}_{10} = J_2 \left[\dot{x}_0^G \quad \dot{y}_0^G \quad \dot{\phi} \right]^T \tag{8.3.13}$$

for the planar robot with prismatic actuators. The matrices J_1 and J_2 are the inverse Jacobian and forward Jacobian of the manipulator, and can be expressed as

$$J_1 = diag\{ \delta_A \quad \delta_B \quad \delta_C \}, \quad J_2 = \begin{bmatrix} \beta_1^A & \beta_2^A & \beta_3^A \\ \beta_1^B & \beta_2^B & \beta_3^B \\ \beta_1^C & \beta_2^C & \beta_3^C \end{bmatrix}, \tag{8.3.14}$$

with

$$\delta_j = \sin\left(\phi - \frac{\pi}{3}\right), \quad \beta_1^j = \sin\left(\phi - \frac{\pi}{3} + \alpha_j\right), \quad \beta_2^j = -\cos\left(\phi - \frac{\pi}{3} + \alpha_j\right)$$
$$\beta_3^j = (x_0^G - x_{00}^j)\cos\left(\phi - \frac{\pi}{3} + \alpha_j\right) + (y_0^G - y_{00}^j)\sin\left(\phi - \frac{\pi}{3} + \alpha_j\right) - \left(\frac{l_0}{\sqrt{3}} + \lambda_{10}^j\right)\cos\left(\phi - \frac{\pi}{3}\right). \tag{8.3.15}$$

On the relative accelerations γ_{10}^j and γ_{32}^j of the robot, new connectivity conditions are obtained by the time derivative of equations in (8.3.12):

$$\begin{aligned} \gamma_{10}^j \vec{u}_i^T q_{10}^T \vec{u}_3 + \gamma_{32}^j \vec{u}_i^T q_{30}^T \vec{u}_3 &= \vec{u}_i^T \ddot{\vec{r}}_0^G - \dot{\phi}^2 \vec{u}_i^T q_{20}^T \tilde{u}_3 \tilde{u}_3 \{ \vec{r}_{32}^j + q_{32}^T \vec{r}_{3G}^j \} \\ &\quad - \ddot{\phi} \vec{u}_i^T q_{20}^T \tilde{u}_3 \{ \vec{r}_{32}^j + q_{32}^T \vec{r}_{3G}^j \} - 2 v_{32}^j \dot{\phi}\, \vec{u}_i^T q_{20}^T \tilde{u}_3 q_{32}^T \vec{u}_3 \quad (i = 1, 2). \end{aligned} \tag{8.3.16}$$

Starting from (8.1.22), the following relations give the angular accelerations $\vec{\varepsilon}_{k0}^j$ and the accelerations $\vec{\gamma}_{k0}^j$ of joints:

$$\begin{aligned} \tilde{\omega}_{k0}^j \tilde{\omega}_{k0}^j + \tilde{\varepsilon}_{k0}^j &= q_{k,k-1} \left\{ \tilde{\omega}_{k-1,0}^j \tilde{\omega}_{k-1,0}^j + \tilde{\varepsilon}_{k-1,0}^j \right\} q_{k,k-1}^T \\ &\quad + \omega_{k,k-1}^j \omega_{k,k-1}^j \tilde{u}_3 \tilde{u}_3 + \varepsilon_{k,k-1}^j \tilde{u}_3 + 2 \omega_{k,k-1}^j q_{k,k-1} \tilde{\omega}_{k-1,0}^j q_{k,k-1}^T \tilde{u}_3 \\ \vec{\varepsilon}_{k0}^j &= q_{k,k-1} \vec{\varepsilon}_{k-1,0}^j + \varepsilon_{k,k-1}^j \vec{u}_3 + \omega_{k,k-1}^j q_{k,k-1} \tilde{\omega}_{k-1,0}^j q_{k,k-1}^T \vec{u}_3, \quad \varepsilon_{21}^A = \ddot{\varphi}, \quad \varepsilon_{\tau,\tau-1}^j = 0 \\ \vec{\gamma}_{k0}^j &= q_{k,k-1} \vec{\gamma}_{k-1,0}^j + q_{k,k-1} \{ \tilde{\omega}_{k-1,0}^j \tilde{\omega}_{k-1,0}^j + \tilde{\varepsilon}_{k-1,0}^j \} \vec{r}_{k,k-1}^j \\ &\quad + 2 v_{k,k-1}^j q_{k,k-1} \tilde{\omega}_{k-1,0}^j q_{k,k-1}^T \vec{u}_3 + \gamma_{k,k-1}^j \vec{u}_3, \quad \gamma_{21}^j = 0, \quad \gamma_{\tau,\tau-1}^j = \ddot{\lambda}_{\tau,\tau-1}^j \\ &\quad (\tau = 1, 3) \quad (k = 1, 2, 3). \end{aligned} \tag{8.3.17}$$

8.3.2 Dynamics Modelling

Three independent mechanical systems acting along three fixed planar directions with the forces $\vec{f}_{10}^{A} = f_{10}^{A}\vec{u}_3, \vec{f}_{10}^{B} = f_{10}^{B}\vec{u}_3, \vec{f}_{10}^{C} = f_{10}^{C}\vec{u}_3$ will control the motion of the moving platform.

Starting from the known external forces acting on the robot as well as the kinematics state of every link, the active forces required in a given motion of the moving platform and some internal reaction forces can be recursively expressed, based on the principle of virtual work. Using the same equivalent graph from Fig. 8.4, this planar robot can artificially by transformed into a set of three independent trees by separating two passive joints for the moving platform. During three independent virtual motions of the robot, the virtual velocities should be consistent with the motions imposed by all kinematical constraints at a given instant in time.

The forces of inertia and the wrench of the external and internal forces applied to an arbitrary rigid body T_k^j can be computed using the above (8.1.23)–(8.1.26) general formulae.

Concerning the first leg A, for example, the characteristic *virtual velocities* are expressed as functions of the pose of the mechanism by the general kinematical equation (8.3.12), where we add the contributions of successive virtual translations during two orthogonal fictitious displacements of intermediary revolute joint A_2, as follows:

$$\begin{aligned} & v_{10A}^{v}\vec{u}_i^T a_{10}^T \vec{u}_3 + v_{21}^{Ayv}\vec{u}_i^T a_{10}^T \vec{u}_2 + v_{21}^{Azv} a_{10}^T \vec{u}_3 \\ & \quad + \omega_{21A}^{v}\vec{u}_i^T a_{20}^T \tilde{u}_3 \{\vec{r}_{32}^{A} + a_{32}^T \vec{r}_{3G}^{A}\} + v_{32A}^{v}\vec{u}_i^T a_{30}^T \vec{u}_3 = \vec{u}_i^T \vec{v}_{0G}^{v} \quad (i = 1, 2). \end{aligned} \tag{8.3.18}$$

Now, let us assume that the robot has successively some virtual motions determined by the following set of velocities:

$$\begin{aligned} & v_{10a}^{Av} = 1, \quad v_{10a}^{Bv} = 0, \quad v_{10a}^{Cv} = 0, \quad v_{21a}^{iyv} = 0, \quad v_{21a}^{izv} = 0 \\ & v_{10a}^{iv} = 0, \quad v_{21a}^{Ayv} = 1, \quad v_{21a}^{Byv} = 0, \quad v_{21a}^{Byv} = 0, \quad v_{21a}^{izv} = 0 \\ & v_{10a}^{iv} = 0, \quad v_{21a}^{iyv} = 0, \quad v_{21a}^{Azv} = 1, \quad v_{21a}^{Bzv} = 0, \quad v_{21a}^{Czv} = 0 \quad (i = A, B, C). \end{aligned} \tag{8.3.19}$$

Total virtual work contributed by the inertia forces and moments of inertia forces $\vec{f}_{ki}^{\,n}, \vec{m}_{ki}^{\,n}$, by the wrench of known external forces $\vec{f}_k^{\,*}, \vec{m}_k^{\,*}$ and by the first active force $\vec{f}_{10}^{A}$ or some internal joint forces, for example, can be written in a compact form, based on the relative virtual velocities. Applying *the equations of the parallel robots dynamics* [18], the following compact relations result

$$f_{10}^{A} = \vec{u}_3^T \{\vec{F}_1^A + \omega_{21a}^{Av}\vec{M}_2^A + v_{32a}^{Av}\vec{F}_3^A + \omega_{21a}^{Bv}\vec{M}_2^B + \omega_{21a}^{Cv}\vec{M}_2^C\} \tag{8.3.20}$$

for the *input force* of the first prismatic actuator, and

$$\begin{aligned} f_{21y}^{A} &= \vec{u}_2^T a_{21}^T \vec{F}_2^A + \vec{u}_3^T \left\{ \omega_{21a}^{Ayv} \vec{M}_2^A + v_{32a}^{Ayv} \vec{F}_3^A + \omega_{21}^{Byv} \vec{M}_2^B + \omega_{21}^{Cyv} \vec{M}_2^C \right\} \\ f_{21z}^{A} &= \vec{u}_3^T a_{21}^T \vec{F}_2^A + \vec{u}_3^T \left\{ \omega_{21a}^{Azv} \vec{M}_2^A + v_{32a}^{Azv} \vec{F}_3^A + \omega_{21}^{Bzv} \vec{M}_2^B + \omega_{21}^{Czv} \vec{M}_2^C \right\} \end{aligned} \quad (8.3.21)$$

for the two orthogonal *joint forces* acting in the internal middle revolute joint A_2.

Relations (8.3.20)–(8.2.21) represent the *complete inverse dynamics model* of the 3-*PRP* planar parallel robot, indicating a high degree of coupling between the connectors caused by the gravitational, Coriolis, tangential and centrifugal accelerations acting upon the system.

As an application let us consider a robot which has the following geometrical and architectural characteristics:

$$x_{0G}^* = 0, \quad y_{0G}^* = 0.025\,\text{m}, \quad \phi^* = \frac{\pi}{12}, \quad l_0 = OA_0 = OB_0 = OC_0 = 0.3\,\text{m}, \quad l = l_0\sqrt{3}$$
$$m_1 = 1\,\text{kg}, \quad m_2 = 0.75\,\text{kg}, \quad m_3 = 3\,\text{kg}, \quad \Delta t = 3\,\text{s}.$$

A program is developed to solve the kinematics and dynamics, assuming that the platform starts at rest from the central configuration, evolving successively a rotation and a vertical translation. The results concern the displacements λ_{10}^j, active powers $P_{10}^j = v_{10}^j f_{10}^j$ required by each actuator, and the internal joint forces f_{21y}^j, f_{21z}^j $(j = A, B, C)$.

First, we consider the *rotation motion* of the moving platform about the z_0 central horizontal axis with variable angular acceleration (Figs. 8.30, 8.31, 8.32 and

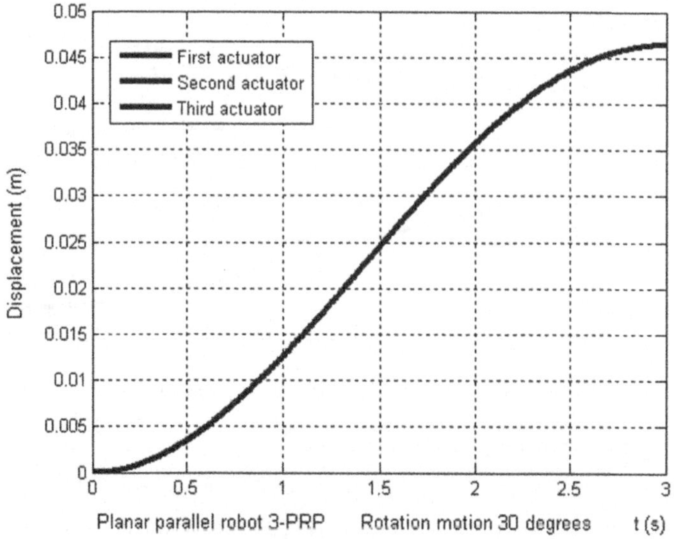

Fig. 8.30 Displacements $\lambda_{10}^A, \lambda_{10}^B, \lambda_{10}^C$ (rotation)

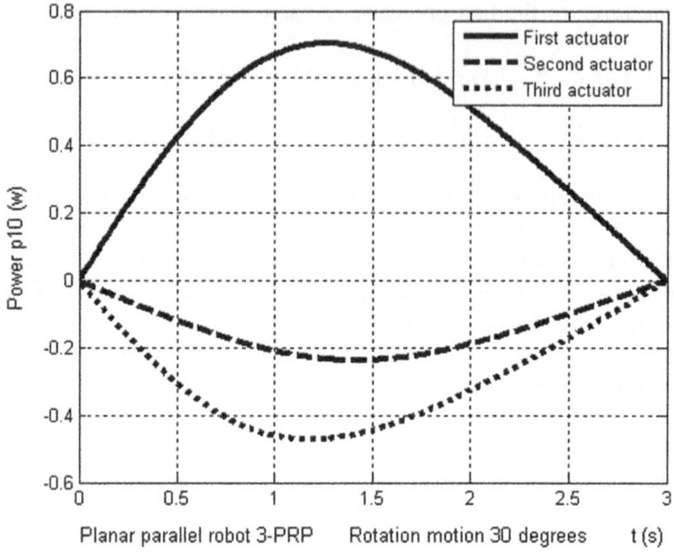

Fig. 8.31 Powers $P_{10}^{A}, P_{10}^{B}, P_{10}^{C}$ of actuators (rotation)

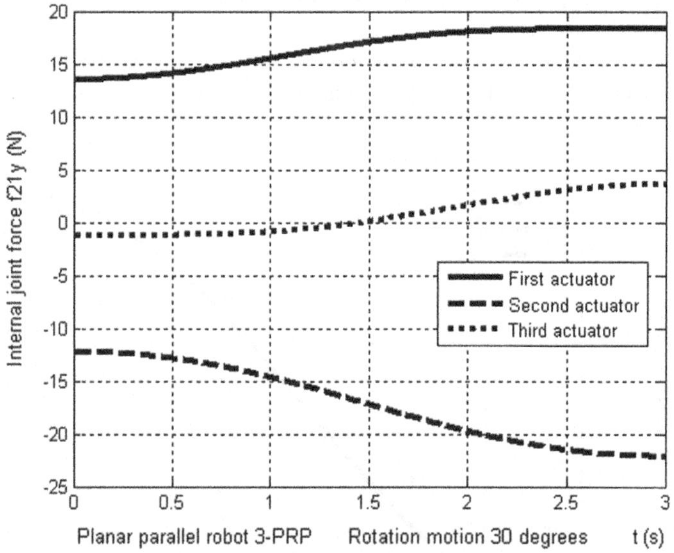

Fig. 8.32 Internal joint forces $f_{21y}^{A}, f_{21y}^{B}, f_{21y}^{C}$ (rotation)

8.33). From Fig. 8.30, we remark that during the rotation of the platform all the displacements of the actuators are identically distributed.

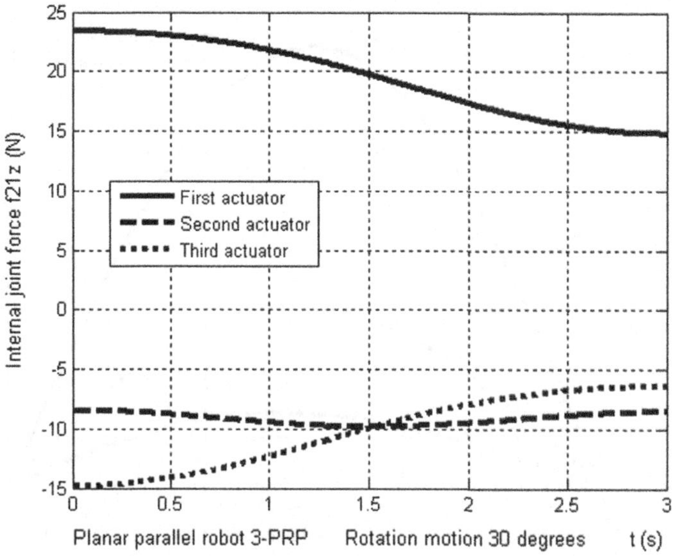

Fig. 8.33 Internal joint forces $f_{21z}^{A}, f_{21z}^{B}, f_{21z}^{C}$ (rotation)

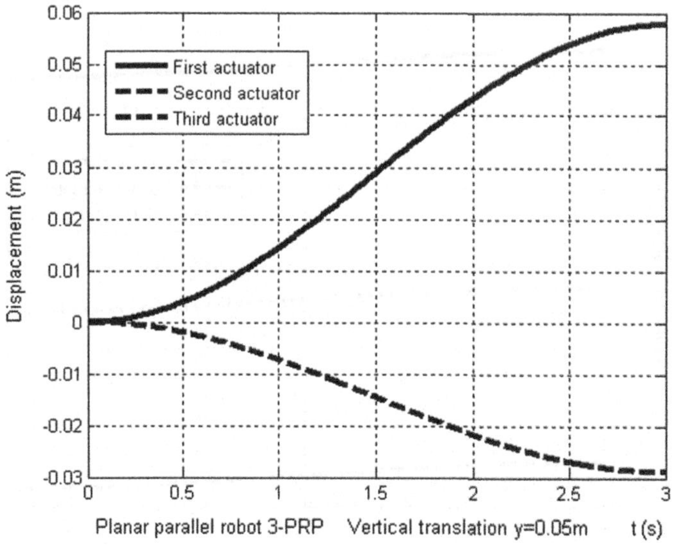

Fig. 8.34 Displacements $\lambda_{10}^{A}, \lambda_{10}^{B}, \lambda_{10}^{C}$ (translation)

If the platform's centre G moves along a *vertical trajectory* without rotation of the platform, the displacements and the input powers of the actuators and also the joint forces are depicted with regard to time (Figs. 8.34, 8.35, 8.36 and 8.37).

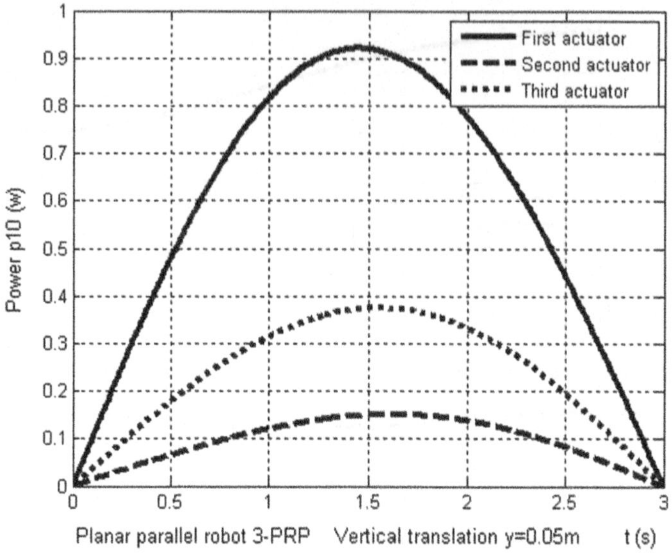

Fig. 8.35 Powers $P_{10}^{A}, P_{10}^{B}, P_{10}^{C}$ of actuators (translation)

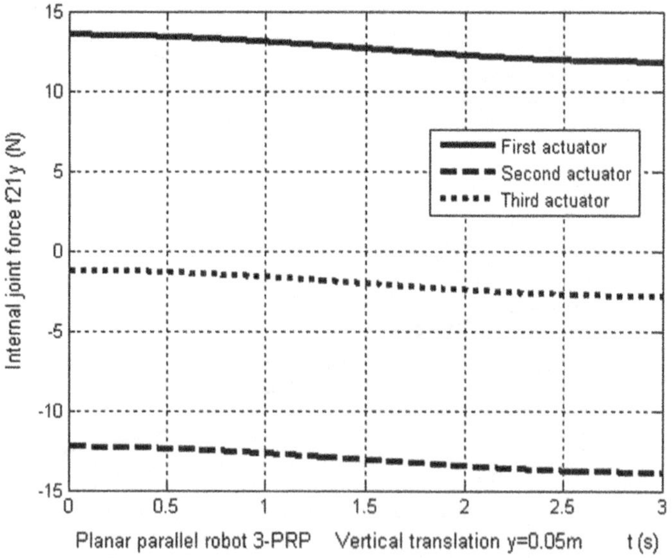

Fig. 8.36 Internal joint forces $f_{21y}^{A}, f_{21y}^{B}, f_{21y}^{C}$ (translation)

We remark that the distribution of displacement depicted in Fig. 8.34 is the same, at any instant, for two of the three actuators.

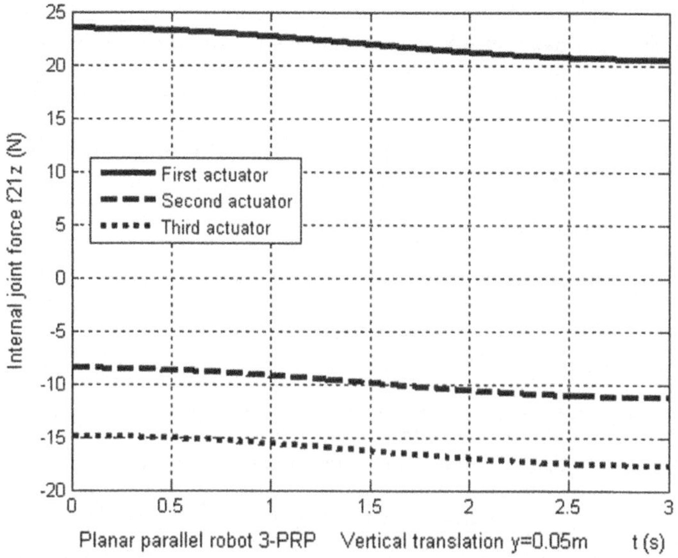

Fig. 8.37 Internal joint forces $f_{21z}^{A}, f_{21z}^{B}, f_{21z}^{C}$ (translation)

The simulation using the program confirms that the current matrix recursive formulation can easily be transformed in a model which is expected successfully to be deployed for automatic robotic control of the parallel hybrid robot and that one of its major advantages is the effectiveness and accuracy of the numerical computation involved.

References

1. Tsai, L.-W.: Robot Analysis: The Mechanics of Serial and Parallel Manipulators. Wiley, New York (1999)
2. Merlet, J.-P.: Parallel Robots. Kluwer Academic, Dordrecht (2000)
3. Aradyfio, D.D., Qiao, D.: Kinematic simulation of novel robotic mechanisms having closed chains. In: Proceedings of ASME Mechanisms Conference, Paper 85-DET-81 (1985)
4. Gosselin, C., Angeles, J.: The optimum kinematic design of a planar three-degree-of-freedom parallel manipulator. ASME J. Mech. Transm. Automat. Des. **110**(1), 35–41 (1988)
5. Pennock, G.R., Kassner, D.J.: Kinematic analysis of a planar eight-bar linkage: application to a platform-type robot. In: Proceedings of ASME Mechanisms Conference, Paper DE-25, pp. 37–43 (1990)
6. Sefrioui, J., Gosselin, C.: On the quadratic nature of the singularity curves of planar three-degree-of-freedom parallel manipulators. Mech. Mach. Theory **30**(4), 533–551 (1995)
7. Mohammadi-Daniali, H., Zsombor-Murray, P., Angeles, J.: Singularity analysis of planar parallel manipulators. Mech. Mach. Theory **30**(5), 665–678 (1995)
8. Williams II, R.L., Reinholtz, C.F.: Closed-form workspace determination and optimization for parallel mechanisms. In: Proceedings of the 20th Biennial ASME Mechanisms Conference, Kissimmee, DE, vol. 5, pp. 341–351 (1988)

9. Yang, G., Chen, W., Chen, I-M.: A geometrical method for the singularity analysis of 3-RRR planar parallel robots with different actuation schemes. In: Proceedings of the IEEE/RSJ International Conference on Intelligent Robots and Systems, Lausanne, pp. 2055–2060 (2002)
10. Bonev, I., Zlatanov, D., Gosselin, C.: Singularity analysis of 3-DOF planar parallel mechanisms via screw theory. J. Mech. Des. **125**(3), 573–581 (2003)
11. Angeles, J.: Fundamentals of Robotic Mechanical Systems: Theory, Methods and Algorithms. Springer, New York (2002)
12. Staicu, S.: Power requirement comparison in the 3-RPR planar parallel robot dynamics. Mech. Mach. Theory **44**(5), 1045–1057 (2009)
13. Staicu, S.: Dynamics analysis of the Star parallel manipulator. Robot. Auton. Syst. **57**(11), 1057–1064 (2009)
14. Wang, J., Gosselin, C.: A new approach for the dynamic analysis of parallel manipulators. Multibody Syst. Dyn. **2**(3), 317–334 (1998)
15. Cheng, G., Shan, X.: Dynamics analysis of a parallel hip-joint simulator with four degree of freedoms (3R1T). Nonlinear Dyn. **70**(4), 2475–2486 (2012)
16. Staicu, S.: Dynamics of the spherical 3-UPS/S parallel mechanism with prismatic actuators. Multibody Syst. Dyn. **22**(2), 115–132 (2009)
17. Li, Y., Staicu, S.: Inverse dynamics of a 3-PRC parallel kinematic machine. Nonlinear Dyn. **67**(2), 1031–1041 (2012)
18. Staicu, S., Zhang, D.: A novel dynamic modelling approach for parallel mechanisms analysis. Robot. Comput.-Integr. Manuf. **24**(1), 167–172 (2008)

Chapter 9
Spatial Parallel Robots

We have already seen that parallel structures have a number of advantages over serial manipulators such as higher payload since this is divided by several legs, has higher accuracy due to none cumulative joint errors, higher structural stiffness, since the load is usually carried by some links and, also, the location of motors are closed to the base. Due to these clear advantages, parallel robots have attracted many researchers and considerable efforts have been devoted to the kinematics and dynamics of these mechanisms. There is a great deal of scope opening up for possible applications of parallel robots in non-traditional field such as like surgical robots and rehabilitation robotics.

Analysis of spatial parallel architectures is complicated due to their closed loop structure and the presence of constraints and singularities. Accurate modelling of the kinematics and dynamics is especially important for high-performance control algorithms and simulations. Also, as the primary field of application for parallel robots is in material handling due to their high dynamic capability, it is desirable that a parallel robot can coordinate and cooperate with other parallel robots in handling not only bulky and heavy objects but also delicate ones like glass. For this designing process, a system that includes accurate trajectory planning, communication between two manipulators and robust force control schemes is very important.

9.1 Dynamics Modelling of Delta Translational Parallel Robot

Kinematics and dynamics analysis of a Delta parallel mechanism which can be used in several applications, including machining tools, is proposed in the following [1]. The success of this type of spatial parallel robot in packaging lines is helping them establish themselves in a new market. The core advantage of this manipulator is the

S. Staicu, *Dynamics of Parallel Robots*, Parallel Robots: Theory and Applications,
https://doi.org/10.1007/978-3-319-99522-9_9

speed of manipulating light and small objects. Whereas, a typical robot arm has to move not only the payload, but also all servomotors in each joint, the only moving part of this mechanism is its frame, which is usually made of lightweight composite materials.

Due to its high speeds, the parallel Delta robots are popular in usage in packaging and pick-and-place operations for relatively light objects, because they can be quite fast, some executing up to 300 picks per minute. Consisting of three arms connected to universal joints at the base, the key design feature is the use of parallelograms in the arms, which restrict the movement of the mobile platform to pure translation with no rotation and accurately maintains the orientation of the end-effector.

The model of the robot has also been improved or modified in accordance with the customer requirements and updated technology. Many companies purchased the license for the Delta parallel manipulator in different specification and gradually manufactured their own robots. Also, many universities or centers of academic studies in the world have come out with their own Delta robots or modified models.

In 1988, Reymond Clavel [2] developed at Lausanne Federal Polytechnic Institute the first prototype of three-arm parallel mechanism driven by motors which are mounted to a base plate: the Delta parallel robot. The arms are attached with revolute joints, which allow the end-effectors to attain three degrees of freedom. Allowing three spatial translations, the manipulator utilizes their symmetry and rigidity to create a simplistic system that can perform pick-and-place operations at high speeds and accuracy on small objects. The parallel manipulator built at the University of Maryland (Tsai and Stamper [3]) employs only revolute joints to constrain the platform's motion to a pure translation.

Knowing the translation motion of the platform, we develop first the inverse kinematics problem and determine the position, velocity and acceleration of every element of the three-degree-of-freedom translational Delta parallel robot with revolute actuators. Further, the inverse dynamic problem is solved using an efficient and fast recursive matrix approach based on the fundamental principle of virtual work. Finally, a comparative study on time-history evolution of the torques of three revolute actuators is analyzed.

9.1.1 Inverse Geometric Model

The mechanism input of the robot is structured into three active revolute joints while the output body is connected to the fixed base through a set of three variable length identical legs. The architecture of one of the three kinematical closed chains of the manipulator is made of an electrical motor, an active revolute joint, and an intermediate planar mechanism that connect four bars parallel two by two, ending with a passive revolute joint connected to the moving platform (Fig. 1.3). Actuation could be achieved with rotational actuators, with or without reduction. Since the actuators are all located in the base, the arms can be made of light composite materials. As a result of this, the moving parts of the Delta robot have a small inertia and this allows for very high speeds and high accelerations.

Due to the special arrangement of the four-bar parallelograms and the position of three active revolute joints at fixed points A_1, B_1, C_1, the mechanism has three translational degrees of freedom.

In the followings we develop by inverse kinematics a means to determine the velocities and accelerations of all moving links of the robot, given the characteristics of the translation motion of the moving platform. Let us locate a fixed Cartesian frame $Ox_0y_0z_0(T_0)$ at the mass centre of the triangular base of side L, about which the Delta parallel manipulator of known geometry moves. The elements of its legs have known size and given mass. To simplify the graphical image of the kinematical scheme of the mechanism, in what follows we will represent the intermediate reference systems by only two axes. The z_k axis is represented for each component element T_k. It is noted that any relative rotation with angle $\varphi_{k,k-1}$ of the body T_k must always be pointed along the direction of the z_k axis.

One of the three active elements of the robot is the input link of the first upside-down leg A (Fig. 9.1). This is a homogenous crank of length $A_1A_2 = l_1$ and mass m_1, which rotates about z_1^A axis with the angle φ_{10}^A, angular velocity ω_{10}^A and angular acceleration ε_{10}^A. The total tensor of inertia of the input link-rotor mounted on this link is $\hat{J}_1$. The centre of the transmission rod $A_3A_5 = l_2$ is denoted by A_2. This link is connected to the frame $A_2x_2^Ay_2^Az_2^A$ (called T_2^A) and has a relative rotation about axis z_2^A with the angle φ_{21}^A, so that $\omega_{21}^A = \dot{\varphi}_{21}^A$ and $\varepsilon_{21}^A = \ddot{\varphi}_{21}^A$. It has the mass m_2 and the central tensor of inertia $\hat{J}_2$.

Two identical and parallel bars A_3A_4 and A_5A_6 with the same length l_3 rotate about two axes z_3^A, z_5^A respectively, with the angle $\varphi_{32}^A = \varphi_{52}^A$. They have also the same mass m_3 and the same tensor of inertia $\hat{J}_3$. The four-bar parallelogram is closed by an element T_4^A identical with T_2^A, but its tensor of inertia is $\hat{J}_4$. This element rotates about z_4^A axis with the relative angle $\varphi_{43}^A = \varphi_{32}^A$. The centre A_7 of the interval between the two revolute joints connects to a moving triangular platform. Having a general translation, this platform is an equilateral triangle of side $l = L - \sqrt{3}l_3 \sin\beta$ and mass m_p. Finally, another translational reference system $Gx_Gy_Gz_G$ is located just at the centre G of the moving platform. Following angles give the initial position of the robot

$$\alpha_A = \frac{\pi}{3}, \quad \alpha_B = \pi, \quad \alpha_C = -\frac{\pi}{3}, \quad \beta_A = \beta_B = \beta_C = \beta = \frac{\pi}{6}. \tag{9.1.1}$$

The angle α_A, for example, is measured from the fixed x_0 axis to the line which connects the centre O at the first actuated joint A_1, while the angle $\beta_A = \beta$ denote the initial inclination of the four-bar parallelogram, which is defined from the extended line of A_1A_2 to the median x_2^A axis.

In the following, we apply the method of successive displacements to geometric analysis of closed-loop chains and we note that a joint variable is the displacement required to move a link from the initial location to the actual position. The variable

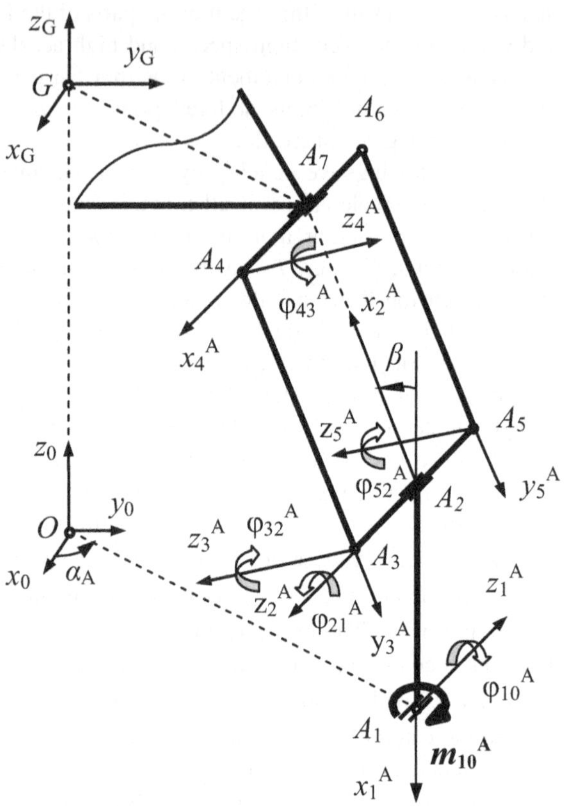

Fig. 9.1 Kinematical scheme of the leg A of upside-down mechanism

angles $\varphi_{k,k-1}$ of rotation about the joint axes z_k are the parameters needed to bring the next link from a reference configuration to the next configuration. We call the matrix $a^{\varphi}_{k,k-1}$, for example, the orthogonal transformation 3×3 matrix of relative rotation with the angle $\varphi^{A}_{k,k-1}$ of link T^{A}_{k} around z^{A}_{k} axis. In the study of the kinematics of robot manipulators, we are interested in deriving a matrix equation relating the location of an arbitrary T_k body to the joint variables. When the change of coordinates is successively considered, the corresponding matrices are multiplied.

Starting from the reference origin O and pursuing the independent legs $A_1A_2A_3A_4$, $B_1B_2B_3B_4$, $C_1C_2C_3C_4$ we obtain following transformation matrices

$$\begin{aligned} &q_{10} = q^{\varphi}_{10}\theta_2\theta_1 a^{j}_{\alpha}, \quad q_{21} = q^{\varphi}_{21} a_{\beta}\theta_3, \quad q_{32} = q^{\varphi}_{32}\theta_2\theta_1, \quad q_{43} = q^{\varphi}_{32}\theta_3 \\ &\quad (q = a, b, c), \quad (j = A, B, C), \end{aligned} \tag{9.1.2}$$

where we denote [4]:

$$q^{\varphi}_{k,k-1} = rot(z, \varphi^j_{k,k-1}) = \begin{bmatrix} \cos\varphi^j_{k,k-1} & \sin\varphi^j_{k,k-1} & 0 \\ -\sin\varphi^j_{k,k-1} & \cos\varphi^j_{k,k-1} & 0 \\ 0 & 0 & 1 \end{bmatrix}$$

$$q_{k0} = \prod_{\sigma=1}^{k} q_{k-\sigma+1,k-\sigma} \quad (k = 1,2,3,4) \tag{9.1.3}$$

$$a^j_{\alpha} = rot(z, \alpha_j), \quad a_{\beta} = rot(z, \beta)$$

$$\theta_1 = rot\left(z, \frac{\pi}{2}\right) = \begin{bmatrix} 0 & 1 & 0 \\ -1 & 0 & 0 \\ 0 & 0 & 1 \end{bmatrix}, \quad \theta_2 = rot\left(y, \frac{\pi}{2}\right) = \begin{bmatrix} 0 & 0 & -1 \\ 0 & 1 & 0 \\ 1 & 0 & 0 \end{bmatrix}$$

$$\theta_3 = rot(y, \pi) = \begin{bmatrix} -1 & 0 & 0 \\ 0 & 1 & 0 \\ 0 & 0 & -1 \end{bmatrix}.$$

For the inverse geometric analysis, the position of an arbitrary end-point is treated as known and the goal is to find the joint variables $\varphi^A_{10}, \varphi^B_{10}, \varphi^C_{10}$ that yield the given location of the tool.

The three rotation angles φ^A_{10}, φ^B_{10}, φ^C_{10} of the actuators A_1, B_1, C_1 are the joint variables that give the input vector $\vec{\varphi}_{10} = \begin{bmatrix} \varphi^A_{10} & \varphi^B_{10} & \varphi^C_{10} \end{bmatrix}^T$ of the instantaneous position of the mechanism. However, the objective of the *inverse geometric problem* is to find the vector $\vec{\varphi}_{10}$ and the position of the mechanism, given three absolute coordinates of the centre G of the platform:x^G_0, y^G_0, z^G_0.

A complex trajectory of the mobile platform could be considered for simulation, to illustrate the dynamic model. Consider, for example, that during three seconds the moving platform remain in same orientation and the motion of the mass centre G along a *rectilinear trajectory* is expressed in the fixed frame $Ox_0y_0z_0$ through the following equations

$$\frac{x^G_0}{x^{G*}_0} = \frac{y^G_0}{y^{G*}_0} = \frac{h - z^G_0}{z^{G*}_0} = 1 - \cos\frac{\pi}{3}t, \quad h = l_1 + l_3\cos\beta, \quad \vec{r}^{\,G}_0 = \begin{bmatrix} x^G_0 & y^G_0 & z^G_0 \end{bmatrix}^T, \tag{9.1.4}$$

where t is the time variable and x^G_0, y^G_0, z^G_0 are evaluated in units of meters.

The input angles φ^j_{10} $(j = A, B, C)$ of the robot and the variables φ^j_{21}, $\varphi^j_{32} = \varphi^j_{43} = \varphi^j_{52}$ will be obtained from the following geometrical conditions

$$\vec{r}^{\,j}_{10} + \sum_{k=1}^{4} q^T_{k0}\vec{r}^{\,j}_{k+1,k} = \vec{r}^{\,G}_0 + \vec{r}^{\,j_7}_G \quad (q = a, b, c), \quad (j = A, B, C) \tag{9.1.5}$$

where one denotes:

$$\begin{aligned} &\vec{r}^{\,j}_{10} = \frac{L}{\sqrt{3}} a^{jT}_{\alpha} \vec{u}_1, \quad \vec{r}^{\,j}_{21} = -l_1 \vec{u}_1, \quad \vec{r}^{\,j}_{32} = \frac{l_2}{2} \vec{u}_3 \\ &\vec{r}^{\,j}_{43} = -l_3 \vec{u}_2, \quad \vec{r}^{\,j}_{54} = -\frac{l_2}{2} \vec{u}_1, \quad \vec{r}^{\,j7}_{G} = \frac{l}{\sqrt{3}} a^{jT}_{\alpha} \vec{u}_1 \\ &\vec{u}_1 = \begin{bmatrix} 1 \\ 0 \\ 0 \end{bmatrix}, \quad \vec{u}_2 = \begin{bmatrix} 0 \\ 1 \\ 0 \end{bmatrix}, \quad \vec{u}_3 = \begin{bmatrix} 0 \\ 0 \\ 1 \end{bmatrix}, \quad \tilde{u}_3 = \begin{bmatrix} 0 & -1 & 0 \\ 1 & 0 & 0 \\ 0 & 0 & 0 \end{bmatrix}. \end{aligned} \tag{9.1.6}$$

Actually, these equations mean that there is only one inverse geometric solution for the manipulator:

$$\begin{aligned} &l_3 \sin \varphi^j_{32} = -x^G_0 \sin \alpha_j + y^G_0 \cos \alpha_j \\ &l_1 \sin \varphi^j_{10} + l_3 \sin(\varphi^j_{10} - \varphi^j_{21} - \beta) \cos \varphi^j_{32} = x^G_0 \cos \alpha_j + y^G_0 \sin \alpha_j - l_3 \sin \beta \\ &l_1 \cos \varphi^j_{10} + l_3 \cos(\varphi^j_{10} - \varphi^j_{21} - \beta) \cos \varphi^j_{32} = z^G_0 \; (j = A, B, C) \end{aligned} \tag{9.1.7}$$

9.1.2 Robot Kinematics

Turning to inverse kinematics, we determine the velocities and the accelerations of all moving components of the robot, given the characteristics of the translation motion of the moving platform. The motions of the component elements of each leg (for example the leg $j = A$) are characterized by the following skew-symmetric matrices [5]:

$$\tilde{\omega}^A_{k0} = a_{k,k-1} \tilde{\omega}^A_{k-1,0} a^T_{k,k-1} + \omega^A_{k,k-1} \tilde{u}_3, \tag{9.1.8}$$

which are *associated* with the absolute angular velocities given by the recursive relations

$$\vec{\omega}^A_{k0} = a_{k,k-1} \vec{\omega}^A_{k-1,0} + \omega^A_{k,k-1} \vec{u}_3, \quad \omega^A_{k,k-1} = \dot{\varphi}^A_{k,k-1}. \tag{9.1.9}$$

General relations (7.9) give the following absolute velocities $\vec{v}^A_{k0}$ of centres of the joints A_k

$$\begin{aligned} &\vec{v}^A_{k0} = a_{k,k-1} \vec{v}^A_{k-1,0} + a_{k,k-1} \tilde{\omega}^A_{k-1,0} \vec{r}^A_{k,k-1} + v^A_{k,k-1} \vec{u}_3, \quad v^A_{k,k-1} = 0. \\ &(k = 1, 2, 3, 4) \end{aligned} \tag{9.1.10}$$

Equations of the geometrical constraints (9.1.5) can be derived with respect to time to obtain the *matrix conditions of connectivity* $\vec{V}_j = [N_j]^{-1}\vec{P}_j$ for the relative angular velocities [6], where the following terms determine the contents of the 3×3 invertible square matrix $[N_j]$ and the column matrix $\vec{P}_j$:

$$\begin{aligned}
&n_{i1}^j = u_i^T q_{10}^T \tilde{u}_3 \{\vec{r}_{21}^j + q_{21}^T q_{32}^T \vec{r}_{43}^j\}, \quad n_{i2}^j = \vec{u}_i^T q_{20}^T \tilde{u}_3 q_{32}^T \vec{r}_{43}^j, \\
&n_{i3}^j = \vec{u}_i^T q_{30}^T \tilde{u}_3 \vec{r}_{43}^j \quad (i = 1, 2, 3) \\
&\vec{P}_j = \begin{bmatrix} \dot{x}_0^G & \dot{y}_0^G & \dot{z}_0^G \end{bmatrix}^T.
\end{aligned} \tag{9.1.11}$$

Finally, the expressions of three relative angular velocities $\omega_{10}^A, \omega_{21}^A, \omega_{32}^A$ along the leg $j = A$ are obtained from the column matrices $\vec{V}_j = \begin{bmatrix} \omega_{10}^j & \omega_{21}^j & \omega_{32}^j \end{bmatrix}^T$ as functions of the platform's *translation velocity*. Same relations (9.1.11) immediately give the *complete* Jacobian matrix of the manipulator, which relates all angular velocities to the actuated joint rates. This matrix is a fundamental element in the analysis of the robot workspace and the finding of the particular configurations of singularities where the manipulator becomes uncontrollable [7].

Rearranging this, the above nine constraint equation (9.1.7) of the Delta robot can be written as follows

$$\begin{aligned}
&(x_0^G \cos\alpha_j + y_0^G \sin\alpha_j - l_1 \sin\varphi_{10}^j - l_3 \sin\beta)^2 + (z_0^G - l_1 \cos\varphi_{10}^j)^2 \\
&\quad + (y_0^G \cos\alpha_j - x_0^G \sin\alpha_j)^2 = l_3^2 \quad (j = A, B, C),
\end{aligned} \tag{9.1.12}$$

where the "zero" position $\vec{r}_0^{0G} = \begin{bmatrix} 0 & 0 & h \end{bmatrix}^T$ corresponds to the joints variables $\vec{\varphi}_{10}^0 = \begin{bmatrix} 0 & 0 & 0 \end{bmatrix}^T$. The derivative with respect to time of above conditions (9.1.12) leads to the matrix equation

$$J_1 \dot{\vec{\varphi}}_{10} = J_2 \dot{\vec{r}}_0^G. \tag{9.1.13}$$

The matrices J_1 and J_2 are, respectively, the inverse and forward Jacobian of the manipulator, which can be expressed as

$$J_1 = diag\{\, \delta_A \quad \delta_B \quad \delta_C \,\}, \quad J_2 = \begin{bmatrix} \beta_1^A & \beta_2^A & \beta_3^A \\ \beta_1^B & \beta_2^B & \beta_3^B \\ \beta_1^C & \beta_2^C & \beta_3^C \end{bmatrix}, \tag{9.1.14}$$

with

$$\begin{aligned}
&\delta_j = l_1 \left[(x_0^G \cos\alpha_j + y_0^G \sin\alpha_j - l_3 \sin\beta) \cos\varphi_{10}^j - z_0^G \sin\varphi_{10}^j \right] \\
&\beta_1^j = x_0^G - (l_1 \sin\varphi_{10}^j + l_3 \sin\beta) \cos\alpha_j \\
&\beta_2^j = y_0^G - (l_1 \sin\varphi_{10}^j + l_3 \sin\beta) \sin\alpha_j \\
&\beta_3^j - z_0^G - l_1 \cos\varphi_{10}^j \quad (j = A, B, C).
\end{aligned} \tag{9.1.15}$$

The three kinds of singularities of the three closed-loop kinematical chains can be determined through the analysis of two Jacobian matrices J_1 and J_2 [8, 9].

Based on Eqs. (9.1.9), (9.1.10) and the conditions of connectivity (9.1.11), in the direct kinematics problem we calculate the velocities of all links in terms of input velocities ω_{10}^{i}. The idea of expressing the velocities and the accelerations of all moving components in terms of those of the actuators was explicitly discussed [10], using the natural orthogonal complement method, where various Jacobian matrices were derived to relate the velocities and accelerations of each body to those of the actuators for numeric computation of the inverse dynamics.

Let us assume that the robot has a virtual motion determined by three input angular velocities $\omega_{10a}^{Av} = 1$, $\omega_{10a}^{Bv} = 0$, $\omega_{10a}^{Cv} = 0$. Characteristic virtual velocities $v_{xa}^{Gv}, v_{ya}^{Gv}, v_{za}^{Gv}, \omega_{21a}^{jv}, \omega_{32a}^{jv}\ (j = A, B, C)$ are expressed at any instant in time as functions of the position of the mechanism by the kinematical constraints equation (9.1.11). Some other relations of connectivity can be obtained if we consider successively that $\omega_{10b}^{Bv} = 1$, $\omega_{10b}^{Cv} = 0$, $\omega_{10b}^{Av} = 0$ and $\omega_{10c}^{Cv} = 1$, $\omega_{10c}^{Av} = 0$, $\omega_{10c}^{Bv} = 0$.

Relative accelerations $\vec{\Gamma}_i = [\varepsilon_{10}^{i}\ \ \varepsilon_{21}^{i}\ \ \gamma_{32}^{i}]^T$ are obtained using new *conditions of connectivity* $\vec{\Gamma}_j = [N_j]^{-1}\vec{W}_j$, where the following terms determine the contents of the column matrix $\vec{W}_j = \dot{\vec{P}}_j - [\dot{N}_j]\vec{V}_j$:

$$\begin{aligned} w_i^j = {} & \vec{u}_i^T \ddot{\vec{r}}_0^G - \omega_{10}^j \omega_{10}^j \vec{u}_i^T q_{10}^T \tilde{u}_3 \tilde{u}_3 \{\vec{r}_{21}^j + q_{21}^T q_{32}^T \vec{r}_{43}^j\} - \omega_{21}^j \omega_{21}^j \vec{u}_i^T q_{20}^T \tilde{u}_3 \tilde{u}_3 q_{32}^T \vec{r}_{43}^j \\ & - \omega_{32}^j \omega_{32}^j \vec{u}_i^T q_{30}^T \tilde{u}_3 \tilde{u}_3 \vec{r}_{43}^j - 2\omega_{10}^j \omega_{21}^j \vec{u}_i^T q_{10}^T \tilde{u}_3 q_{21}^T \tilde{u}_3 q_{32}^T \vec{r}_{43}^j \\ & - 2\omega_{10}^j \omega_{32}^j \vec{u}_i^T q_{10}^T \tilde{u}_3 q_{21}^T q_{32}^T \tilde{u}_3 \vec{r}_{43}^j - 2\omega_{21}^j \omega_{32}^j \vec{u}_i^T q_{20}^T \tilde{u}_3 q_{32}^T \tilde{u}_3 r_{43}^T. \\ & (i = 1, 2, 3) \end{aligned} \tag{9.1.16}$$

Starting from (9.1.9) and (9.1.10), the following recursive relations give the angular accelerations $\vec{\varepsilon}_{k0}^{A}$ and the accelerations $\vec{\gamma}_{k0}^{A}$ of joints A_k

$$\begin{aligned} \vec{\varepsilon}_{k0}^{A} &= a_{k,k-1}\vec{\varepsilon}_{k-1,0}^{A} + \varepsilon_{k,k-1}^{A}\vec{u}_3 + \omega_{k,k-1}^{A} a_{k,k-1}\tilde{\omega}_{k-1,0}^{A} a_{k,k-1}^{T}\vec{u}_3, \quad \varepsilon_{k,k-1}^{A} = \ddot{\varphi}_{k,k-1}^{A} \\ \vec{\gamma}_{k0}^{A} &= a_{k,k-1}\vec{\gamma}_{k-1,0}^{A} + a_{k,k-1}\{\tilde{\omega}_{k-1,0}^{A}\tilde{\omega}_{k-1.0}^{A} + \tilde{\varepsilon}_{k-1,0}^{A}\}\vec{r}_{k,k-1}^{A} \\ &\quad + 2v_{k,k-1}^{A} a_{k,k-1}\tilde{\omega}_{k-1,0}^{A} a_{k,k-1}^{T}\vec{u}_3 + \gamma_{k,k-1}^{A}\vec{u}_3, \quad \gamma_{k,k-1}^{A} = 0. \end{aligned} \tag{9.1.17}$$

The matrix conditions of connectivity $\vec{V}_j = [N_j]^{-1}\vec{P}_j$ and $\vec{\Gamma}_j = [N_j]^{-1}\{\dot{\vec{P}}_j - [\dot{N}_j]\vec{V}_j\}$ are the *inverse kinematics model* of the Delta parallel robot.

9.1.3 Inverse Robot Dynamics

In the context of designing of a suitable strategy for accurate real-time control, ignoring friction forces and taking into account the gravitational effects, the key issue in dynamics analysis for the parallel robot is to establish an inverse dynamic model, so that the required actuator torques can be computed, where the time history of a desired trajectory of the end-effector is given. There are three methods for this, which can provide the same result concerning these actuating moments. The first one is using the Newton-Euler classic procedure [7, 11–13], the second one uses Lagrange's equations and multipliers formalism [3, 14, 15], and the third one is based on the fundamental principle of virtual work [16–18].

In what follows, the dynamics analysis of the Delta robot is examined by two efficient methods, namely the principle of virtual work and Lagrange formulation. Furthermore, these approaches are presented to formulate the dynamic equations of the parallel manipulator into a closed form, by which the dynamic matrices are more tractable and dynamics verification becomes possible.

9.1.3.1 Principle of Virtual Work

Within the inverse dynamic problem, the principle of virtual work is used in order to establish some recursive matrix relations for the torques of three active systems.

Three independent electric motors A_1, B_1, C_1 that generate three moments $\vec{m}_{10}^A = m_{10}^A \vec{u}_3$, $\vec{m}_{10}^B = m_{10}^B \vec{u}_3$, $\vec{m}_{10}^C = m_{10}^C \vec{u}_3$ have fixed directions of the axes z_1^A, z_1^B, z_1^C and control the motion of the platform $A_5B_5C_5$.

The force of inertia and the torque of inertia forces acting on an arbitrary rigid body T_k^A, for example,

$$\begin{aligned} \vec{f}_k^{inA} &= -m_k^A \left[\vec{\gamma}_{k0}^A + \left(\tilde{\omega}_{k0}^A \tilde{\omega}_{k0}^A + \tilde{\varepsilon}_{k0}^A\right) \vec{r}_k^{CA}\right], \\ \vec{m}_k^{inA} &= -\left[m_k^A \tilde{r}_k^{CA} \vec{\gamma}_{k0}^A + \hat{J}_k^A \vec{\varepsilon}_{k0}^A + \tilde{\omega}_{k0}^A \hat{J}_k^A \vec{\omega}_{k0}^A\right] \end{aligned} \tag{9.1.18}$$

are computed with respect to the centre of joint A_k. On the other hand, the wrench of two vectors $\vec{f}_k^{*A}$ and $\vec{m}_k^{*A}$ evaluates the influence of the weight $m_k^A g$ and of other external and internal forces possible applied to the same element T_k^A, for example:

$$\vec{f}_k^{*A} = -g m_k^A a_{k0} \vec{u}_3, \quad \vec{m}_k^{*A} = -g m_k^A \tilde{r}_k^{CA} a_{k0} \vec{u}_3. \tag{9.1.19}$$

Knowing the kinematic state of every link as well as the external forces acting on the robot, in the following one can apply the principle of virtual work to solve an inverse dynamic problem. Using the *equivalent graph* from Fig. 9.2, this parallel robot can artificially be transformed into a set of seven independent trees by separating the joints A_6, B_6, C_6 and three passive joints A_7, B_7, C_7 for a moving platform.

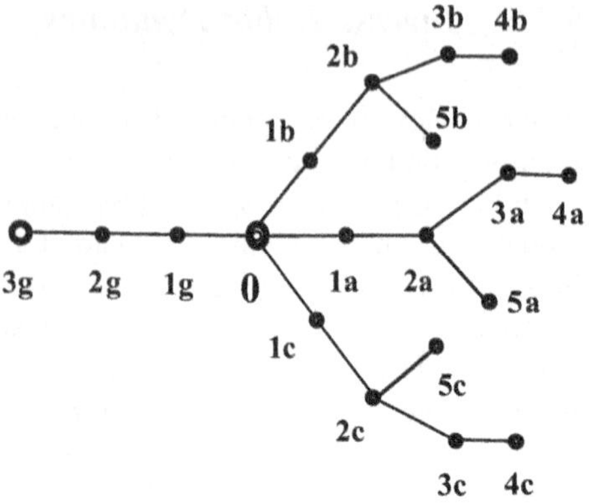

Fig. 9.2 Seven open trees of the parallel robot

Starting from *the fundamental equations of the parallel robots dynamics* (7.18), the following expression results for the first active torque

$$\begin{aligned} m_{10}^{A} = \vec{u}_3^{T}\{ & \vec{M}_1^{A} + \omega_{21a}^{Av}\vec{M}_2^{A} + \omega_{32a}^{Av}(\vec{M}_3^{A} + \vec{M}_4^{A} + \vec{M}_5^{A}) \\ & + \omega_{21a}^{Bv}\vec{M}_2^{B} + \omega_{32a}^{Bv}(\vec{M}_3^{B} + \vec{M}_4^{B} + \vec{M}_5^{B}) \\ & + \omega_{21a}^{Cv}\vec{M}_2^{C} + \omega_{32a}^{Cv}(\vec{M}_3^{C} + \vec{M}_4^{C} + \vec{M}_5^{C})\} \\ & + v_{xa}^{Gv}F_x^{G} + v_{ya}^{Gv}F_y^{G} + v_{za}^{Gv}F_z^{G} \end{aligned} \tag{9.1.20}$$

Using the notations

$$\begin{aligned} & \vec{F}_{k0}^{j} = -\vec{f}_k^{inj} - \vec{f}_k^{*j}, \quad \vec{M}_{k0}^{j} = -\vec{m}_k^{inj} - \vec{m}_k^{*j} \\ & F_x^{G} = m_p\ddot{x}_0^{G}, \quad F_y^{G} = m_p\ddot{y}_0^{G}, \quad F_z^{G} = m_p(g + \ddot{z}_0^{G}), \end{aligned} \tag{9.1.21}$$

significant recursive relations generate the vectors

$$\begin{aligned} & \vec{F}_k^{j} = \vec{F}_{k0}^{j} + q_{k+1,k}^{T}\vec{F}_{k+1}^{j}, \quad \vec{M}_k^{j} = \vec{M}_{k0}^{i} + q_{k+1,k}^{T}\vec{M}_{k+1}^{j} + \tilde{r}_{k+1,k}^{j}q_{k+1,k}^{T}\vec{F}_{k+1}^{j} \\ & (j = A, B, C)\ (k = 1, 2, 3, 4). \end{aligned} \tag{9.1.22}$$

Especially, in the joints A_2, B_2 and C_2 we have following wrenches

$$\vec{F}_2^{j} = \vec{F}_{20}^{j} + q_{32}^{T}(\vec{F}_3^{j} + \vec{F}_5^{j}), \quad \vec{M}_2^{j} = \vec{M}_{20}^{j} + q_{32}^{T}(\vec{M}_3^{j} + \vec{M}_5^{j}) + \tilde{r}_{32}^{j}q_{32}^{T}\vec{F}_3^{j} + \tilde{r}_{52}^{j}q_{32}^{T}\vec{F}_5^{j}. \tag{9.1.23}$$

The relations (9.1.20)–(9.1.22) represent just the inverse *dynamics model* of Delta translational parallel robot. We note that the Coriolis and centrifugal forces coupling and the gravitational actions are considered in these explicit equations.

9.1.3.2 Lagrange Equations

A solution of the dynamics problem of the Delta parallel manipulator can be developed based on Lagrange's equations for a mechanical system with constraints. The generalized coordinates of the robot are represented by 12 parameters $s_1 = x_0^G$, $s_2 = y_0^G$, $s_3 = z_0^G$, $s_4 = \varphi_{10}^A$, $s_5 = \varphi_{21}^A$, $s_6 = \varphi_{32}^A$, $s_7 = \varphi_{10}^B$, $s_8 = \varphi_{21}^B$, $s_9 = \varphi_{32}^B$, $s_{10} = \varphi_{10}^C$, $s_{11} = \varphi_{21}^C$, $s_{12} = \varphi_{32}^C$.

Lagrange equations with their nine multipliers $\lambda_1, \lambda_2, \ldots, \lambda_9$ will be expressed by 12 differential relations

$$\frac{d}{dt}\left\{\frac{\partial L}{\partial \dot{s}_l}\right\} - \frac{\partial L}{\partial s_l} = \Phi_l + \sum_{v=1}^{9} \lambda_v c_{vl} \quad (l = 1, 2, \ldots, 12), \tag{9.1.24}$$

which contain following 12 generalized forces

$$\begin{aligned} &\Phi_1 = 0, \quad \Phi_2 = 0, \quad \Phi_3 = 0, \quad \Phi_4 = m_{10}^A, \quad \Phi_5 = 0, \quad \Phi_6 = 0 \\ &\Phi_7 = m_{10}^B, \quad \Phi_8 = 0, \quad \Phi_9 = 0, \quad \Phi_{10} = m_{10}^C, \quad \Phi_{11} = 0, \quad \Phi_{12} = 0. \end{aligned} \tag{9.1.25}$$

Using the relations (9.1.11), a total of nine kinematical conditions of constraint are composed

$$\sum_{l=1}^{12} c_{vl}\dot{s}_l = 0 \quad (v = 1, 2, \ldots, 9). \tag{9.1.26}$$

Components of the general expression of the Lagrange function $L = L_p + \sum_{\mu=1}^{4} (L_\mu^A + L_\mu^B + L_\mu^C)$ are expressed as analytical functions of the generalized coordinates and their first derivatives with respect to time:

$$\begin{aligned} L_p &= \frac{1}{2} m_p (\dot{x}_0^{G2} + \dot{y}_0^{G2} + \dot{z}_0^{G2}) - m_p g z_0^G \\ L_1^j &= \frac{1}{2} \vec{\omega}_{10}^{jT} \hat{J}_1^j \vec{\omega}_{10}^j - m_1^j g \vec{u}_3^T (\vec{r}_{10}^j + q_{10}^T \vec{r}_1^{Cj}) \\ L_2^j &= \frac{1}{2} m_2^j \vec{v}_{20}^{jT} \vec{v}_{20}^j + \frac{1}{2} \vec{\omega}_{20}^{jT} \hat{J}_2^j \vec{\omega}_{20}^j - m_2^j g \vec{u}_3^T (\vec{r}_{10}^j + q_{10}^T \vec{r}_{21}^j) \\ L_3^j &= 2\left\{\frac{1}{2} m_3^j \vec{v}_{30}^{jT} \vec{v}_{30}^j + \frac{1}{2} \vec{\omega}_{30}^{jT} \hat{J}_3^j \vec{\omega}_{30}^j + m_3^j \vec{v}_{30}^{jT} \tilde{\omega}_{30}^j \vec{r}_3^{Cj} \right. \\ &\quad \left. - m_3^j g \vec{u}_3^T (\vec{r}_{10}^j + q_{10}^T \vec{r}_{21}^j + q_{20}^T \vec{r}_{32}^j + q_{30}^T \vec{r}_3^{Cj})\right\} \\ L_4^j &= \frac{1}{2} m_4^j \vec{v}_{40}^{jT} \vec{v}_{40}^j + \frac{1}{2} \vec{\omega}_{40}^{jT} \hat{J}_4^j \vec{\omega}_{40}^j + m_4^j \vec{v}_{40}^{jT} \tilde{\omega}_{40}^j \vec{r}_4^{Cj} \\ &\quad - m_{4}^j g \vec{u}_3^T (\vec{r}_{10}^j + q_{10}^T \vec{r}_{21}^j + q_{20}^T \vec{r}_{32}^j + q_{30}^T \vec{r}_{43}^j + q_{40}^T \vec{r}_4^{Cj}) \\ q &= (a, b, c), \quad j = (A, B, C). \end{aligned} \tag{9.1.27}$$

A long calculus of the derivatives with respect to time $\frac{d}{dt}\left\{\frac{\partial L}{\partial \dot{s}_l}\right\}(l = 1, 2, \ldots, 12)$ of the above functions leads to an algebraic system of 12 relations, starting from the relations (9.1.8)–(9.1.10) and using the first derivatives of orthogonal transformation matrices $q_{k,k-1}$:

$$\begin{aligned} &\dot{q}_{k,k-1} = \dot{\varphi}^j_{k,k-1}\tilde{u}_3^T q_{k,k-1}, \quad \dot{q}^T_{k,k-1} = \dot{\varphi}^j_{k,k-1} q^T_{k,k-1}\tilde{u}_3, \\ &\frac{\partial q_{k,k-1}}{\partial \varphi^j_{k,k-1}} = \tilde{u}_3^T q_{k,k-1}, \quad \frac{\partial q^T_{k,k-1}}{\partial \varphi^j_{k,k-1}} = q^T_{k,k-1}\tilde{u}_3. \end{aligned} \tag{9.1.28}$$

The kinematics constraint relations (9.1.26) are adjoined to the system of differential equations delivered by Lagrange formulation. In the direct or inverse dynamics problem, *after elimination of nine multipliers*, we finally obtain the same expressions (9.1.21) for the six input torques $m^A_{10}, m^B_{10}, m^C_{10}$ required by the revolute actuators.

9.1.4 Simulation Procedure

The procedure for solving the inverse dynamics of the Delta parallel robot by using the principle of virtual work can be summarised in several basic steps.

1. For a period of $\Delta t = 3$ s, it is assumed that the time history of the moving platform is specified in terms of the position of the centre G from Eq. (9.1.4). The relations (9.1.7) give the evolution of joint variables φ^j_{10}, φ^j_{21}, $\varphi^j_{32} = \varphi^j_{43}$, φ^j_{54}, $(j = A, B, C)$.
2. Using the relations (9.1.2), (9.1.3), we compute the transformation matrices of three legs A, B, C: q^j_{10}, q^j_{21}, q^j_{32}, q^j_{43} and $q^j_{20} = q^j_{21}q^j_{10}$, $q^j_{30} = q^j_{32}q^j_{20}$, $q^j_{40} = q^j_{43}q^j_{30}$ $(q = a, b, c)$.
3. Determine the velocity and the acceleration of all links by performing the inverse kinematics analysis in terms of prescribed velocities $\dot{x}^G_0$, $\dot{y}^G_0$, $\dot{z}^G_0$ and accelerations $\ddot{x}^G_0$, $\ddot{y}^G_0$, $\ddot{z}^G_0$ of the moving platform. Specifically, for every leg, from the conditions of connectivity (9.1.11), (9.1.16) we compute the relative angular velocities $\omega^j_{10}, \omega^j_{21}, \omega^j_{32}, \omega^j_{43}, \omega^j_{54}$ and the relative angular accelerations $\varepsilon^j_{10}, \varepsilon^j_{21}, \varepsilon^j_{32}, \varepsilon^j_{43}, \varepsilon^j_{54}$.
4. The input velocities ω^j_{10} could be easily verified through the relation (9.1.13), based on the inverse and forward conventional Jacobian matrices of the manipulator.
5. Using the same Eq. (9.1.11), where we introduce $\omega^{Av}_{10a} = 1$, $\omega^{Bv}_{10a} = 0$, $\omega^{Cv}_{10a} = 0$, for example, we compute the virtual characteristic velocities of each element of the robot. Other sets of virtual velocities are obtained if we consider in each case that $\omega^{Bv}_{10b} = 1$, $\omega^{Cv}_{10c} = 1$.

6. We compute the velocity and the acceleration of the centre of mass as well as the angular velocity and the angular acceleration of each link using (9.1.8)–(9.1.10) and (9.1.17).
7. Artificially, decompose the manipulator into several open-loop chains by cutting it open at six passive revolute joints. An interesting system is the open serial chain which includes only the moving platform.
8. For every link and platform we determine the inertia force (9.1.18) and the resulting force (excluding the actuator force) exerted to the rigid body T_k^j, from the relations (9.1.19).
9. For every link and platform, we determine the moment of inertia forces (9.1.18) and the resulting moment (excluding the actuator torque) exerted at the joint A_k, from the relations (9.1.19).
10. We find the input torque m_{10}^A and the power $P_{10}^A = \omega_{10}^A m_{10}^A$ from Eq. (9.1.2). Analogous relations give the torques m_{10}^B, m_{10}^C and the powers P_{10}^B, P_{10}^C of the other two actuators.

As an application, let us consider a robot which has the following architectural characteristics:

$$x_0^{G*} = 0.04\,\mathrm{m}, \quad y_0^{G*} = 0.06\,\mathrm{m}, \quad z_0^{G*} = 0.08\,\mathrm{m}$$

$$L = 1.25\,\mathrm{m}, \quad l_1 = 0.25\,\mathrm{m}, \quad l_2 = l_4 = 0.05\,\mathrm{m}, \quad l_3 = l_6 = 0.55\,\mathrm{m}$$

$$m_1 = 1.5\,\mathrm{kg}, \quad m_2 = m_4 = 0.05\,\mathrm{kg}, \quad m_3 = m_6 = 1\,\mathrm{kg},$$

$$m_p = 5\,\mathrm{kg}, \quad J_m = 0.02\,\mathrm{kg\,m^2}, \quad \Delta t = 3\,\mathrm{s}$$

$$\hat{J}_1 = \begin{bmatrix} J_m/2 & & \\ & m_1 l_1^2/3 + J_m/2 & \\ & & m_1 l_1^2/3 + J_m \end{bmatrix},$$

$$\hat{J}_2 = \begin{bmatrix} 0.0002 & & \\ & 0.0002 & \\ & & 0.0001 \end{bmatrix}$$

$$\hat{J}_3 = \begin{bmatrix} 0.15 & & \\ & 0.02 & \\ & & 0.15 \end{bmatrix},$$

$$\hat{J}_4 = \begin{bmatrix} 0.0001 & & \\ & 0.0004 & \\ & & 0.0004 \end{bmatrix}.$$

Based on the algorithm established in this simulation procedure, a computer program was developed to solve the inverse dynamics of the Delta parallel robot, using MATLAB software, assuming that the platform starts at rest from a central configuration and translates along a given direction for a period of three seconds. Furthermore, at the initial location, the platform is located 0.725 m lower than the fixed base, namely $t = 0$: $x_0^G = 0, y_0^G = 0$ $z_0^G = 0.725$ m.

Fig. 9.3 Input torques of actuators (vertical translation)

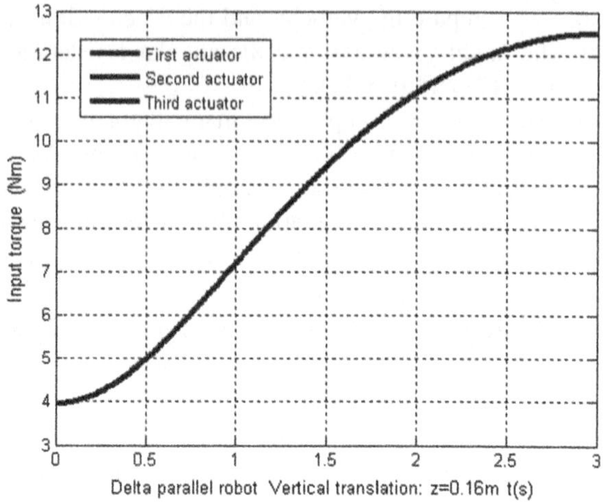

The time-history of input torques m_{10}^{A}, m_{10}^{B}, m_{10}^{C} and input powers P_{10}^{A}, P_{10}^{B}, P_{10}^{C} of three actuators are graphically plotted during the evolution of the platform.

For the first example, the moving platform moves along the z_0 *vertical direction* with variable acceleration, while all the other positional parameters remain equal to zero. As can be seen from Figs. 9.3 and 9.4, it is proved to be true that all actuating torques and powers are constantly equal to one another.

If the platform's centre G moves along a *rectilinear trajectory*, the torques exerted by the actuators A_1, B_1, C_1 and the required powers are also calculated by the program and plotted against time (Figs. 9.5 and 9.6). When the moving platform is moving towards to the fixed base, the legs become more horizontally oriented, coupled therefore with an increase for the actuating torques.

Fig. 9.4 Powers of actuators (vertical translation)

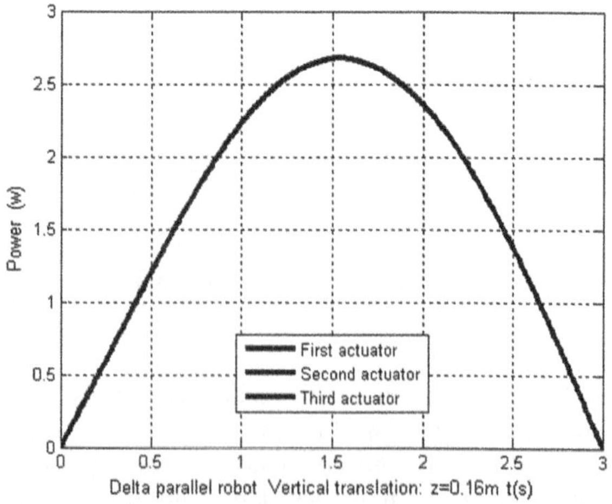

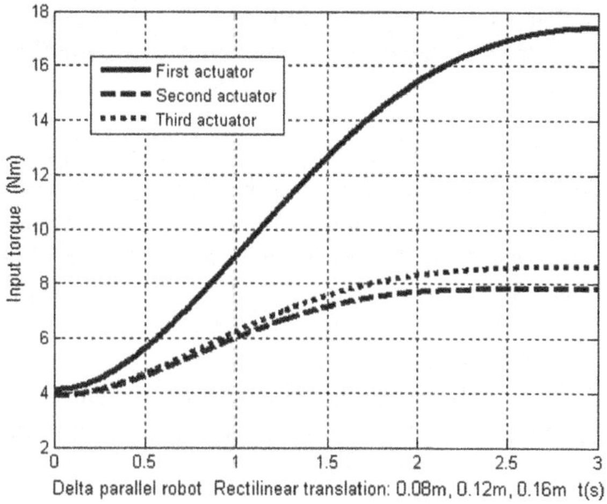

Fig. 9.5 Input torques of actuators (rectilinear translation)

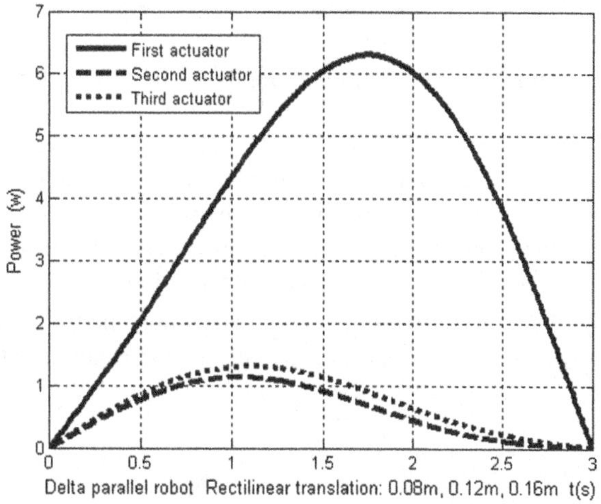

Fig. 9.6 Powers of actuators (rectilinear translation)

9.1.5 Comparative Analysis

The analytical inverse dynamics [1, 3] of the University of Maryland manipulator are solved using the Lagrange equations of second kind, which are written in terms of a set of three redundant coordinates x_0^G, y_0^G, z_0^G. The number of generalized coordinates x_0^G, y_0^G, z_0^G, φ_{10}^A, φ_{10}^B, φ_{10}^C exceeds the number of degrees of freedom of the moving platform. Three scalar constraint equations containing only the above six coordinates are easily obtained from the general vector conditions (9.1.5). However, generally, this formalism leads to a cumbersome expression for the Lagrange equations, due to the complex kinematics of the manipulator.

The simulation through the MATLAB program confirms that one of the major advantages of the current matrix recursive formulation is a shorter processing time of numerical computation in comparison with the approach based on the Lagrange equations.

To *simplify* the analysis, the assumption was made that the mass m_3 of each connecting rod in the upper arm assembly is divided evenly and concentrated at its two endpoints. The masses m_2 and m_4 of two transmission bars are taken to be *negligible* with respect to the mass of other links. Within these *simplifying assumptions*, one obtains a set of six equations [1] from which the three Lagrange multipliers $\lambda_1, \lambda_2, \lambda_3$ and the actuator torques τ_1, τ_2, τ_3 are determined.

A comparative analysis of two sets of torques m_{10}^A, m_{10}^B, m_{10}^C, obtained in the present study using the principle of virtual work and τ_1, τ_2, τ_3 in applying Lagrange's equations of the second kind is illustrated graphically through the relative coefficients of error

$$err\, m_{10}^j = \frac{m_{10}^j - \tau_i}{m_{10}^j} \times 100, \quad (i = 1, 2, 3) \ (j = A, B, C). \tag{9.1.29}$$

Finally, one obtains the graphs of the three relative coefficients of error $err\, m_{10}^j$ (Fig. 9.7).

The simplified analysis is the consequence outcome of expressing of the geometrical constraint conditions in its reduced form (9.1.13), which are obtained from the observation that the distance between A_3 and A_4 is always equal to the length l_3 of the connecting rod of the upper arm. Moreover, the conventional Jacobian matrices (9.1.14) can relate only the input velocities ω_{10}^i to the velocity of the moving platform and therefore avoid absolutely the presence of the velocities ω_{21}^i, ω_{32}^i, ω_{43}^i, ω_{54}^i. Consequently, the transformation between the end-effector output

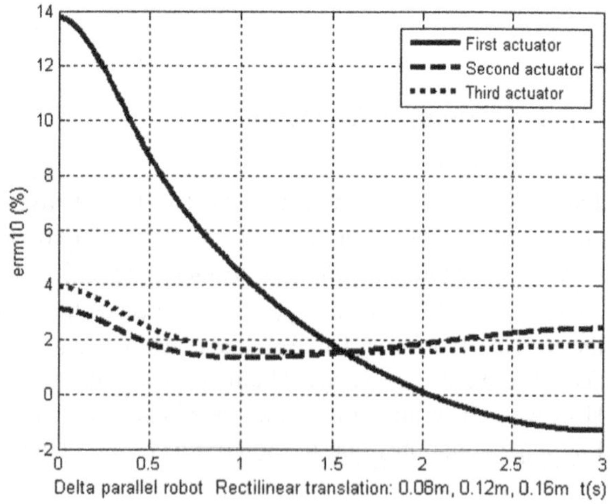

Fig. 9.7 Relative error coefficients of three actuators

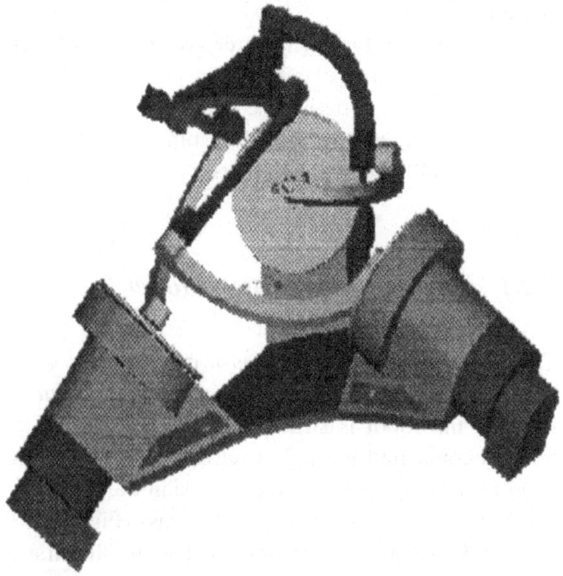

Fig. 9.8 Agile Wrist spherical parallel robot

forces and the active joint torques is governed, into the *absence* of gravitational effects, by these conventional matrices.

We note that the Newton-Euler method leads to a large number of equations with unknowns including even the connecting forces in the joints. Thus, in the dynamics of the Delta robot, a one-hundred-and-eight-equations system, that must be solved, would result.

9.2 Inverse Dynamics of Agile Wrist Spherical Parallel Robot

Spherical parallel manipulators provide high stiffness and accuracy, due to their mechanical arrangement in which all the actuators are fixed to the base and the end-effector is supported by three kinematic chains.

Parallel spherical wrists are used in technical applications where it is desired to orient a rigid body at high speed. Accuracy and precision in the execution of the task of the wrist are essential, since the robot is intended to operate on various objects; here positioning errors of the tool could end in costly damage. Examples of such applications include the orientation of a camera at high velocity [19].

The kinematics and dynamics of spherical parallel structures have been studied extensively over the last two decades. When a good dynamic performance and a precise positioning of the moving platform under high load are required, the dynamic model of the robot is important for its automatic control. Gosselin and

Angeles [20] analysed extensively the direct kinematics and singularity loci of the Agile Wrist prototype with three concurrent rotations.

In that following, we focus on an efficient and fast recursive matrix method, which is adopted to derive the kinematic model and the inverse dynamics equations of a three-degree-of-freedom symmetric spherical parallel robot with revolute actuators and three identical legs.

9.2.1 Inverse Kinematics Model

The spherical manipulator has a construction with the axes of all nine revolute joints concurring in a common *center O of rotation* of the device. The mechanism input of the robot is structured into three active revolute joints while the output body is connected through a set of legs with identical topology [20].

Let $Ox_0y_0z_0(T_0)$ be a fixed Cartesian frame, about which the manipulator moves. It has three legs of known size and mass (Fig. 9.9). Each of the legs of this robot is a serial 3-$\underline{R}RR$ symmetric spherical wrist: the first joint couples the base with the proximal link, the second joint couples the proximal link with the distal link, and the last one couples the distal link with the end-effector (Fig. 9.8).

The first element T_1 of leg $A(\alpha_A = 0)$, one of the three driving components of the robot, is called the *proximal link*. It is a homogenous rod, rotating about the axis Oz_1^A with the angular velocity $\omega_{10}^A = \ddot{\varphi}_{10}^A$ and the angular acceleration $\varepsilon_{10}^A = \ddot{\varphi}_{10}^A$. It

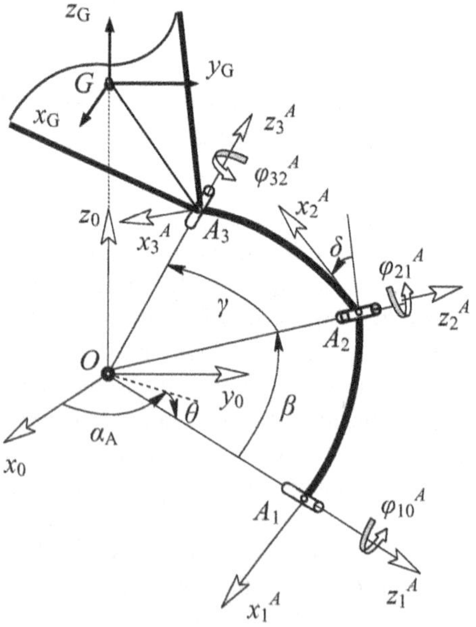

Fig. 9.9 Kinematical scheme of leg *A* of the mechanism

has the radius r, mass m_1 and tensor of inertia $\hat{J}_1$. The *distal link* T_2 of the same radius r is connected to the $A_2x_2^Ay_2^Az_2^A$ frame and has a relative rotation with the angle φ_{21}^A and angular velocity $\omega_{21}^A = \dot{\varphi}_{21}^A$; it has the mass m_2 and the tensor of inertia $\hat{J}_2$. The platform of the robot is a three-armed star of length $l = r\sqrt{(2 + \sin^2\delta)/3}$, mass m_3 and tensor of inertia $\hat{J}_3$, which rotates with an angular velocity $\omega_{32}^A = \dot{\varphi}_{32}^A$ with respect to the neighboring body (Fig. 9.8).

The wrist architecture consists of two main elements, the *base* and the *moving platform*, which is free to undergo arbitrary rotations with respect to the center of three identical serial legs, each of these composed of two links coupled to each other by means of a revolute joint.

The link angles of the manipulator are assumed to be the same on each of the legs connecting the end-effector to the base and are noted β for the proximal link and γ for the distal link. Given are the geometry of the base and the initial position of the manipulator arms through four angles $\alpha_A = 0$, $\alpha_B = 2\pi/3$, $\alpha_C = -2\pi/3$, $\delta = \pi/6$ and the functions $\sin\theta = 1/\sqrt{3}$, $\cos\theta = \sqrt{2/3}$. In particular, this serial wrist is called orthogonal, i.e., its neighboring revolute axes lay out at $\beta = \pi/2$ and $\gamma = \pi/2$. Consider the rotation angles $\varphi_{10}^A, \varphi_{10}^B, \varphi_{10}^C$ of the three actuators A_1, B_1, C_1 as the variables that give the instantaneous position of the mechanism. However, in an inverse geometric problem, it can be considered that the Euler angles $\alpha_1, \alpha_2, \alpha_3$ give the absolute orientation of the mobile platform.

Starting from the reference origin O and pursuing the independent legs $OA_1A_2A_3$, $OB_1B_2B_3$, $OC_1C_2C_3$, we obtain the following transformation matrices [21]

$$\begin{gathered} q_{10} = p_{10}^{\varphi}\theta_1 a_\theta a_\alpha^j, \quad q_{21} = q_{21}^{\varphi} a_\delta \theta_\beta \theta_2, \quad q_{32} = q_{32}^{\varphi}\theta_\gamma \\ (q = a, b, c) \;\; (j = A, B, C) \end{gathered} \tag{9.2.1}$$

with the notations

$$\begin{aligned} & q_{k,k-1}^{\varphi} = rot(z, \varphi_{k,k-1}^j), \quad q_{k0} = \prod_{\sigma=1}^{k} q_{k-\sigma+1,k-\sigma} \;\; (k = 1, 2, 3) \\ & \theta_1 = rot\left(y, \frac{\pi}{2}\right), \quad \theta_2 = rot(z, \pi), \quad a_\alpha^j = rot(z, \alpha_j), \\ & a_\delta = rot(z, \delta), \quad a_\theta = rot(y, \theta) \\ & \theta_\beta = rot\left(y, \beta = \frac{\pi}{2}\right), \quad \theta_\gamma = rot\left(y, \gamma = \frac{\pi}{2}\right). \end{aligned} \tag{9.2.2}$$

The three rotation angles $\varphi_{10}^A, \varphi_{10}^B, \varphi_{10}^C$ of the actuators A_1, B_1, C_1 are the joint variables that give the input vector $\vec{\varphi}_{10} = \begin{bmatrix} \varphi_{10}^A & \varphi_{10}^B & \varphi_{10}^C \end{bmatrix}^T$ of the position of the mechanism. The objective of the inverse geometric problem is to find the vector $\vec{\varphi}_{10}$ and the position of the mechanism with the given orientation of the moving platform.

The absolute motion of the platform is a general rotation around the fixed point O, which is supposed to be known and expressed by the three Euler angles

$$\alpha_l = \alpha_l^* \left(1 - \cos\frac{\pi}{3}t\right), \quad (l = 1, 2, 3). \tag{9.2.3}$$

Representing the orientation of the platform in the fixed frame, three relative rotations around the movable axes Ox, Oy, Oz successively lead to the commonly known rotation matrix $R = R_{32}R_{21}R_{10}$, where

$$R_{10} = rot(x, \alpha_1) = \begin{bmatrix} 1 & 0 & 0 \\ 0 & \cos\alpha_1 & \sin\alpha_1 \\ 0 & -\sin\alpha_1 & \cos\alpha_1 \end{bmatrix}$$
$$R_{21} = rot(y, \alpha_2) = \begin{bmatrix} \cos\alpha_2 & 0 & -\sin\alpha_2 \\ 0 & 1 & 0 \\ \sin\alpha_2 & 0 & \cos\alpha_2 \end{bmatrix} \tag{9.2.4}$$
$$R_{32} = rot(z, \alpha_3) = \begin{bmatrix} \cos\alpha_3 & \sin\alpha_3 & 0 \\ -\sin\alpha_3 & \cos\alpha_3 & 0 \\ 0 & 0 & 1 \end{bmatrix}.$$

The geometric conditions of rotation for the moving platform are given by the following identities

$$q_{30}^{\circ T} q_{30} = R, \quad q_{30}^{\circ} = q_{30}(t=0) = \theta_1 a_\delta \theta_1 \theta_2 \theta_1 a_\theta a_\alpha^j$$
$$(q = a, b, c), \quad (j = A, B, C). \tag{9.2.5}$$

From these relations, one obtains the real time evolution of nine characteristic angles φ_{10}^j, φ_{21}^j, φ_{32}^j, $(j = A, B, C)$. These equations mean that there is only one inverse geometric solution with the leg A, for example:

$$\sin\varphi_{10}^A(A_{31}\sin\alpha_A - A_{32}\cos\alpha_A) + \cos\varphi_{10}^A[(A_{31}\cos\alpha_A + A_{32}\sin\alpha_A)\sin\theta + A_{33}\cos\theta] = 0$$
$$\cos(\varphi_{21}^A + \delta) = -(A_{31}\cos\alpha_A + A_{32}\sin\alpha_A)\cos\theta + A_{33}\sin\theta$$
$$\sin\varphi_{32}^A[\sin\varphi_{10}^A\cos(\varphi_{21}^A + \delta)\cos\theta + \sin(\varphi_{21}^A + \delta)\sin\theta] + \cos\varphi_{32}^A\cos\varphi_{10}^A\cos\theta + A_{13} = 0, \tag{9.2.6}$$

with the notations $A_{13} = \vec{u}_1^T a_{30}^{\circ} R\,\vec{u}_3$, $A_{31} = \vec{u}_3^T a_{30}^{\circ} R\,\vec{u}_1$, $A_{32} = \vec{u}_3^T a_{30}^{\circ} R\,\vec{u}_2$, $A_{33} = \vec{u}_3^T a_{30}^{\circ} R\,\vec{u}_3$.

Kinematic analysis must generate analytical expressions that relate the motion of the connectors to the motion of the platform. Supposing as known the spherical motion of the platform by the relations (9.2.3), we develop the inverse kinematical problem and determine the characteristic velocities $\vec{v}_{k0}^j$, $\vec{\omega}_{k0}^j$ and accelerations $\vec{\gamma}_{k0}^j$, $\vec{\varepsilon}_{k0}^j$

of each of the moving links. The kinematics of compounding elements of the first leg A, for example, are characterized by three skew-symmetric matrices

$$\tilde{\omega}_{k0}^{A} = a_{k,k-1}\tilde{\omega}_{k-1,0}^{A}a_{k,k-1}^{T} + \omega_{k,k-1}^{A}\tilde{u}_3 \quad (k = 1, 2, 3), \tag{9.2.7}$$

which are *associated* with the absolute angular velocities given by the recurrence relations

$$\vec{\omega}_{k0}^{A} = a_{k,k-1}\vec{\omega}_{k-1,0}^{A} + \omega_{k,k-1}^{A}\vec{u}_3, \quad \omega_{k,k-1}^{A} = \dot{\varphi}_{k,k-1}^{A}, \tag{9.2.8}$$

and by the following absolute velocities of centres of joints A_k:

$$\vec{v}_{k0}^{A} = a_{k,k-1}\vec{v}_{k-1,0}^{A} + a_{k,k-1}\tilde{\omega}_{k-1,0}^{A}\vec{r}_{k,k-1}^{A}. \tag{9.2.9}$$

Equations of geometrical constraints (9.2.5) can be derived with respect to time to obtain the following *matrix conditions of connectivity* established for the relative angular velocities [22]:

$$\begin{aligned} &\omega_{10}^{A}\vec{u}_i^{T}a_{10}^{T}\vec{u}_3 + \omega_{21}^{A}\vec{u}_i^{T}a_{20}^{T}\vec{u}_3 + \omega_{32}^{A}\vec{u}_i^{T}a_{30}^{T}\vec{u}_3 \\ &\quad = \vec{u}_i^{T}\{\dot{\alpha}_1 R_{10}^{T}\vec{u}_1 + \dot{\alpha}_2 R_{10}^{T}R_{21}^{T}\vec{u}_2 + \dot{\alpha}_3 R_{10}^{T}R_{21}^{T}R_{32}^{T}\vec{u}_3\} \quad (i = 1, 2, 3), \end{aligned} \tag{9.2.10}$$

where $\vec{u}_1$, $\vec{u}_2$, $\vec{u}_3$ are three orthogonal unit vectors. These relations give the angular velocities $\omega_{10}^{A}, \omega_{21}^{A}, \omega_{32}^{A}$ of the elements of first leg $OA_1A_2A_3$, as functions of the angular velocities $\dot{\alpha}_1, \dot{\alpha}_2, \dot{\alpha}_3$ of the end-effector. The expression of a *complete* Jacobian matrix of the robot can be written in an invariant form from (9.2.10).

Let us assume now that the robot has a *virtual motion* defined by the angular velocities $\omega_{10a}^{Av} = 1$, $\omega_{10a}^{Bv} = 0$, $\omega_{10a}^{Cv} = 0$. Characteristic virtual velocities expressed as function of the position of the robot are given by the kinematical constraint equation (9.2.10). Other relations of connectivity at the level of virtual velocities can be obtained if one considers successively that $\omega_{10b}^{Bv} = 1$, $\omega_{10b}^{Cv} = 0$, $\omega_{10b}^{Av} = 0$ and $\omega_{10c}^{Cv} = 1$, $\omega_{10c}^{Av} = 0$, $\omega_{10c}^{Bv} = 0$.

As for the relative angular accelerations ε_{10}^{A}, ε_{21}^{A}, ε_{32}^{A} these are given by some other *conditions of connectivity*, obtained by deriving the relations (9.2.10), as follows

$$\begin{aligned} &\varepsilon_{10}^{A}\vec{u}_i^{T}a_{10}^{T}\vec{u}_3 + \varepsilon_{21}^{A}\vec{u}_i^{T}a_{20}^{T}\vec{u}_3 + \varepsilon_{32}^{A}\vec{u}_i^{T}a_{30}^{T}\vec{u}_3 \\ &\quad = \vec{u}_i^{T}\{\ddot{\alpha}_1 R_{10}^{T}\vec{u}_1 + \ddot{\alpha}_2 R_{10}^{T}R_{21}^{T}\vec{u}_2 + \ddot{\alpha}_3 R_{10}^{T}R_{21}^{T}R_{32}^{T}\vec{u}_3 \\ &\qquad + \dot{\alpha}_1\dot{\alpha}_2 R_{10}^{T}\tilde{u}_1 R_{21}^{T}\vec{u}_2 + \dot{\alpha}_2\dot{\alpha}_3 R_{19}^{T}R_{21}^{T}\tilde{u}_2 R_{3}^{T}\vec{u}_3 + \dot{\alpha}_3\dot{\alpha}_1 R_{10}^{T}\tilde{u}_1 R_{21}^{T}R_{32}^{T}\vec{u}_3 \\ &\qquad - \omega_{10}^{A}\omega_{21}^{A}a_{10}^{T}\tilde{u}_3 a_{21}^{T}\vec{u}_3 - \omega_{21}^{A}\omega_{32}^{A}a_{20}^{T}\tilde{u}_3 a_{32}^{T}\vec{u}_3 - \omega_{32}^{A}\omega_{10}^{A}a_{10}^{T}\tilde{u}_3 a_{21}^{T}a_{32}^{T}\vec{u}_3\}, \end{aligned} \tag{9.2.11}$$

where $\tilde{u}_1, \tilde{u}_2, \tilde{u}_3$ are three skew-symmetric matrices *associated* with the unit vectors. If the other two chains B and C of the manipulator are pursued, analogous relations can then be obtained. Finally, recursive general relations (7.11) give the angular accelerations $\vec{\varepsilon}_{k0}^{A}$ and the linear accelerations $\vec{\gamma}_{k0}^{A}$ of joints A_k. Relations (9.2.10) and (9.2.11) constitute *the inverse kinematics model* of the 3-$\underline{R}RR$ symmetric spherical parallel mechanism.

9.2.2 Dynamics Equations

Since spherical parallel robots are mostly used in applications involving high-speed motion, the dynamics of this kind of robot have a very important effect on the level of actuator torques required. Dynamics analysis determines the relationships between the externally applied loads, gravity, actuator forces and the motion of the platform.

In order to establish some recursive matrix relations for the torques of the three actuators, in the inverse dynamic problem we use the principle of virtual work. The electric motors A_1, B_1, C_1 that generate three torques of moments $\vec{m}_{10}^{A} = m_{10}^{A}\vec{u}_3$, $\vec{m}_{10}^{B} = m_{10}^{B}\vec{u}_3$, $\vec{m}_{10}^{C} = m_{10}^{C}\vec{u}_3$ about the directions z_1^A, z_1^B, z_1^C, control the motion of the legs of the robot.

The wrench $\vec{f}_k^{inA}$, $\vec{m}_k^{inA}$ of the inertia forces acting on the rigid body T_k^A and the wrench of two vectors $\vec{f}_k^{*A}$, $\vec{m}_k^{*A}$, evaluating the influence of the action of the weight $m_k^A\vec{g}$ and of other external and internal forces applied to the same element T_k^A, for example, are determined with respect to the centre of joint A_k, pursuing the same vector expressions (9.1.18) and (9.1.19) give above. At this stage, only the torques m_{10}^{j} applied at the proximal links are of interest.

Using the same equivalent graph from Fig. 8.4, for the three independent trees of the spherical robot, we apply *the fundamental equations of the parallel robots dynamics* (7.18) to the obtain following relation

$$m_{10}^{A} = \vec{u}_3^{T}\{\vec{M}_1^{A} + \omega_{21a}^{Av}\vec{M}_2^{A} + \omega_{32a}^{Av}\vec{M}_3^{A} + \omega_{21a}^{Bv}\vec{M}_{20}^{B} + \omega_{21a}^{Cv}\vec{M}_{20}^{C}\}, \tag{9.2.12}$$

where the vectors $\vec{F}_k^{A}, \vec{M}_k^{A}$ are expressed by the above relations (9.1.22).

The explicit equation (9.2.12) represents only the *inverse dynamics model* of an Agile Wrist spherical parallel robot. This dynamics approach developed here is completely general and can be used for any spherical manipulator with revolute or prismatic actuators.

The dynamics modelling of Wang and Gosselin [23] is obtained using Newton-Euler equations of dynamic equilibrium, which are written for each of seven moving rigid bodies. The assumptions employed here are that the joints connecting the distal links to the end-effector cannot transmit torques and consequently that they are equivalent to *spherical* joints. Since the mechanism is over-constrained, based on the principle of virtual work, this novel approach can eliminate any assumption concerning the influence of the non-working forces applied to the distal links by the end-effector. The Newton-Euler method, developed by Khalil and Ibrahim [24] in inverse dynamics of the Delta robot and the Gough-Stewart platform, consists of applying the free-body diagram (*FBD*) procedure for each leg connecting the end-effector, where *all* components of joint forces and the actuated forces are unknown.

Consider, for example, a spherical manipulator that has the following mechanical characteristics

$$\begin{gathered} m_1^A = m_1^B = m_1^C = m_2^A = m_2^B = m_2^C = 0.14\,\mathrm{kg}, \\ m_3 = 0.7\,\mathrm{kg}, \quad r = 0.074\,\mathrm{m}, \quad \Delta t = 3\,\mathrm{s}, \end{gathered}$$

$$\alpha_1^* = \frac{5\pi}{180}, \quad \alpha_2^* = \frac{7\pi}{180}, \quad \alpha_3^* = \frac{10\pi}{180}.$$

Pursuing a simulation procedure, we find the actuator torque m_{10}^A and the power $P_{10}^A = \omega_{10}^A m_{10}^A$ from Eq. (9.2.12). Analogous relations give the powers P_{10}^B, P_{10}^C of the other two actuators. A *comparative study* of the time-history evolution of the powers P_{10}^A, P_{10}^B and P_{10}^C, developed by the actuators during the evolution of the platform, are shown in two computational hypotheses: *complete* dynamic model and a *simplified* dynamic model in which the mass m_2^i and inertia tensor $\hat{J}_2^i$ of the *distal link* are *negligible*. The computational accuracy is now compared.

Following three examples, a computer program of simulation in MATLAB was developed to solve the inverse dynamics of the Agile Wrist spherical parallel robot. In the first example, we consider the *rotation motion* of the moving platform about the Gz_G axis with variable angular acceleration, when the platform starts at rest from a central configuration and moves for a period of three seconds. From Fig. 9.10 it is proved to be true that the actuating powers of three actuators are permanently equal to one another.

If the platform rotates simultaneously about Gx_G, Gy_G axes, the powers required by the actuators A_1, B_1, C_1 are calculated and plotted versus time as follows: P_{10}^A (Fig. 9.11), P_{10}^B (Fig. 9.12) and P_{10}^C (Fig. 9.13). In the third example, the moving platform has a *general rotation* about its centre G. The powers developed by the actuators are given in the graphs P_{10}^A (Fig. 9.14), P_{10}^B (Fig. 9.15) and P_{10}^C (Fig. 9.16).

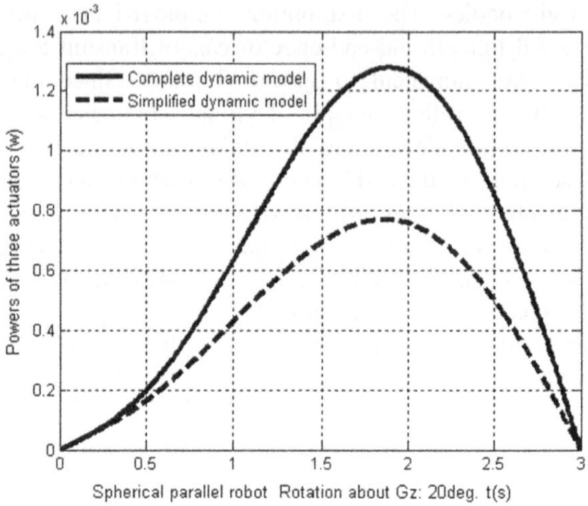

Fig. 9.10 Powers $P_{10}^{A}, P_{10}^{B}, P_{10}^{C}$ of three actuators

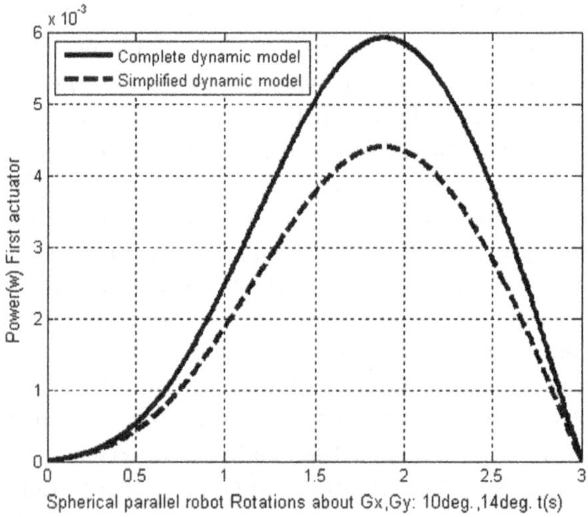

Fig. 9.11 Power P_{10}^{A} of first actuator (rotation)

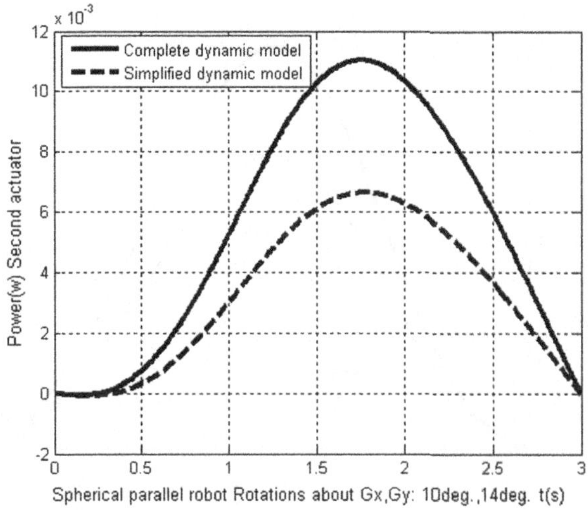

Fig. 9.12 Power P_{10}^{B} of second actuator (rotation)

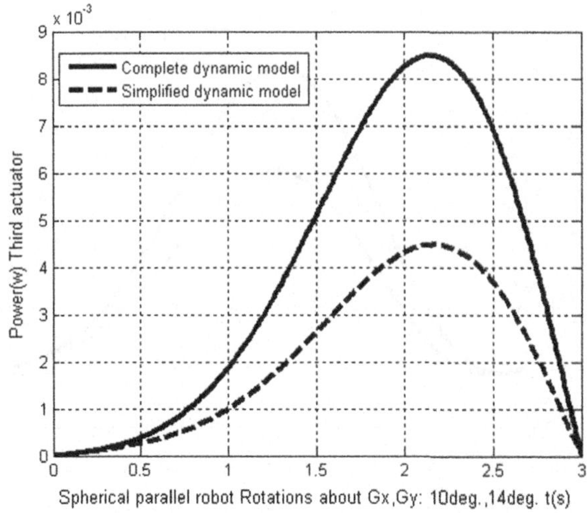

Fig. 9.13 Power P_{10}^{C} of third actuator (rotation)

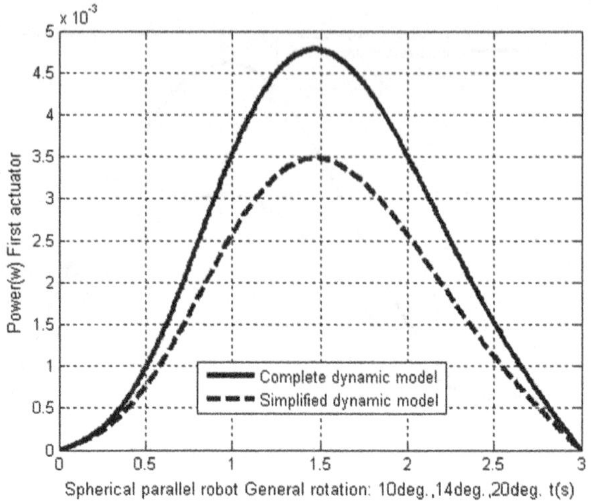

Fig. 9.14 Power P_{10}^{A} of first actuator (general rotation)

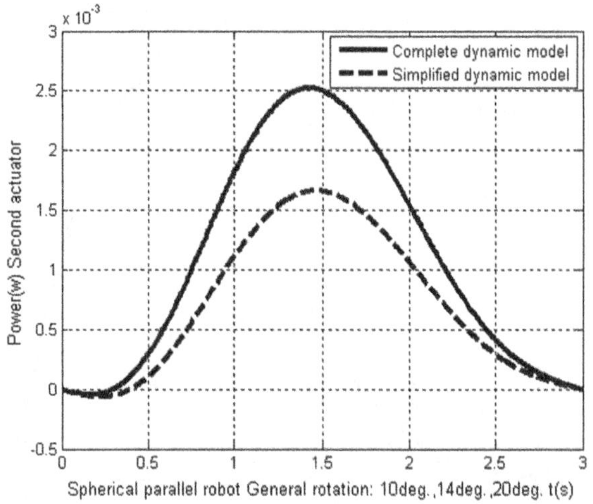

Fig. 9.15 Power P_{10}^{B} of second actuator (general rotation)

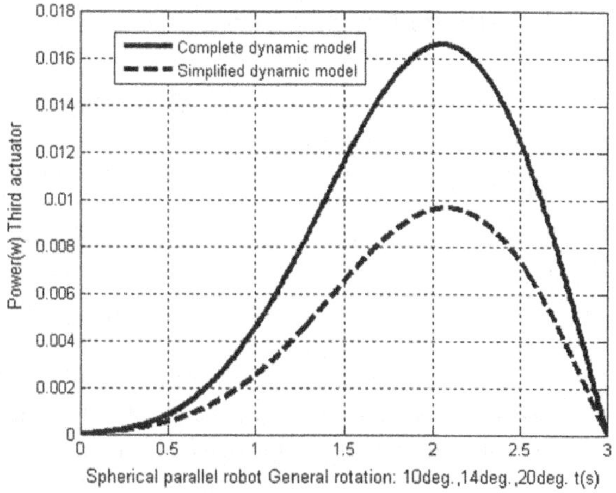

Fig. 9.16 Power P^C_{10} of third actuator (general rotation)

9.3 Dynamics of the 6-6 Stewart Parallel Manipulator

The first octahedral hexapod as a symmetrical arrangement of a six-degrees-of-freedom machine, built by E. Gough, as far back in time as 1950, was fully operational in 1954, but the design for industrial purposes can be dated to 1965, when D. Stewart implemented a six-linear jack system for use as a universal tire-testing machine. Some years later, the next historic reference comes from 1969, when Stewart built and published the design of a similar platform robot with six telescoping legs for use as a flight simulator [25, 26] (Fig. 1.1), which had a tremendous impact on subsequent development in the field of parallel robots. This device consists of a platform raised by six prismatic actuators, which allow manipulating of the position and the orientation of this platform with six degrees of freedom: three translational and three rotational.

Although parallel robots can take on a large variety of forms, we focus here on platform manipulators, with architecture similar to that of fight simulators. In fight simulators, the pose of the moving platform is controlled by hydraulic actuators that vary the distance between the centres of the spherical and the universal joints of the corresponding legs. The analysis of parallel manipulators is usually implemented through analytical methods in classical mechanics, in which the projection and resolution of vector equations on the reference axes are written in a considerable number of cumbersome, scalar relations and the solutions are rendered by large scale computations together with time-consuming computer codes [27, 28].

A recursive matrix method, already implemented in the inverse dynamics of planar parallel robots, is applied to this spatial 6-DOF mechanism. It has been proved to reduce the number of equations and computation operations significantly by using a set of matrices for kinematics and dynamics models. The Stewart-Gough platform is a well-known example of a parallel robot, where each of the six legs is a *UPS* structure, consisting of rigid links sequentially connected by universal, prismatic and spherical joints, with the mobile prismatic joint actuated.

9.3.1 *Kinematics Modelling*

A spatial six-degree-of-freedom parallel manipulator, which can be employed in several applications including machine tools, is proposed in the following study. Since the pneumatic joints can easily achieve high accuracy and heavy loads, many 3-DOF or 6-DOF parallel mechanisms use the actuated prismatic joints.

The 6-6 Stewart platform is a six-degrees-of-freedom spatial parallel mechanism in which a moving platform is connected together to a fixed base through six extensible legs from universal and spherical joints. Each leg is made up of a cylinder and a piston that are connected together by a prismatic joint. The upper end of each leg is connected to the platform by a spherical joint, whereas the lower end is connected to the base by a universal joint. Hydraulic or pneumatic actuators are used to vary the lengths of the prismatic joints and to control the location of the platform. Changing permanently the position and the orientation of its end-effectors, practically, the degree-of-freedom value $F = 6$ of the parallel mechanism, predicted by the Grübler-Kutzbach mobility criterion (3.36), is equal to the degrees of freedom associated with all the moving links $v = 36$ minus the total number of independent constraint relations $l = 30$ imposed by the joints.

The mechanism of the manipulator consists of six chains, including six variable lengths with identical topology, all connecting the fixed base to the moving platform. We suppose a circular moving platform symbolically represented by six pairs of spherical joints A_4, B_4, C_4, D_4, E_4, F_4 and a fixed base represented by another six universal joints located at the points A_1, B_1, C_1, D_1, E_1, F_1 (Fig. 9.17). Since the parallel manipulator is an assemblage of links and joints, this can be symbolised in a more abstract form known as equivalent graph representation, using the associated graph to represent the topology of the mechanism (Fig. 9.18). In the kinematical graph representation, with *five independent loops*, the links are denoted by vertices, the prismatic joints by thick edges, the revolute and spherical joints by thin edges and, finally, the fixed link 0 by two small concentric circles.

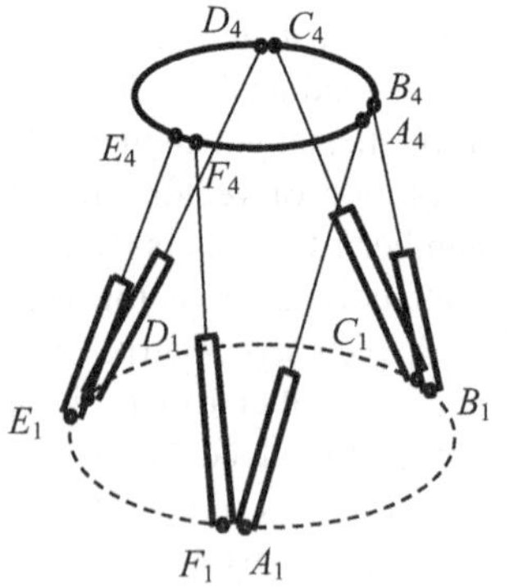

Fig. 9.17 The 6-6 Stewart parallel manipulator

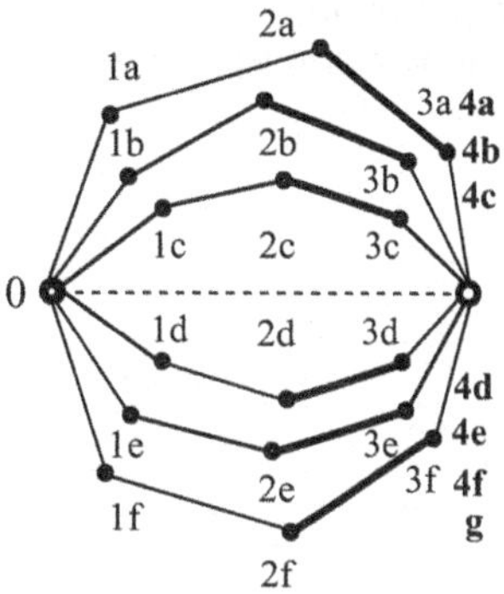

Fig. 9.18 Associated graph of the mechanism

In what follows, we consider that the moving platform is initially located at a *central configuration*, where the moving platform is not rotated with respect to the fixed base and the mass centre G is at an elevation $OG = h$ above the centre of the fixed base. We use a Cartesian coordinate system $Ox_0y_0z_0(T_0)$ to the fixed base with its origin located at the centre O of fixed platform, the Oz_0 axis perpendicular to the base and the Ox_0 axis pointing to the midpoint line linking the points F_1, A_1. Another coordinate central frame $Gx_Gy_Gz_G$ could be linked at the centre G of the moving platform only.

One of the six identical active legs (for example leg $A \equiv A_1A_2A_3A_4$) consists of a little cross of a universal fixed Hooke joint, characterized by a mass m_1 and a tensor of inertia $\hat{J}_1$, linked at the frame $A_1x_1^Ay_1^Az_1^A$ which has the angular velocity $\omega_{10}^A = \dot{\varphi}_{10}^A$ and the angular acceleration $\varepsilon_{10}^A = \dot{\omega}_{10}^A$ connected at a moving cylinder $A_2x_2^Ay_2^Az_2^A$ of length l_2, mass m_2 and tensor of inertia $\hat{J}_2$, having a relative rotation around $A_2z_2^A$ axis with the angle φ_{21}^A, so that $\omega_{21}^A = \dot{\varphi}_{21}^A$, $\varepsilon_{21}^A = \dot{\omega}_{21}^A$. An actuated prismatic joint is (as is also a piston) linked to the $A_3x_3^Ay_3^Az_3^A$ frame, having a relative displacement λ_{32}^A, velocity $v_{32}^A = \dot{\lambda}_{32}^A$ and acceleration $\gamma_{32}^A = \dot{v}_{32}^A$. It has the length l_3, mass m_3 and tensor of inertia $\hat{J}_3$. Finally, a ball-joint or a spherical joint A_4 is attached at the edge of the moving platform, which can be schematised as a circle of radius r_p, mass m_p and central tensor of inertia $\hat{J}_p$ (Fig. 9.19).

The six active sliders are start from the same initial relative position $A_2A_3 = B_2B_3 = \cdots = F_2F_3 = l_1$, when the angle of initial inclination β and the angles of orientation of universal and spherical joints have constant values

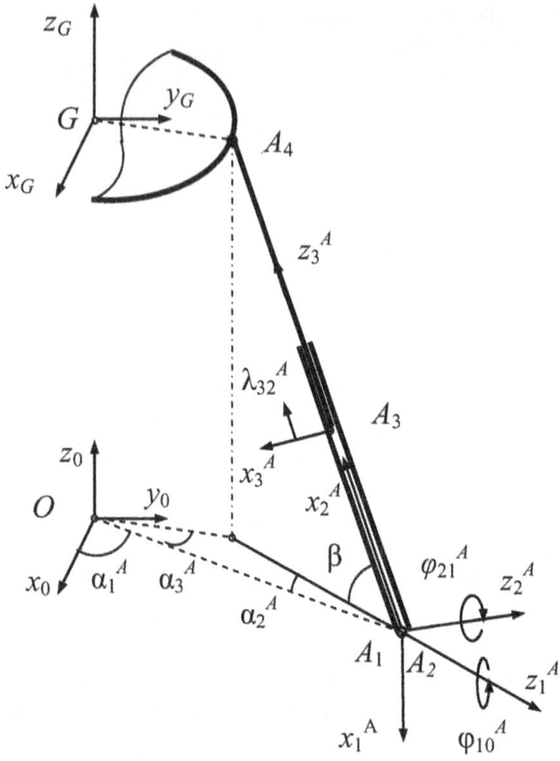

Fig. 9.19 Kinematics scheme of first leg A of the mechanism

$$
\begin{aligned}
&\alpha_1^A = \alpha_0, \quad \alpha_1^B = \frac{2\pi}{3} - \alpha_0, \quad \alpha_1^C = \frac{2\pi}{3} + \alpha_0\\
&\alpha_1^D = -\frac{2\pi}{3} - \alpha_0, \quad \alpha_1^E = -\frac{2\pi}{3} + \alpha_0, \quad \alpha_1^F = -\alpha_0\\
&\alpha_2^A = \alpha_2^C = \alpha_2^E = -\frac{\pi}{3} - \alpha_0, \quad \alpha_2^B = \alpha_2^D = \alpha_2^F = \frac{\pi}{3} + \alpha_0\\
&\alpha_3^A = \alpha_3^C = \alpha_3^E = \frac{\pi}{3} - 2\alpha_0, \quad \alpha_3^B = \alpha_3^D = \alpha_3^F = -\frac{\pi}{3} + 2\alpha_0.\\
&\quad h = l_0(\cos\alpha_0 - \sqrt{3}\sin\alpha_0)tg\beta = (l_1 + l_3)\sin\beta.\\
&\quad \alpha_j = \alpha_1^j + \alpha_2^j \ \ (j = A, B, C, D, E, F)
\end{aligned}
\tag{9.3.1}
$$

Starting from the origin O and pursuing successively 6 independent legs $OA_1A_2A_3, OB_1B_2B_3, \ldots, OF_1F_2F_3$, we obtain the following matrices of transformation

$$
\begin{aligned}
&q_{10} = q_{10}^{\varphi}\theta_1 a_{2\alpha}^j a_{1\alpha}^j, \quad q_{21} = q_{21}^{\varphi} a_{\beta}\theta_1\theta_2, \quad q_{32} = \theta_1, \quad q_{20} = q_{21}q_{10}, \quad q_{30} = q_{32}q_{20}\\
&(q = a, b, c, d, e, f) \ \ (j = A, B, C, D, E, F),
\end{aligned}
\tag{9.3.2}
$$

where we denote the matrices

$$\begin{aligned} q^{\varphi}_{\tau,\tau-1} &= rot(z, \varphi^{j}_{\tau,\tau-1}) \ \ (\tau = 1,2) \\ \theta_1 &= rot(y, \pi/2), \quad \theta_2 = rot(z, \pi/2), \quad a_{\beta} = rot(z, \beta) \\ a^{j}_{\chi\alpha} &= rot(z, \alpha^{j}_{\chi}) \ \ (\chi = 1,2,3). \end{aligned} \tag{9.3.3}$$

For the Stewart-Gough platform, in which the leg lengths can be adjusted by actuating the prismatic joints, the inverse kinematics are unique and straightforward to obtain, whereas the forward kinematics will have multiple solutions. Six displacements $\lambda^{A}_{32}, \lambda^{B}_{32}, \lambda^{C}_{32}, \lambda^{D}_{32}, \lambda^{E}_{32}, \lambda^{F}_{32}$ of the active links are the input variables that can give the instantaneous position of the mechanism. In the inverse geometric problem, it can be considered that the coordinates $x^{G}_{0}, y^{G}_{0}, z^{G}_{0}$ of the mass center G of the moving platform and the other three Euler angles $\alpha_1, \alpha_2, \alpha_3$, characterizing the sequence of concurrent rotations around its center G, give the position of the parallel mechanism. Since all rotations take place successively about the moving coordinate axes, the general rotation matrix $R = R_{32}R_{21}R_{10}$ of the moving platform from $Ox_0y_0z_0$ to $Gx_Gy_Gz_G$ reference system is obtained by multiplying the three relative basic matrices of transformation

$$R_{10} = a_1\theta_1, \quad R_{21} = a_2\theta_1\theta_2, \quad R_{32} = a_3\theta_2\theta_1\theta_2, \tag{9.3.4}$$

where

$$a_{\sigma} = rot(z, \alpha_{\sigma}) = \begin{bmatrix} \cos\alpha_{\sigma} & \sin\alpha_{\sigma} & 0 \\ -\sin\alpha_{\sigma} & \cos\alpha_{\sigma} & 0 \\ 0 & 0 & 1 \end{bmatrix}, \quad (\sigma = 1,2,3). \tag{9.3.5}$$

Consider, for example, that during three seconds the coordinates of the platform's centre G and the angles of orientation $\alpha_1, \alpha_2, \alpha_3$ can describe in fixed frame $Ox_0y_0z_0$ for the general absolute motion of the moving platform through the following analytical functions

$$\frac{x^{G}_{0}}{x^{G*}_{0}} = \frac{y^{G}_{0}}{y^{G*}_{0}} = \frac{h - z^{G}_{0}}{z^{G*}_{0}} = \frac{\alpha_{\sigma}}{\alpha^{*}_{\sigma}} = 1 - \cos\frac{\pi}{3}t, \quad (\sigma = 1,2,3). \tag{9.3.6}$$

A set of 18 independent variables $\varphi^{A}_{10}, \varphi^{A}_{21}, \lambda^{A}_{32}, \ldots, \varphi^{F}_{10}, \varphi^{F}_{21}, \lambda^{F}_{32}$ will be determined through 18 scalar equations which are easily generated by some vector-loop constraint conditions

$$\vec{r}^{\,j}_{10} + \sum_{k=1}^{3} q^{T}_{k0}\vec{r}^{\,j}_{k+1,k} = \vec{r}^{\,G}_{0} + R^{T}\vec{r}^{\,j_4}_{G}, \tag{9.3.7}$$

where

$$\vec{r}_{10}^{\,j} = l_0 a_{1\alpha}^{jT} \vec{u}_1, \vec{r}_{21}^{\,j} = \vec{0}, \quad \vec{r}_{32}^{\,j} = (l_1 + \lambda_{32}^j)\vec{u}_1, \quad \vec{r}_{43}^{\,j} = l_3 \vec{u}_3, \quad \vec{r}_G^{\,j4} = r_p a_{1\alpha}^{jT} a_{3\alpha}^{jT} \vec{u}_1$$
$$\vec{r}_0^{\,G} = \begin{bmatrix} x_0^G & y_0^G & z_0^G \end{bmatrix}^T (q = a, b, c, d, e, f) \ (j = A, B, C, D, E, F)$$
$$\vec{u}_1 = \begin{bmatrix} 1 & 0 & 0 \end{bmatrix}^T, \quad \vec{u}_2 = \begin{bmatrix} 0 & 1 & 0 \end{bmatrix}^T, \ \vec{u}_3 = \begin{bmatrix} 0 & 0 & 1 \end{bmatrix}^T. \tag{9.3.8}$$

Actually, the vector equation (9.3.7) mean that there is only one inverse geometric solution for the manipulator:

$$\varphi_{10}^j = a\tan(\mu_2^j/\mu_3^j), \quad \varphi_{21}^j = -\beta + A\tan 2(\mu_3^j/\cos\varphi_{10}^j, -\mu_1^j), \ \lambda_{32}^j = L_j - l_1 - l_3,$$

with the following notations

$$\mu_1^j = (x_0^G - x_{10}^j)\cos\alpha_j + (y_0^G - y_{10}^j)\sin\alpha_j + \cos\alpha_j \vec{u}_1^T R^T \vec{r}_G^{\,j4} + \sin\alpha_j \vec{u}_2^T R^T \vec{r}_G^{\,j4}$$
$$\mu_2^j = (x_0^G - x_{10}^j)\sin\alpha_j - (y_0^G - y_{10}^j)\cos\alpha_j + \sin\alpha_j \vec{u}_1^T R^T \vec{r}_G^{\,j4} - \cos\alpha_j \vec{u}_2^T R^T \vec{r}_G^{\,j4}$$
$$\mu_3^j = z_0^G + \vec{u}_3^T R^T \vec{r}_G^{\,j4}, \quad L_j = \sqrt{(\mu_1^j)^2 + (\mu_2^j)^2 + (\mu_3^j)^2}.$$

We develop, by inverse kinematics, a means to determine the absolute velocities and accelerations of all moving components of the robot, given the characteristics of the general motion of the moving platform by the relations (9.3.6). The angular velocity $\vec{\omega}_p$ of the moving platform and the velocity $\dot{\vec{r}}_0^{\,G}$ of its centre G are expressed

$$\vec{\omega}_p = \dot{\alpha}_1 R_{32} R_{21} \vec{u}_3 + \dot{\alpha}_2 R_{32} \vec{u}_3 + \dot{\alpha}_3 \vec{u}_3, \quad \dot{\vec{r}}_0^{\,G} = R^T \vec{v}_p^{\,G} = \begin{bmatrix} \dot{x}_0^G & \dot{y}_0^G & \dot{z}_0^G \end{bmatrix}^T. \tag{9.3.9}$$

The motion of the compounding elements of six legs is characterized by the following relative velocities of joints and angular velocities

$$\vec{v}_{\tau,\tau-1}^{\,j} = \vec{0}, \quad \vec{v}_{32}^{\,j} = \dot{\lambda}_{32}^j \vec{u}_3, \quad \vec{\omega}_{\tau,\tau-1}^{\,j} = \omega_{\tau,\tau-1}^j \vec{u}_3, \quad \vec{\omega}_{32}^{\,j} = \vec{0}, \quad (\tau = 1, 2) \tag{9.3.10}$$

To describe the absolute kinematical state of each body of the leg j, we compute the linear velocity $\vec{v}_{k0}^{\,j}$, the angular velocity $\vec{\omega}_{k0}^{\,j}$ and the *associate skew-symmetric matrix* $\tilde{\omega}_{k0}^j$ in terms of the vectors of the preceding body, using a recursive approach:

$$\vec{v}_{k0}^{\,j} = q_{k,k-1} \vec{v}_{k-1,0}^{\,j} + q_{k,k-1} \tilde{\omega}_{k-1,0}^j \vec{r}_{k,k-1}^{\,j} + v_{k,k-1}^j \vec{u}_3 \quad (k = 1, 2, 3)$$
$$\vec{\omega}_{k0}^{\,j} = q_{k,k-1} \vec{\omega}_{k-1,0}^{\,j} + \omega_{k,k-1}^j \vec{u}_3, \quad \tilde{\omega}_{k0}^j = q_{k,k-1} \tilde{\omega}_{k-1,0}^j q_{k,k-1}^T + \omega_{k,k-1}^j \tilde{u}_3 \tag{9.3.11}$$

Starting from the derivatives of geometrical constraints (9.3.7) and using the skew-symmetric matrix

$$\tilde{\omega}_p = \dot{\alpha}_1 R_{32} R_{21} \tilde{u}_3 R_{21}^T R_{32}^T + \dot{\alpha}_2 R_{32} \tilde{u}_3 R_{32}^T + \dot{\alpha}_3 \tilde{u}_3 \tag{9.3.12}$$

associated with the angular velocities $\vec{\omega}_p$ of the moving platform, we obtain the *matrix conditions of connectivity* [29] and the relative velocities $\vec{V}_j = \begin{bmatrix} \omega_{10}^j & \omega_{21}^j & v_{32}^j \end{bmatrix}^T$:

$$\vec{V}_j = [N_j]^{-1} \vec{P}_j, \tag{9.3.13}$$

where the following terms determine the contents of 3×3 invertible square matrix $[N_j]$ and the column matrix $\vec{P}_j$:

$$\begin{aligned} & n_{i1}^j = u_i^T q_{10}^T \tilde{u}_3 q_{21}^T \vec{r}_{42}^{\,j}, \quad n_{i2}^j = \vec{u}_i^T q_{20}^T \tilde{u}_3 \vec{r}_{42}^{\,j}, \quad n_{i3}^j = \vec{u}_i^T q_{20}^T \vec{u}_1, \quad \vec{r}_{42}^{\,j} = \vec{r}_{32}^{\,j} + q_{32}^T \vec{r}_{43}^{\,j}, \\ & p_i^j = \vec{u}_i^T \dot{\vec{r}}_0^{\,G} + \vec{u}_i^T R^T \tilde{\omega}_p \vec{r}_G^{\,j4} (i = 1, 2, 3)\ (j = A, B, C, D, E, F). \end{aligned} \tag{9.3.14}$$

Given as per (9.3.14), the *complete* Jacobian matrix is a fundamental prerequisite for the analysis of the robot workspace and the particular configurations of singularities where the spatial mechanism becomes uncontrollable [30, 31].

Based on different conditions of connectivity and the square matrix

$$\begin{aligned} \tilde{\omega}_p \tilde{\omega}_p + \tilde{\varepsilon}_p = {} & \ddot{\alpha}_1 R_{32} R_{21} \tilde{u}_3 R_{21}^T R_{32}^T + \ddot{\alpha}_2 R_{32} \tilde{u}_3 R_{32}^T + \ddot{\alpha}_3 \tilde{u}_3 \\ & + \dot{\alpha}_1^2 R_{32} R_{21} \tilde{u}_3 \tilde{u}_3 R_{21}^T R_{32}^T + \dot{\alpha}_2^2 R_{32} \tilde{u}_3 \tilde{u}_3 R_{32}^T + \dot{\alpha}_3^2 \tilde{u}_3 \tilde{u}_3 \\ & + 2 \dot{\alpha}_1 \dot{\alpha}_2 R_{32} R_{21} \tilde{u}_3 R_{21}^T \tilde{u}_3 R_{32}^T + 2 \dot{\alpha}_2 \dot{\alpha}_3 R_{32} \tilde{u}_3 R_{32}^T \tilde{u}_3 \\ & + 2 \dot{\alpha}_3 \dot{\alpha}_1 R_{32} R_{21} \tilde{u}_3 R_{21}^T R_{32}^T \tilde{u}_3, \end{aligned} \tag{9.3.15}$$

the expressions of relative accelerations $\vec{\Gamma}_j = [\varepsilon_{10}^j \ \ \varepsilon_{21}^j \ \ \gamma_{32}^j]^T$ are obtained from the column matrix

$$\vec{\Gamma}_j = [N_j]^{-1} \vec{W}_j, \tag{9.3.16}$$

where the following terms determine the contents of column matrices $\vec{W}_j = \dot{\vec{P}}_j - [\dot{N}_j] \vec{V}_j$:

$$\begin{aligned} w_i^j = {} & \vec{u}_i^T \ddot{\vec{r}}_0^{\,G} + \vec{u}_i^T R^T \{\tilde{\omega}_p \tilde{\omega}_p + \tilde{\varepsilon}_p\} \vec{r}_G^{\,j4} - \omega_{10}^j \omega_{10}^j \vec{u}_i^T q_{10}^T \tilde{u}_3 \tilde{u}_3 q_{21}^T \vec{r}_{42}^{\,j} \\ & - \omega_{21}^j \omega_{21}^j \vec{u}_i^T q_{20}^T \tilde{u}_3 \tilde{u}_3 \vec{r}_{42}^{\,j} - 2 \omega_{10}^j \omega_{21}^j \vec{u}_i^T q_{10}^T \tilde{u}_3 q_{21}^T \tilde{u}_3 \vec{r}_{42}^{\,j} \\ & - 2 \omega_{10}^j v_{32}^j \vec{u}_i^T q_{10}^T \tilde{u}_3 q_{21}^T \vec{u}_1 - 2 \omega_{21}^j v_{32}^j \vec{u}_i^T q_{20}^T \tilde{u}_3 \vec{u}_1 \\ & (i = 1, 2, 3) \ \ (j = A, B, C, D, E, F). \end{aligned} \tag{9.3.17}$$

The above characteristic velocities and accelerations of this manipulator can rapidly be verified if the vector relations delivered by the derivatives with respect to time of the constraining conditions (9.3.7) will successively be pre-multiplied by $\vec{u}_1^T q_{20}$, $\vec{u}_2^T q_{20}$ and $\vec{u}_3^T q_{20}$, respectively, as follows:

$$
\begin{aligned}
&L_j\omega_{10}^j \sin(\varphi_{21}^j+\beta) = -\vec{u}_3^T q_{20}\vec{v}_0^{j4}, L_j\omega_{21}^j = \vec{u}_2^T q_{20}\vec{v}_0^{j4}, v_{32}^j = \vec{u}_1^T q_{20}\vec{v}_0^{j4} \\
&L_j\varepsilon_{10}^j \sin(\varphi_{21}^j+\beta) = -\vec{u}_3^T q_{20}\vec{\gamma}_0^{j4} - 2L_j\omega_{10}^j\omega_{21}^j\cos(\varphi_{21}^j+\beta) - 2\omega_{10}^j v_{32}^j \sin(\varphi_{21}^j+\beta) \\
&L_j\varepsilon_{21}^j = \vec{u}_2^T q_{20}\vec{\gamma}_0^{j4} - L_j(\omega_{10}^j)^2\sin(\varphi_{21}^j+\beta)\cos(\varphi_{21}^j+\beta) - 2\omega_{21}^j v_{32}^j \\
&\gamma_{32}^j = \vec{u}_1^T q_{20}\vec{\gamma}_0^{j4} - L_j(\omega_{10}^j)^2\sin^2(\varphi_{21}^j+\beta) + L_j(\omega_{21}^j)^2
\end{aligned}
$$

where $\vec{v}_0^{j4} = \dot{\vec{r}}_0^G + R^T\tilde{\omega}_p\vec{r}_G^{j4}$, $\vec{\gamma}_0^{j4} = \ddot{\vec{r}}_0^G + R^T(\tilde{\omega}_p\tilde{\omega}_p + \tilde{\varepsilon}_p)\vec{r}_G^{j4}$.

The following relations give the linear accelerations $\vec{\gamma}_{k0}^j$, the angular accelerations $\vec{\varepsilon}_{k0}^j$ and the *significant matrices* $\tilde{\omega}_{k0}^j\tilde{\omega}_{k0}^j + \tilde{\varepsilon}_{k0}^j$ of links $k = 1, 2, 3$ from an arbitrary leg j:

$$
\begin{aligned}
\vec{\gamma}_{k0}^j &= q_{k,k-1}\vec{\gamma}_{k-1,0}^j + q_{k,k-1}\{\tilde{\omega}_{k-1,0}^j\tilde{\omega}_{k-1.0}^j + \tilde{\varepsilon}_{k-1,0}^j\}\vec{r}_{k,k-1}^j \\
&\quad + 2v_{k,k-1}^j q_{k,k-1}\tilde{\omega}_{k-1,0}^j q_{k,k-1}^T\vec{u}_3 + \gamma_{k,k-1}^j\vec{u}_3 \\
\vec{\varepsilon}_{k0}^j &= \dot{\vec{\omega}}_{k0}^j = q_{k,k-1}\vec{\varepsilon}_{k-1,0}^j + \varepsilon_{k,k-1}^j\vec{u}_3 + \omega_{k,k-1}^j q_{k,k-1}\tilde{\omega}_{k-1,0}^j q_{k,k-1}^T\vec{u}_3 \\
\tilde{\omega}_{k0}^j\tilde{\omega}_{k0}^j + \dot{\tilde{\omega}}_{k0}^j &= q_{k,k-1}\{\tilde{\omega}_{k-1,0}^j\tilde{\omega}_{k-1,0}^j + \tilde{\varepsilon}_{k-1,0}^j\}q_{k,k-1}^T \\
&\quad + \omega_{k,k-1}^j\omega_{k,k-1}^j\tilde{u}_3\tilde{u}_3 + \varepsilon_{k,k-1}^j\tilde{u}_3 + 2\omega_{k,k-1}^j q_{k,k-1}\tilde{\omega}_{k-1,0}^j q_{k,k-1}^T\tilde{u}_3 \\
&\quad (q = a, b, c, d, e, f)\ \ (j = A, B, C, D, E, F).
\end{aligned}
\tag{9.3.18}
$$

9.3.2 *Inverse Dynamics Modelling*

The inverse dynamic analysis of the Stewart manipulator has been done using Lagrange equations and the Newton-Euler method [32]. Starting from recursive equations based on the Newton-Euler procedure, numerical solutions have been developed and used for the dynamics analysis. Dasgupta and Mruthyunjaya [12] used the Newton-Euler approach to develop closed-form dynamic equations of the Stewart manipulator, taking into account all dynamic and gravity effects as well as viscous friction at joints. Geng [33] and Tsai [1] presented an algorithm to solve the inverse dynamics for a Stewart platform-type using Lagrange formalism. The Lagrange formulation is well structured and can be expressed in closed form, but a large amount of symbolic computation is needed to find partial derivatives of Lagrange's function, the analytical calculi involved being too long for each scheme of the manipulator.

We will gain an appreciation of the principle of virtual work in dynamics, which is used for obtaining recursive dynamic equations and to have a better understanding of the basic joint-force relationships in the analysis of interconnected bodies.

9.3.2.1 Principle of Virtual Work

For higher payloads, the links of the parallel manipulator must be stiffer and the actuators stronger. This will increase the inertial parameters, which can increase the dynamic effects on the system.

Knowing the position and kinematics state of every link as well as the external forces acting on the robot, we derive some explicit recursive matrix equations for the inverse dynamic problem in order to obtain the input forces and powers required by all prismatic actuators. The equations of motion allow the designer to examine the effects of different factors upon the dynamic behaviour of the system, such as geometry, link sizes and inertial properties. The explicit equations also are required for designing control systems that will improve the dynamic performance of the parallel robot.

In the inverse dynamic analysis of the parallel robot, the desired motion is specified and the forces required to generate this motion are determined. So, the inverse dynamic model is the tool that a designer would use to determine whether a given Stewart robot or a proposed design is capable of generating a desired motion. This analysis can also be used to determine the effects of changing platform specifications upon the required forces, a determination which is also useful in the design process.

Some independent pneumatic or hydraulic systems, but also electrical prismatic actuators with very good dynamic behaviour and positioning accuracy, that generate six active forces $\vec{f}_{32}^{j} = f_{32}^{j}\vec{u}_3$, can be used to vary the lengths of the prismatic joints and to control the location of the platform and the motion of the six moving pistons of the legs. The parallel robot can artificially be transformed into a set of *seven independent trees* subject to the constraint—comprising also the *principal kinematical chain OG*—which connect the moving platform. This is possible by imaginarily cutting six spherical joints $A_4, B_4, C_4, D_4, E_4, F_4$ for the moving platform and taking the effect of this into account by introducing the corresponding constraint conditions (Fig. 9.18).

The force of inertia and the moment of inertia forces (9.1.18) of an arbitrary rigid body T_k^A, for example, are determined with respect to the centre of joint A_k. On the other hand, the wrench of two vectors (9.1.19) evaluates the influence of the action of the weight $m_k^A\vec{g}$ on the same element T_k^A of the manipulator.

Pursuing the first leg A, two known *significant recursive relations* generate the vectors

$$\vec{F}_k^A = \vec{F}_{k0}^A + a_{k+1,k}^T\vec{F}_{k+1}^A, \quad \vec{M}_k^A = \vec{M}_{k0}^A + a_{k+1,k}^T\vec{M}_{k+1}^A + \tilde{r}_{k+1,k}^A a_{k+1,k}^T\vec{F}_{k+1}^A \tag{9.3.19}$$

with the notations

$$\vec{F}_{k0}^{A} = -\vec{f}_{k}^{in\,A} - \vec{f}_{k}^{*A}, \quad \vec{M}_{k0}^{A} = -\vec{m}_{k}^{in\,A} - \vec{m}_{k}^{*A}. \tag{9.3.20}$$

As example, starting from (9.3.19), we develop a set of six recursive relations:

$$\vec{F}_{3}^{A} = \vec{F}_{30}^{A}, \quad \vec{F}_{2}^{A} = \vec{F}_{20}^{A} + a_{32}^{T}\vec{F}_{3}^{A}, \quad \vec{F}_{1}^{A} = \vec{F}_{10}^{A} + a_{21}^{T}\vec{F}_{2}^{A}$$

$$\vec{M}_{3}^{A} = \vec{M}_{30}^{A}, \quad \vec{M}_{2}^{A} = \vec{M}_{20}^{A} + a_{32}^{T}\vec{M}_{3}^{A} + \tilde{r}_{32}^{A}a_{32}^{T}\vec{F}_{3}^{A}, \quad \vec{M}_{1}^{A} = \vec{M}_{10}^{A} + a_{21}^{T}\vec{M}_{2}^{A} + \tilde{r}_{21}^{A}a_{21}^{T}\vec{F}_{2}^{A}. \tag{9.3.21}$$

The characteristic *relative virtual velocities* $\vec{v}_{k,k-1}^{v} = v_{k,k-1}^{v}\vec{u}_{3}$, $\vec{\omega}_{k,k-1}^{v} = \omega_{k,k-1}^{v}\vec{u}_{3}$ can be expressed as functions of the pose of the robot by the matrix equations of connectivity (9.3.13). Let us assume that the mechanism has successively *six independent* virtual motions determined by the following sets of velocities:

$$\begin{aligned} &v_{32a}^{Av} = 1, \quad v_{32a}^{jv} = 0\ (j \neq A); \quad v_{32b}^{Bv} = 1, \quad v_{32b}^{jv} = 0\ (j \neq B); \quad v_{32c}^{Cv} = 1, \quad v_{32c}^{jv} = 0\ (j \neq C) \\ &v_{32d}^{Dv} = 1, \quad v_{32d}^{jv} = 0\ (j \neq D); \quad v_{32e}^{Ev} = 1, \quad v_{32e}^{jv} = 0\ (j \neq E); \quad v_{32f}^{Fv} = 1, \quad v_{32f}^{jv} = 0\ (j \neq F). \end{aligned} \tag{9.3.22}$$

Applying the explicit form of the *equations of parallel robots dynamics* [6], the following compact matrix relation is obtained

$$\begin{aligned} f_{32}^{A} = \vec{u}_{3}^{T}\{&\vec{F}_{3}^{A} + \omega_{10a}^{Av}\vec{M}_{1}^{A} + \omega_{21a}^{Av}\vec{M}_{2}^{A} + \omega_{10a}^{Bv}\vec{M}_{1}^{B} + \omega_{21a}^{Bv}\vec{M}_{2}^{B} + \omega_{10a}^{Cv}\vec{M}_{1}^{C} + \omega_{21a}^{Cv}\vec{M}_{2}^{C} \\ &+ \omega_{10a}^{Dv}\vec{M}_{1}^{D} + \omega_{21a}^{Dv}\vec{M}_{2}^{D} + \omega_{10a}^{Ev}\vec{M}_{1}^{E} + \omega_{21a}^{Ev}\vec{M}_{2}^{E} + \omega_{10a}^{Fv}\vec{M}_{1}^{F} + \omega_{21a}^{Fv}\vec{M}_{2}^{F} \\ &+ v_{10a}^{Gv}\vec{F}_{1}^{G} + v_{21a}^{Gv}\vec{F}_{2}^{G} + v_{32a}^{Gv}\vec{F}_{3}^{G} + \omega_{43a}^{Gv}\vec{M}_{4}^{G} + \omega_{54a}^{Gv}\vec{M}_{5}^{G} + \omega_{65a}^{Gv}\vec{M}_{6}^{G}\} \end{aligned} \tag{9.3.23}$$

for the *input force* required by the first active prismatic joint A_3 along the axis $A_3z_3^A$.

Based on the *principle of virtual work*, the relations (9.3.19)–(9.3.23) represent the *inverse dynamics model* of the 6-6 Stewart parallel robot, which can be transformed into a model for automatic command.

In the following, we were able to apply the *Newton-Euler procedure* to establish the set of analytical equations for each compounding rigid body of a prototype parallel robot in a real application. These equations give all connecting forces in the external and internal joints. Several relations from the general system of equations could, if required, constitute verification for the input forces or active torques obtained by the method based on the principle of virtual work. Now, we can calculate the friction forces and the friction torques in the joints, based on the friction coefficients and the maximum of the connecting forces in the joints. We apply again the above explicit equations, where the contribution of the virtual work of *internal friction forces* in joints is added. The new active torques and input forces are compared to the values obtained in the first calculus.

9.3.2.2 Lagrange Equations

The generalized coordinates of the 6-6 Stewart parallel robot, in a Lagrange formalism, are represented by 24 parameters $s_1 = x_0^G$, $s_2 = y_0^G$, $s_3 = z_0^G$, $s_4 = \alpha_1$, $s_5 = \alpha_2$, $s_6 = \alpha_3$, $s_7 = \varphi_{10}^A$, $s_8 = \varphi_{21}^A$, $s_9 = \lambda_{32}^A$, $s_{10} = \varphi_{10}^B$, $s_{11} = \varphi_{21}^B$, $s_{12} = \lambda_{32}^B$, ..., $s_{22} = \varphi_{10}^F$, $s_{23} = \varphi_{21}^F$, $s_{24} = \lambda_{32}^F$.

Lagrange's equations with 18 multipliers $\lambda_1, \lambda_2, \ldots, \lambda_{18}$ will be expressed by 24 differential relations

$$\frac{d}{dt}\left\{\frac{\partial L}{\partial \dot{s}_l}\right\} - \frac{\partial L}{\partial s_l} = \Phi_l + \sum_{v=1}^{18} \lambda_v c_{vl} \quad (l = 1, 2, \ldots, 24), \tag{9.3.24}$$

which contain following 24 generalized forces $\Phi_1 = 0$, $\Phi_2 = 0$, $\Phi_3 = 0$, $\Phi_4 = 0$, $\Phi_5 = 0$, $\Phi_6 = 0$, $\Phi_7 = 0$, $\Phi_8 = 0$, $\Phi_9 = f_{32}^A$, $\Phi_{10} = 0$, $\Phi_{11} = 0$, $\Phi_{12} = f_{32}^B$, ..., $\Phi_{22} = 0$, $\Phi_{23} = 0$, $\Phi_{24} = f_{32}^F$.

The number of unknowns is 42. Because the relations (9.3.24) deliver only 24 equations, in order to be able to solve this system one needs to have additional 18 new equations, which are the nonlinear algebraic constraint equations. A number of 18 kinematical conditions of constraint are given by the relations (9.3.13):

$$\sum_{l=1}^{24} c_{vl}\dot{s}_l = 0 \quad (v = 1, 2, \ldots, 18). \tag{9.3.25}$$

If the kinematic constraints are linearly independent, 18 independent coordinates can be selected in such a manner that the matrix C of the coefficients c_{vl} be a non-singular matrix.

The components of the general expression of the Lagrange function $L = L_p + \sum_{\mu=1}^{3} (L_\mu^A + L_\mu^B + \cdots + L_\mu^F)$ are expressed as analytical functions of the generalized coordinates and their first derivatives with respect to time:

$$\begin{aligned}
L_p &= \frac{1}{2} m_p(\dot{x}_0^{G2} + \dot{y}_0^{G2} + \dot{z}_0^{G2}) + \frac{1}{2}\vec{\omega}_p^T \hat{J}_p \vec{\omega}_p - m_p g z_0^G \\
L_1^j &= \frac{1}{2}\vec{\omega}_{10}^{jT} \hat{J}_1^j \vec{\omega}_{10}^j \\
L_2^j &= \frac{1}{2}\vec{\omega}_{20}^{jT} \hat{J}_2^j \vec{\omega}_{20}^j - m_2^j g \vec{u}_3^T (\vec{r}_{10}^j + q_{10}^T \vec{r}_{21}^j + q_{20}^T \vec{r}_2^{Cj}) \\
L_3^j &= \frac{1}{2} m_3^j \vec{v}_{30}^{jT} \vec{v}_{30}^j + \frac{1}{2}\vec{\omega}_{30}^{jT} \hat{J}_3^j \vec{\omega}_{30}^j + m_3^j \vec{v}_{30}^{jT} \tilde{\omega}_{30}^j \vec{r}_3^{Cj} \\
&\quad - m_3^j g \vec{u}_3^T (\vec{r}_{10}^j + q_{10}^T \vec{r}_{21}^j + q_{20}^T \vec{r}_{32}^j + q_{30}^T \vec{r}_3^{Cj}) \\
&\quad (q = a, b, c, d, e, f) \ (j = A, B, C, D, E, F).
\end{aligned} \tag{9.3.26}$$

Based on the relations (9.3.10), (9.3.11), we can calculate the partial derivatives of the Lagrange functions, using the above first derivatives (9.1.28) of orthogonal transformation matrices $q_{k,k-1}$:

$$
\begin{aligned}
&\frac{\partial L_p}{\partial x_0^G} = 0, \quad \frac{\partial L_p}{\partial y_0^G} = 0, \quad \frac{\partial L_p}{\partial z_0^G} = -m_p g \\
&\frac{\partial L_p}{\partial \dot{x}_0^G} = m_p \dot{x}_0^G, \quad \frac{\partial L_p}{\partial \dot{y}_0^G} = m_p \dot{y}_0^G, \quad \frac{\partial L_p}{\partial \dot{z}_0^G} = m_p \dot{z}_0^G \\
&\frac{\partial L_p}{\partial \alpha_1} = 0, \quad \frac{\partial L_p}{\partial \dot{\alpha}_1} = \vec{u}_3^T R_{21}^T R_{32}^T \hat{J}_p \vec{\omega}_p \\
&\frac{\partial L_p}{\partial \alpha_2} = \dot{\alpha}_1 \vec{u}_3^T R_{21}^T \tilde{u}_3 R_{32}^T \hat{J}_p \vec{\omega}_p, \quad \frac{\partial L_p}{\partial \dot{\alpha}_2} = \vec{u}_3^T R_{32}^T \hat{J}_p \vec{\omega}_p \\
&\frac{\partial L_p}{\partial \alpha_3} = \dot{\alpha}_1 \vec{u}_3^T R_{21}^T R_{32}^T \tilde{u}_3 \hat{J}_p \vec{\omega}_p + \dot{\alpha}_2 \vec{u}_3^T R_{32}^T \tilde{u}_3 \hat{J}_p \vec{\omega}_p, \quad \frac{\partial L_p}{\partial \dot{\alpha}_3} = \vec{u}_3^T \hat{J}_p \vec{\omega}_p \\
&\frac{\partial L_1^j}{\partial \varphi_{10}^j} = 0, \quad \frac{\partial L_1^j}{\partial \dot{\varphi}_{10}^j} = \vec{u}_3^T \hat{J}_1^j \vec{\omega}_{10}^j \\
&\frac{\partial L_2^j}{\partial \varphi_{10}^j} = -m_2^j g \vec{u}_3^T q_{10}^T \tilde{u}_3 q_{21}^T \vec{r}_2^{\,Cj}, \quad \frac{\partial L_2^j}{\partial \dot{\varphi}_{10}^j} = \vec{u}_3^T q_{21}^T \hat{J}_2^j \vec{\omega}_{20}^j \\
&\frac{\partial L_2^j}{\partial \varphi_{21}^j} = \dot{\varphi}_{10}^j \vec{u}_3^T q_{21}^T \tilde{u}_3 \hat{J}_2^j \vec{\omega}_{20}^j - m_2^j g \vec{u}_3^T q_{20}^T \tilde{u}_3 \vec{r}_2^{\,Cj}, \quad \frac{\partial L_2^j}{\partial \dot{\varphi}_{21}^j} = \vec{u}_3^T \hat{J}_2^j \vec{\omega}_{20}^j \\
&\frac{\partial L_3^j}{\partial \varphi_{10}^j} = -m_3^j g \vec{u}_3^T q_{10}^T \tilde{u}_3 q_{21}^T \{\vec{r}_{32}^{\,j} + q_{32}^T \vec{r}_3^{\,Cj}\} \\
&\frac{\partial L_3^j}{\partial \dot{\varphi}_{10}^j} = m_3^j \vec{u}_3^T q_{21}^T \tilde{r}_{32}^{ji} q_{32}^T \{\vec{v}_{30}^j + \tilde{\omega}_{30}^j \vec{r}_3^{\,Cj}\} + \vec{u}_3^T q_{21}^T q_{32}^T \{m_3^j \tilde{r}_3^{\,Cj} \vec{v}_{30}^j + \hat{J}_3^j \vec{\omega}_{30}^j\} \\
&\frac{\partial L_3^j}{\partial \varphi_{21}^j} = m_3^j \dot{\varphi}_{10}^j \vec{u}_3^T q_{21}^T \tilde{u}_3 \tilde{r}_{32}^j q_{32}^T \{\vec{v}_{30}^j + \tilde{\omega}_{30}^j \vec{r}_3^{\,Cj}\} + \dot{\varphi}_{10}^j \vec{u}_3^T q_{21}^T \tilde{u}_3 q_{32}^T \{m_3^j \tilde{r}_3^{\,Cj} \vec{v}_{30}^j \\
&\qquad + \hat{J}_3^j \vec{\omega}_{30}^j\} - m_3^j g \vec{u}_3^T q_{20}^T \tilde{u}_3 \{\vec{r}_{32}^{\,j} + q_{32}^T \vec{r}_3^{\,Cj}\} \\
&\frac{\partial L_3^j}{\partial \dot{\varphi}_{21}^j} = m_3^j \vec{u}_3^T \tilde{r}_{32}^j q_{32}^T \{\vec{v}_{30}^j + \tilde{\omega}_{30}^j \vec{r}_3^{\,Cj}\} + \vec{u}_3^T q_{32}^T \{m_3^j \tilde{r}_3^{\,Cj} \vec{v}_{30}^j + \hat{J}_3^j \vec{\omega}_{30}^j\} \\
&\frac{\partial L_3^j}{\partial \lambda_{32}^j} = m_3^j \vec{\omega}_{20}^{jT} \tilde{u}_1 q_{32}^T \{\vec{v}_{30}^j + \tilde{\omega}_{30}^j \vec{r}_3^{\,Cj}\} - m_3^j g \vec{u}_3^T q_{20}^T \vec{u}_1, \quad \frac{\partial L_3^j}{\partial \dot{\lambda}_{32}^j} = m_3^j \vec{u}_3^T \{\vec{v}_{30}^j + \tilde{\omega}_{30}^j \vec{r}_3^{\,Cj}\}.
\end{aligned}
\tag{9.3.27}
$$

Performing the derivatives with respect to time $\frac{d}{dt}\left\{\frac{\partial L}{\partial \dot{s}_l}\right\}$ $(l = 1, 2, \ldots, 24)$, a complex system of 24 relations leads to the same expressions (9.3.23) for the six input forces $f_{32}^A, f_{32}^B, \ldots, f_{32}^F$ required by the six prismatic actuators, after elimination of the 18 multipliers.

The dynamic simulation for testing some robot designs will enable the designer to gain some basic insight of the dynamic behaviour of the system and its relation to different system parameters, such as geometry and inertia. We assume that the manipulator starts at rest from a central configuration, when the moving platform is assumed to be located $h = 1\,\mathrm{m}$ lower than the fixed base, namely $t = 0$: $x_0^G = 0$, $y_0^G = 0, z_0^G = 1\,\mathrm{m}$.

As application, we consider a manipulator, which has the following mechanical and architectural characteristics

$$x_0^{G*} = 0, \quad y_0^{G*} = 0, \quad z_0^{G*} = 0.15\,\mathrm{m}, \quad \alpha_1^* = 0, \quad \alpha_2^* = 0, \quad \alpha_3^* = \frac{\pi}{12}, \quad \alpha_0 = \frac{\pi}{36}$$

$$m_1 = 0.25\,\mathrm{kg}, \quad m_2 = 1\,\mathrm{kg}, \quad m_3 = 1\,\mathrm{kg}, \quad m_p = 15\,\mathrm{kg}$$

$$l_0 = OA_1 = 1\,\mathrm{m}, \quad l_2 = 1\,\mathrm{m}, \quad l_3 = 1\,\mathrm{m}, \quad GA_4 = l_4 = 1\,\mathrm{m}, \quad OG = h = 1\,\mathrm{m}$$

$$g = 9.81\,\mathrm{ms}^{-2}, \quad \Delta t = 3\,\mathrm{s}, \quad l_0(\cos\alpha_0 - \sqrt{3}\sin\alpha_0)tg\beta = h$$

$$\hat{J}_1 = \begin{bmatrix} 0.001 & & \\ & 0.001 & \\ & & 0.002 \end{bmatrix}, \quad \hat{J}_2 = \begin{bmatrix} 0.01 & & \\ & 0.325 & \\ & & 0.325 \end{bmatrix}$$

$$\hat{J}_3 = \begin{bmatrix} 0.325 & & \\ & 0.325 & \\ & & 0.01 \end{bmatrix}, \quad \hat{J}_p = \begin{bmatrix} 3.75 & & \\ & 3.75 & \\ & & 7.5 \end{bmatrix}.$$

MATLAB software for a computer program was developed to solve the inverse kinematics and dynamics of the 6-6 Stewart parallel robot. Considering the weights of compounding elements as passive external forces acting on the spatial manipulator, the time-history evolution of the input forces and powers required by all six pneumatic or hydraulic active systems are shown for a period of three seconds of the platform's motion.

When the platform moves along the *vertical* z_0 direction with variable acceleration, we remark from Figs. 9.20 and 9.21 that all input forces and powers are constantly equal to one another.

For the second example, we consider the *rotation* of the platform *about* z_0 axis with the variable acceleration $\ddot{\alpha}_3$. During this rotation, the actuators B, D, F only are working in an active regime with increasing forces and positive values of the absorbed powers (Figs. 9.22 and 9.23). The recursive matrix relations (9.3.23) represent a set of explicit equations of the dynamics simulation which, in a context of automatic command, can easily be transformed into a robust model for computerized control of the spatial 6-6 Stewart parallel manipulator.

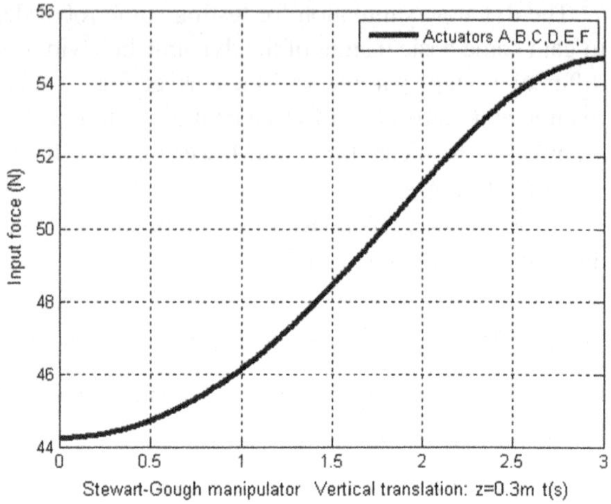

Fig. 9.20 Input forces of six actuators (translation)

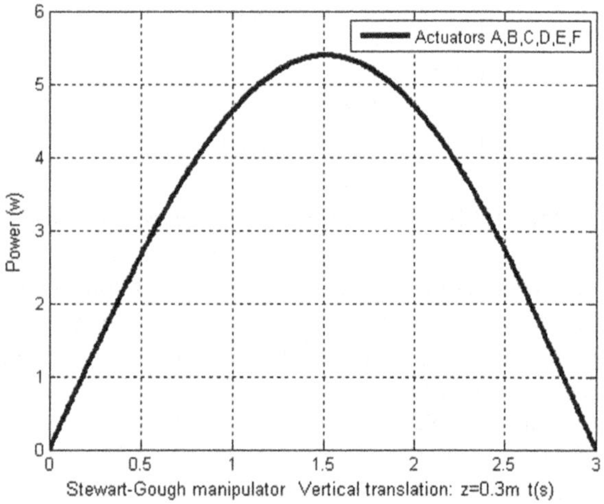

Fig. 9.21 Powers of six actuators (translation)

9.4 Internal Joint Forces in Dynamics of a 3-RPS Parallel Manipulator

Inverse and forward kinematics problem for this type of manipulator were already analyzed [34, 35]. Because the robot motions in the translation and orientation degrees of freedom are *interconnected*, this makes the investigations here more complicated in comparison to other manipulators with a different spatial structure [6, 36].

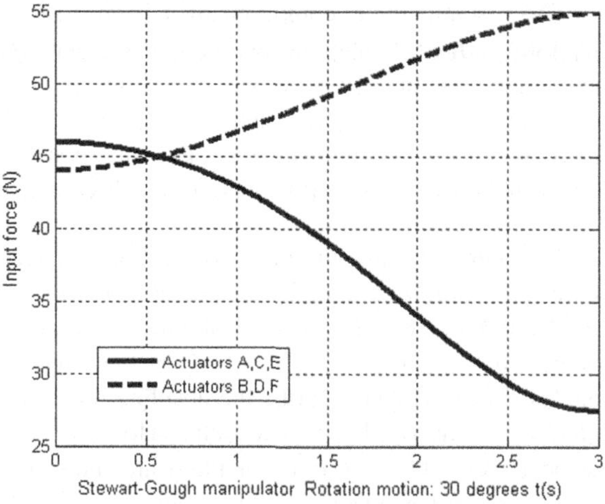

Fig. 9.22 Input forces of six actuators (rotation)

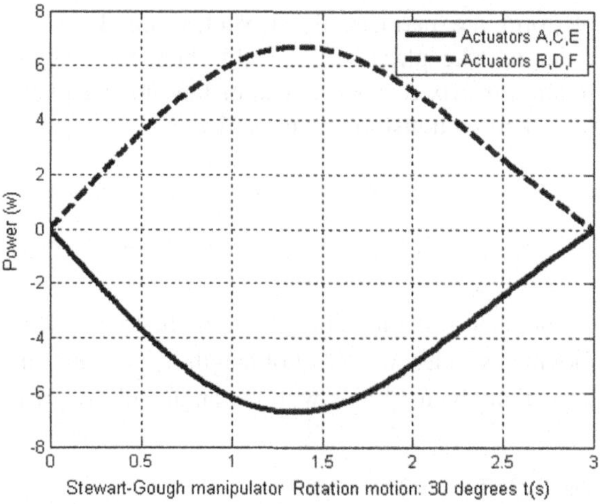

Fig. 9.23 Powers of six actuators (rotation)

A recursive method is developed, using a set of matrices for calculating the input powers and the *reaction internal joint forces* in the inverse dynamics of the three-degrees-of-freedom 3-$\underline{R}$PS spatial parallel mechanism.

9.4.1 *Kinematic Analysis*

Already well-known in the mechanisms community, the special symmetrical parallel architecture 3-$\underline{R}$PS consists of a moving platform, a fixed base and three limbs of identical structure.

The manipulator is composed of a fixed base $A_1B_1C_1$ and the upper moving platform $A_3B_3C_3$, which are two equilateral triangles with $L = l_0\sqrt{3}$ and $l = r\sqrt{3}$ the lengths of the sides, respectively. The extensible legs connect the moving platform by spherical joints and the base by means of revolute joints, their axes being parallel to the opposite edge. Each leg is made up of a cylinder and a piston connected together by a prismatic joint. Three actuated revolute joints of the legs drive this spatial manipulator.

The frame $Ox_0y_0z_0$ with its z_0 vertical axis is attached at the centre O of a fixed base. The platform is initially located at a *central configuration*, since the centre G of the frame $Gx_Gy_Gz_G$ is fixed at $OG = h$ elevation (Fig. 9.24). *Six variables*, namely three coordinates x_0^G, y_0^G, z_0^G of the mass center and three classical Euler angles α_1, α_2, α_3 can describe the absolute position and orientation of the moving platform in the fixed reference frame. However, the mechanism is only a 3-DOF device; therefore, only three of them are independent. Here, z_0^G, α_1, α_2 will be chosen as independent variables and x_0^G, y_0^G, α_3 are parameters of *parasitic* motions. The three parasitic motions from the six commonly-known motions of the platform are permanently dependent on three independent variables. Since all rotations take place successively about the moving coordinate axes, the general rotation matrix $R = R_{32}R_{21}R_{10}$ of the moving platform is obtained by multiplying three known transformation matrices

$$\begin{gathered} R_{10} = a_1\theta_1, \quad R_{21} = a_2\theta_1\theta_2, \quad R_{32} = a_3\theta_2\theta_1\theta_2 \\ \theta_1 = rot\left(y, \frac{\pi}{2}\right), \quad \theta_2 = rot\left(z, \frac{\pi}{2}\right), \quad a_\sigma = rot(z, \alpha_\sigma), \quad (\sigma = 1, 2, 3). \end{gathered} \tag{9.4.1}$$

Inclined of the initial angle β, the first leg A consists of a fixed revolute joint, a moving cylinder $A_1x_1^Ay_1^Az_1^A$ of length l_1, mass m_1 and tensor of inertia $\hat{J}_1$, which has a rotation about z_1^A axis with the angle φ_{10}^A. A prismatic joint is also a piston with

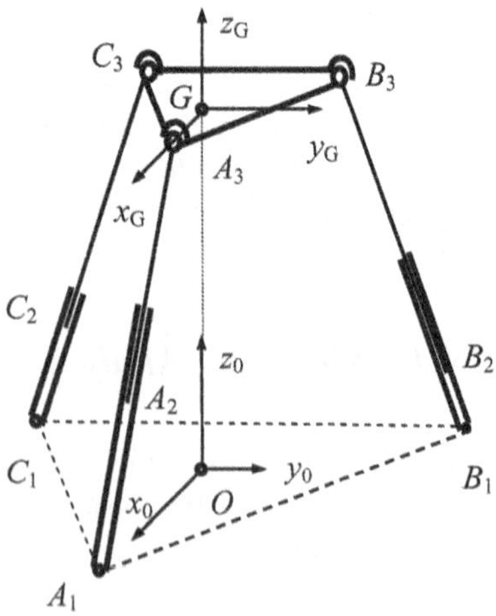

Fig. 9.24 The 3-RPS parallel manipulator

the length l_2, the mass m_2 and the tensor of inertia $\hat{J}_2$ linked at the $A_2x_2^Ay_2^Az_2^A$ frame, having a relative translation motion with the displacement λ_{21}^A. Finally, a spherical joint A_3 is introduced to a planar moving platform, which is schematised as an equilateral triangle with edge length $l = r\sqrt{3}$, mass m_p and tensor of inertia $\hat{J}_p$ (Fig. 9.25).

At the central configuration, we also consider that all legs are initially extended at equal length $\lambda_0 + l_2 = h/\cos\beta = (l_0 - r)/\sin\beta$ and that the angles of orientation of fixed pivots are given by $\alpha_A = 0, \alpha_B = 2\pi/3\ \alpha_C = -2\pi/3$.

Pursuing successively each leg along three distinct ways $OA_1A_2A_3$,$OB_1B_2B_3$, $OC_1C_2C_3$, we obtain the following matrices of transformation [6]

$$p_{10} = p_{10}^{\varphi}a_{\beta}\theta_3 a_{\alpha}^j, \quad p_{21} = \theta_3^T, \quad p_{20} = p_{21}p_{10}, \quad (p = a, b, c) \quad (j = A, B, C) \quad (9.4.2)$$

with the notations

$$p_{10}^{\varphi} = rot(z, \varphi_{10}^i), \quad a_{\alpha}^j = rot(z, \alpha_j), \quad a_{\beta} = rot(z, \beta), \quad \theta_3 = rot\left(x, \frac{\pi}{2}\right). \quad (9.4.3)$$

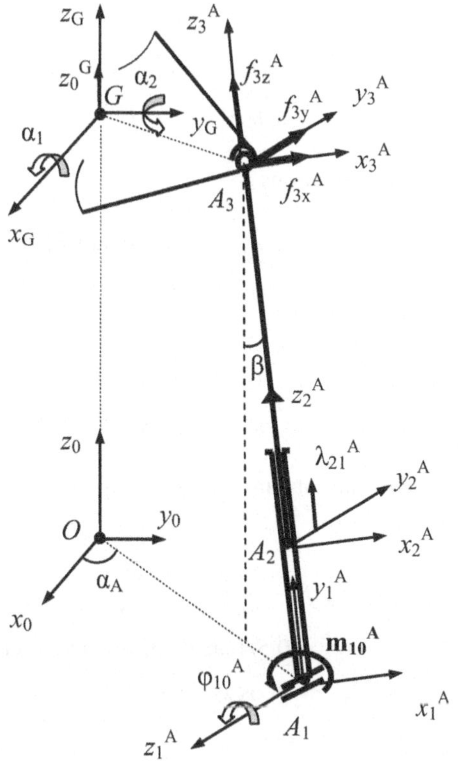

Fig. 9.25 Kinematical scheme of first leg A of the mechanism

We consider that the position of the 3-RPS mechanism is completely controlled in the inverse geometric problem through the independent variable z_0^G and the angles of rotation α_1, α_2. Six independent variables $\varphi_{10}^A, \lambda_{21}^A, \varphi_{10}^B, \lambda_{21}^B, \varphi_{10}^C, \lambda_{21}^C$ and three parameters α_3, x_0^G, y_0^G ca be determined by following vector-loop equations

$$\vec{r}_{10}^{\,j} + \sum_{k=1}^{2} p_{k0}^T \vec{r}_{k+1,k}^{\,j} = \vec{r}_0^{\,G} + R^T \vec{r}_G^{\,j3} \quad (p = a, b, c) \; (j = A, B, C), \tag{9.4.4}$$

with the notations

$$\begin{gathered} \vec{r}_{10}^{\,j} = l_0 a_\alpha^{jT} \vec{u}_1, \quad \vec{r}_{21}^{\,j} = (\lambda_0 + \lambda_{21}^j)\vec{u}_2, \quad \vec{r}_{32}^{\,j} = l_2 \vec{u}_3, \quad \vec{r}_G^{\,j3} = r a_\alpha^{jT} \vec{u}_1 \\ \vec{u}_1 = \begin{bmatrix} 1 \\ 0 \\ 0 \end{bmatrix}, \quad \vec{u}_2 = \begin{bmatrix} 0 \\ 1 \\ 0 \end{bmatrix}, \quad \vec{u}_3 = \begin{bmatrix} 0 \\ 0 \\ 1 \end{bmatrix}, \quad \tilde{u}_3 = \begin{bmatrix} 0 & -1 & 0 \\ 1 & 0 & 0 \\ 0 & 0 & 0 \end{bmatrix}, \quad \vec{r}_G^{\,j3} = \begin{bmatrix} x_G^{j3} \\ y_G^{j3} \\ z_G^{j3} \end{bmatrix}. \end{gathered} \tag{9.4.5}$$

These vector equations mean that there is only one inverse geometric solution for the manipulator:

$$\begin{aligned} -(\lambda_0 + l_2 + \lambda_{21}^j)\cos\alpha_j \sin(\varphi_{10}^j + \beta) &= x_0^G - x_{10}^j + q_1^j \\ -(\lambda_0 + l_2 + \lambda_{21}^j)\sin\alpha_j \sin(\varphi_{10}^j + \beta) &= y_0^G - y_{10}^j + q_2^j \\ (\lambda_0 + l_2 + \lambda_{21}^j)\cos(\varphi_{10}^j + \beta) &= z_0^G + q_3^j \quad (j = A, B, C), \end{aligned} \tag{9.4.6}$$

with the notations

$$\begin{aligned} q_1^j &= x_G^{j3}\cos\alpha_2\cos\alpha_3 - y_G^{j3}\cos\alpha_2\sin\alpha_3 \\ q_2^j &= x_G^j(\sin\alpha_1\sin\alpha_2\cos\alpha_3 + \cos\alpha_1\sin\alpha_3) \\ &\quad + y_G^{j3}(-\sin\alpha_1\sin\alpha_2\sin\alpha_3 + \cos\alpha_1\cos\alpha_3) \\ q_3^j &= x_G^{j3}(-\cos\alpha_1\sin\alpha_2\cos\alpha_3 + \sin\alpha_1\sin\alpha_3) \\ &\quad + y_G^{j3}(\cos\alpha_1\sin\alpha_2\sin\alpha_3 + \sin\alpha_1\cos\alpha_3). \end{aligned} \tag{9.4.7}$$

We can obtain the following commonly known *significant functions* from Eq. (9.4.6):

$$\begin{aligned} \tan\alpha_3 &= -\frac{\sin\alpha_1\sin\alpha_2}{\cos\alpha_1 + \cos\alpha_2} \\ x_0^G &= 0.5r(\sin\alpha_1\sin\alpha_2\sin\alpha_3 - \cos\alpha_1\cos\alpha_3 + \cos\alpha_2\cos\alpha_3) \\ y_0^G &= r\cos\alpha_2\sin\alpha_3. \end{aligned} \tag{9.4.8}$$

In kinematics, the motion of the elements of every leg (e.g. leg A) is characterized by the velocities $\vec{v}_{\tau 0}^{A}$, the accelerations $\vec{\gamma}_{\tau 0}^{A}$ of the joints A_{τ}, the angular velocities $\vec{\omega}_{\tau 0}^{A}$ and the angular accelerations $\vec{\varepsilon}_{\tau 0}^{A}$ as follows

$$\begin{aligned}
\vec{v}_{\tau 0}^{A} &= a_{\tau,\tau-1}\vec{v}_{\tau-1,0}^{A} + a_{\tau,\tau-1}\tilde{\omega}_{\tau-1,0}^{A}\vec{r}_{\tau,\tau-1}^{A} + v_{\tau,\tau-1}^{A}\vec{u}_3, \quad v_{10}^{A} = 0, \ v_{21}^{A} = \dot{\lambda}_{21}^{A} \\
\vec{\gamma}_{\tau 0}^{A} &= a_{\tau,\tau-1}\vec{\gamma}_{\tau-1,0}^{A} + a_{\tau,\tau-1}\{\tilde{\omega}_{\tau-1,0}^{A}\tilde{\omega}_{\tau-1.0}^{A} + \tilde{\varepsilon}_{\tau-1,0}^{A}\}\vec{r}_{\tau,\tau-1}^{A} \\
&\quad + 2v_{\tau,\tau-1}^{A}a_{\tau,\tau-1}\tilde{\omega}_{\tau-1,0}^{A}a_{\tau,\tau-1}^{T}\vec{u}_3 + \gamma_{\tau,\tau-1}^{A}\vec{u}_3, \quad \gamma_{10}^{A} = 0, \ \gamma_{21}^{A} = \ddot{\lambda}_{21}^{A} \\
\vec{\omega}_{\tau 0}^{A} &= a_{\tau,\tau-1}\vec{\omega}_{\tau-1,0}^{A} + \omega_{\tau,\tau-1}^{A}\vec{u}_3, \quad \omega_{10}^{A} = \dot{\varphi}_{10}^{A}, \ \omega_{21}^{A} = 0 \\
\vec{\varepsilon}_{\tau 0}^{A} &= a_{\tau,\tau-1}\vec{\varepsilon}_{\tau-1,0}^{A} + \varepsilon_{\tau,\tau-1}^{A}\vec{u}_3 + \omega_{\tau,\tau-1}^{A}a_{\tau,\tau-1}\tilde{\omega}_{\tau-1,0}^{A}a_{\tau,\tau-1}^{T}\vec{u}_3, \quad \varepsilon_{10}^{A} = \ddot{\varphi}_{10}^{A}, \ \varepsilon_{21}^{A} = 0 \\
&\quad (\tau = 1, 2)
\end{aligned} \tag{9.4.9}$$

In terms of the angular velocities $\dot{\alpha}_1, \dot{\alpha}_2$ and the vertical velocity $\dot{z}_0^G$ we compute the angular velocities $\dot{\alpha}_3, \omega_{10}^{A}, \omega_{10}^{B}, \omega_{10}^{C}$ and the linear velocities $\dot{x}_0^G, \dot{y}_0^G$, starting from following *matrix conditions of connectivity* [21]

$$\begin{aligned}
&\omega_{10}^{A}\vec{u}_i^{T}a_{10}^{T}\tilde{u}_3\vec{r}_{31}^{A} + v_{21}^{A}\vec{u}_i^{T}a_{10}^{T}\vec{u}_2 = \vec{u}_i^{T}\dot{\vec{r}}_0^{G} + \vec{u}_i^{T}R^{T}\tilde{\omega}_p\vec{r}_G^{A_3}, \quad \vec{r}_{31}^{A} = \vec{r}_{21}^{A} + a_{21}^{T}\vec{r}_{32}^{A} \\
&\qquad (i = 1, 2, 3)
\end{aligned} \tag{9.4.10}$$

where

$$\tilde{\omega}_p = \dot{\alpha}_1 R_{32}R_{21}\tilde{u}_3R_{21}^{T}R_{32}^{T} + \dot{\alpha}_2 R_{32}\tilde{u}_3R_{32}^{T} + \dot{\alpha}_3\tilde{u}_3 \tag{9.4.11}$$

denotes the skew-symmetric matrix associated with the angular velocity $\vec{\omega}_p$ of the moving platform.

Concerning the first leg A, the characteristic *virtual velocities* are expressed as functions of the position of the mechanism by the kinematical constraint equation (9.4.10), where we add the contributions of some virtual translations during three successive fictitious displacements of the distal spherical joint A_3 chosen in the directions x_3^A, y_3^A and z_3^A, respectively, as follows:

$$\begin{aligned}
&\omega_{10}^{Av}\vec{u}_i^{T}a_{10}^{T}\tilde{u}_3\vec{r}_{31}^{A} + v_{21}^{Av}\vec{u}_i^{T}a_{10}^{T}\vec{u}_2 + v_{3x}^{Av}\vec{u}_i^{T}a_{20}^{T}\vec{u}_1 + v_{3y}^{Av}\vec{u}_i^{T}a_{20}^{T}\vec{u}_2 + v_{3z}^{Av}\vec{u}_i^{T}a_{20}^{T}\vec{u}_3 = \\
&\vec{u}_i^{T}\vec{v}_0^{Gv} + \vec{u}_i^{T}R^{T}\tilde{\omega}_p^{v}\vec{r}_G^{A_3}, \quad (i = 1, 2, 3).
\end{aligned} \tag{9.4.12}$$

Let us assume that the robot has successively some *virtual motions* determined by the following four sets of velocities concerning the first leg A, for example:

$$
\begin{aligned}
&\omega_{10a}^{Av} = 1, \quad \omega_{10a}^{Bv} = 0, \quad \omega_{10a}^{Cv} = 0, \quad v_{3xa}^{jv} = 0, \quad v_{3ya}^{jv} = 0, \quad v_{3za}^{jv} = 0 \\
&\omega_{10a}^{jv} = 0, \quad v_{3xa}^{Av} = 1, \quad v_{3xa}^{Bv} = 0, \quad v_{3xa}^{Cv} = 0, \quad v_{3ya}^{jv} = 0, \quad v_{3za}^{jv} = 0 \\
&\omega_{10a}^{jv} = 0, \quad v_{3xa}^{jv} = 0, \quad v_{3ya}^{Av} = 1, \quad v_{3ya}^{Bv} = 0, \quad v_{3ya}^{Cv} = 0, \quad v_{3za}^{jv} = 0 \\
&\omega_{10a}^{jv} = 0, \quad v_{3xa}^{jv} = 0, \quad v_{3ya}^{jv} = 0, \quad v_{3za}^{Av} = 1, \quad v_{3za}^{Bv} = 0, \quad v_{3za}^{Cv} = 0. \\
&\quad (j = A, B, C)
\end{aligned}
\tag{9.4.13}
$$

These virtual velocities are required to be fed into the computation of the virtual work of all the forces applied to the component elements of the mechanism.

As for the angular accelerations $\ddot{\alpha}_3, \varepsilon_{10}^A, \varepsilon_{10}^B, \varepsilon_{10}^C$ and the linear accelerations $\ddot{x}_0^G$, $\ddot{y}_0^G, \gamma_{21}^A, \gamma_{21}^B, \gamma_{21}^C$, the derivatives with respect to time of Eq. (9.4.10) and the square matrix (9.3.15) give new conditions of connectivity

$$
\begin{aligned}
&\varepsilon_{10}^A \vec{u}_i^T a_{10}^T \tilde{u}_3 \vec{r}_{31}^A + \gamma_{21}^A \vec{u}_i^T a_{10}^T \vec{u}_2 \\
&\quad = \vec{u}_i^T \ddot{\vec{r}}_0^G + \vec{u}_i^T R_{30}^T \{\tilde{\omega}_p \tilde{\omega}_p + \dot{\tilde{\omega}}_p\} \vec{r}_G^{A_3} \\
&\qquad - \omega_{10}^A \omega_{10}^A \vec{u}_i^T a_{10}^T \tilde{u}_3 \tilde{u}_3 \vec{r}_{31}^A - 2\omega_{10}^A v_{21}^A \vec{u}_i^T a_{10}^T \tilde{u}_3 \vec{u}_2. \\
&\qquad (i = 1, 2, 3)
\end{aligned}
\tag{9.4.14}
$$

9.4.2 Dynamics Modelling

Knowing the position and kinematic state of each link as well as the external forces acting on the 3-RPS parallel mechanism, we apply first the principle of virtual work for the inverse dynamic problem in order to establish some definitive recursive matrix relations concerning the time-history evolution for input torques of the three actuators, but also for some internal reaction forces in joints.

9.4.2.1 Principle of Virtual Work

The spatial evolution of the moving platform is controlled by three electric motors A_1, B_1, C_1 that generate three couples of moments $\vec{m}_{10}^A = m_{10}^A \vec{u}_3$, $\vec{m}_{10}^B = m_{10}^B \vec{u}_3$, $\vec{m}_{10}^C = m_{10}^C \vec{u}_3$.

The force of inertia and the resulting moment of inertia forces of a rigid body T_k^A

$$
\begin{aligned}
\vec{f}_k^{in\,A} &= -m_k^A [\vec{\gamma}_{k0}^A + (\tilde{\omega}_{k0}^A \tilde{\omega}_{k0}^A + \tilde{\varepsilon}_{k0}^A) \vec{r}_k^{CA}], \\
\vec{m}_k^{in\,A} &= -[m_k^A \tilde{r}_k^{CA} \vec{\gamma}_{k0}^A + \hat{J}_k^A \vec{\varepsilon}_{k0}^A + \tilde{\omega}_{k0}^A \hat{J}_k^A \vec{\omega}_{k0}^A]
\end{aligned}
\tag{9.4.15}
$$

are computed with respect to the centre of joint A_k. On the other hand, the wrench of two vectors $\vec{f}_k^{*A} = -g m_k^A a_{k0} \vec{u}_3$ and $\vec{m}_k^{*A} = -g m_k^A \tilde{r}_k^{CA} a_{k0} \vec{u}_3$ evaluates the action of the weight $m_k^A \vec{g}$ on same element T_k^A, for example. The above recursive relations (9.3.19) can generate the vectors $\vec{F}_k^A$ and $\vec{M}_k^A$.

Using the same equivalent graph from Fig. 8.19 for four independent trees of the spatial robot, we can apply *the fundamental equations of the parallel robots dynamics* (7.18) to obtain compact matrix relations for the *torque* of first active revolute joint

$$
\begin{aligned}
m_{10}^{A} = \vec{u}_3^{T}\{ & \vec{M}_1^{A} + v_{21a}^{Av}\vec{F}_2^{A} + v_{21a}^{Bv}\vec{F}_2^{B} + v_{21a}^{Cv}\vec{F}_2^{C} \\
& + v_{10a}^{Gv}\vec{F}_1^{G} + v_{21a}^{Gv}\vec{F}_2^{G} + v_{32a}^{Gv}\vec{F}_3^{G} + \omega_{10a}^{Gv}\vec{M}_1^{G} + \omega_{21a}^{Gv}\vec{M}_2^{G} + \omega_{32a}^{Gv}\vec{M}_3^{G}\}
\end{aligned}
\tag{9.4.16}
$$

and for one of three *internal joint forces* $f_{3x}^{A}, f_{3y}^{A}, f_{3z}^{A}$ acting along three orthogonal axes

$$
\begin{aligned}
f_3^{A} = \vec{u}_3^{T}\{ & v_{21a}^{Av}\vec{F}_2^{A} + v_{21a}^{Bv}\vec{F}_2^{B} + v_{21a}^{Cv}\vec{F}_2^{C} \\
& + v_{10a}^{Gv}\vec{F}_1^{G} + v_{21a}^{Gv}\vec{F}_2^{G} + v_{32a}^{Gv}\vec{F}_3^{G} + \omega_{10a}^{Gv}\vec{M}_1^{G} + \omega_{21a}^{Gv}\vec{M}_2^{G} + \omega_{32a}^{Gv}\vec{M}_3^{G}\},
\end{aligned}
\tag{9.4.17}
$$

where the virtual velocities v_{21a}^{Av}, v_{21a}^{Bv}, v_{21a}^{Cv}, v_{10a}^{Gv}, v_{21a}^{Gv}, v_{32a}^{Gv}, ω_{10a}^{Gv}, ω_{21a}^{Gv}, ω_{32a}^{Gv} are computed successively starting from the conditions (9.4.12) and (9.4.13).

Partial verification of the above (9.4.17) results can quickly be obtained, establishing some equations based on the Newton-Euler procedure, as follows:

$$
\begin{aligned}
f_{3x}^{A} &= m_{10}^{A} + (0.5m_1l_1 + m_2d_{21}^{A})g\sin(\varphi_{10}^{A} + \beta) - m_2(d_{21}^{A})^2\ddot{\varphi}_{10}^{A} \\
&\quad - 2m_2d_{21}^{A}\dot{\varphi}_{10}^{A}\dot{\lambda}_{21}^{A} - (J_1^{z} + J_2^{y} - 0.25m_2l_2^2)\ddot{\varphi}_{10}^{A} \\
f_{3z}^{A} &= m_2d_{21}^{A}\dot{\varphi}_{10}^{2} - m_2g\cos(\varphi_{10}^{A} + \beta) - m_2\ddot{\lambda}_{21}^{A}, \quad d_{21}^{A} = 0.5l_2 + \lambda_0 + \lambda_{21}^{A}.
\end{aligned}
\tag{9.4.18}
$$

9.4.2.2 Lagrange Equations

A solution to the dynamics problem of a 3-RPS parallel robot can be developed based on the Lagrange equations of the second kind. The generalized coordinates are represented by 12 parameters $s_1 = x_0^G$, $s_2 = y_0^G$, $s_3 = z_0^G$, $s_4 = \alpha_1$, $s_5 = \alpha_2$, $s_6 = \alpha_3$, $s_7 = \varphi_{10}^A$, $s_8 = \lambda_{21}^A$, $s_9 = \varphi_{10}^B$, $s_{10} = \lambda_{21}^B$, $s_{11} = \varphi_{10}^C$, $s_{12} = \lambda_{21}^C$.

Lagrange's equations with their nine multipliers $\lambda_1, \lambda_2, \ldots, \lambda_9$ will be expressed by 12 differential relations

$$
\frac{d}{dt}\left\{\frac{\partial L}{\partial \dot{s}_l}\right\} - \frac{\partial L}{\partial s_l} = \Phi_l + \sum_{v=1}^{9}\lambda_v c_{vl} \quad (l = 1, 2, \ldots, 12),
\tag{9.4.19}
$$

which contain following 12 generalized forces $\Phi_1 = 0$, $\Phi_2 = 0$, $\Phi_3 = 0$, $\Phi_4 = 0$, $\Phi_5 = 0$, $\Phi_6 = 0$, $\Phi_7 = m_{10}^A$, $\Phi_8 = 0$, $\Phi_9 = m_{10}^B$, $\Phi_{10} = 0$, $\Phi_{11} = m_{10}^C$, $\Phi_{12} = 0$.

A total of nine kinematical conditions of constraint are given by the relations (9.4.10):

$$\sum_{l=1}^{12} c_{\nu l}\dot{s}_l = 0 \quad (\nu = 1, 2, \ldots, 9). \tag{9.4.20}$$

The components of the general expression of the Lagrange function $L = L_p + \sum_{\mu=1}^{2}(L_\mu^A + L_\mu^B + L_\mu^C)$ are given as analytical functions of the generalized coordinates and their first derivatives with respect to time:

$$\begin{aligned}
L_p &= \frac{1}{2}m_p(\dot{x}_0^{G2} + \dot{y}_0^{G2} + \dot{z}_0^{G2}) + \frac{1}{2}\vec{\omega}_p^T \hat{J}_p \vec{\omega}_p - m_p g z_0^G \\
L_1^j &= \frac{1}{2}\vec{\omega}_{10}^{jT}\hat{J}_1^j\vec{\omega}_{10}^j - m_1^j g \vec{u}_3^T\{\vec{r}_{10}^j + p_{10}^T \vec{r}_1^{Cj}\} \\
L_2^j &= \frac{1}{2}m_2^j \vec{v}_{20}^{jT}\vec{v}_{20}^j + \frac{1}{2}\vec{\omega}_{20}^{jT}\hat{J}_2^j\vec{\omega}_{20}^j + m_2^j \vec{v}_{20}^{jT}\tilde{\omega}_{20}^j \vec{r}_2^{Cj} - m_2^{ji} g \vec{u}_3^T\{\vec{r}_{10}^j + p_{10}^T \vec{r}_{21}^j + p_{20}^T \vec{r}_2^{Cj}\} \\
&\quad j = (A, B, C), \;\; p = (a, b, c).
\end{aligned} \tag{9.4.21}$$

Angular velocities, joint velocities, skew-symmetric matrices associated with the angular velocities and first derivatives of orthogonal matrices $p_{\tau,\tau-1}$ are expressed as follows:

$$\begin{aligned}
&\vec{\omega}_{10}^j = \dot{\varphi}_{10}^j \vec{u}_3, \quad \vec{\omega}_{20}^j = p_{21}\vec{\omega}_{10}^j, \quad \tilde{\omega}_{10}^j = \dot{\varphi}_{10}^j \tilde{u}_3, \quad \tilde{\omega}_{20}^j = p_{21}\tilde{\omega}_{10}^j p_{21}^T \\
&\vec{v}_{10}^j = \vec{0}, \quad \vec{v}_{20}^j = p_{21}\tilde{\omega}_{10}^j \vec{r}_{21}^j + \dot{\lambda}_{21}^j \vec{u}_3, \quad j = (A, B, C) \\
&\dot{p}_{\tau,\tau-1} = \dot{\varphi}_{\tau,\tau-1}^i \tilde{u}_3^T p_{\tau,\tau-1}, \quad \dot{p}_{\tau,\tau-1}^T = \dot{\varphi}_{\tau,\tau-1}^i p_{\tau,\tau-1}^T \tilde{u}_3 \\
&\frac{\partial p_{\tau,\tau-1}}{\partial \varphi_{\tau,\tau-1}^i} = \tilde{u}_3^T p_{\tau,\tau-1}, \quad \frac{\partial p_{\tau,\tau-1}^T}{\partial \varphi_{\tau,\tau-1}^i} = p_{\tau,\tau-1}^T \tilde{u}_3, \quad p = (a, b, c) \;\; (\tau = 1, 2).
\end{aligned} \tag{9.4.22}$$

The calculus of the derivatives with respect to time $\frac{d}{dt}\left\{\frac{\partial L}{\partial \dot{s}_l}\right\}$ $(l = 1, 2, \ldots, 12)$ of some of the above functions leads to a system of 12 relations. In the direct or inverse dynamics problem, after elimination of the nine multipliers, finally we obtain the same expressions (9.4.16) for the three input torques.

In the conditions of a dynamic simulation, furthermore, we suppose that during three seconds following analytical functions can describe the general absolute motion of the moving platform

$$\frac{\alpha_1}{\alpha_1^*} = \frac{\alpha_2}{\alpha_2^*} = \frac{h - z_0^G}{z_0^{G*}} = 1 - \cos\frac{\pi}{3}t. \tag{9.4.23}$$

As an application let us consider a manipulator which has the following architectural and mechanical characteristics

$$z_0^{G*} = 0.1\,\text{m}, \quad \alpha_1^* = \alpha_2^* = \frac{\pi}{18}, \quad l_0 = 1\,\text{m}, \quad r = 0.5\,\text{m}, \quad l = r\sqrt{3}, \quad h = 1.2\,\text{m}$$
$$\lambda_0 = 0.35\,\text{m}, \quad l_1 = 0.7\,\text{m}, \quad m_1 = m_2 = 0.5\,\text{kg}, \quad m_p = 5\,\text{kg}, \quad \Delta t = 3\,\text{s}.$$

Based on the computation of the input powers $P_{10}^j = \omega_{10}^j m_{10}^j$ and the orthogonal internal joint forces $f_{3x}^j, f_{3y}^j, f_{3z}^j$, a computer program in MATLAB was developed to solve the inverse dynamics.

For the first example, when the platform moves along the vertical z_0 direction, all powers P_{10}^j and internal joint forces f_{3x}^j, f_{3z}^j are permanently equal to one another and the second joint force vanishes during this symmetrical evolution (Figs. 9.26, 9.27 and 9.28).

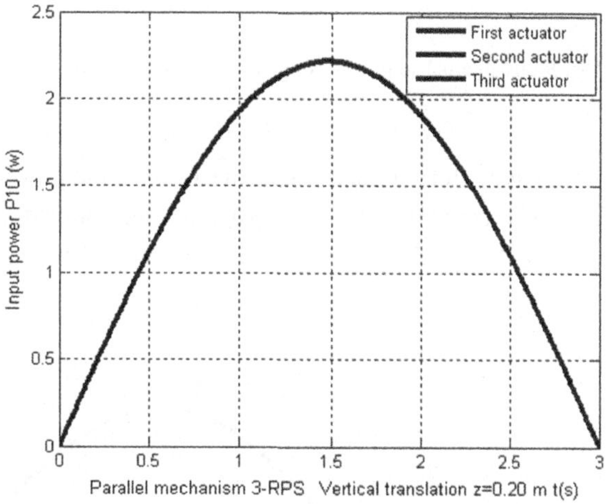

Fig. 9.26 Powers P_{10}^j of three actuators (translation)

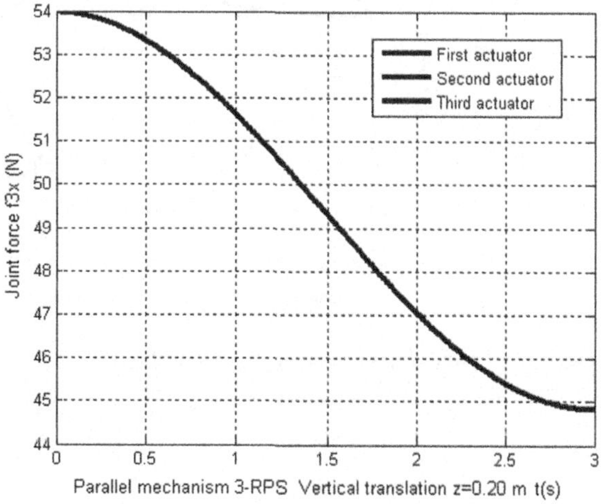

Fig. 9.27 Internal joint forces f_{3x}^j from legs (translation)

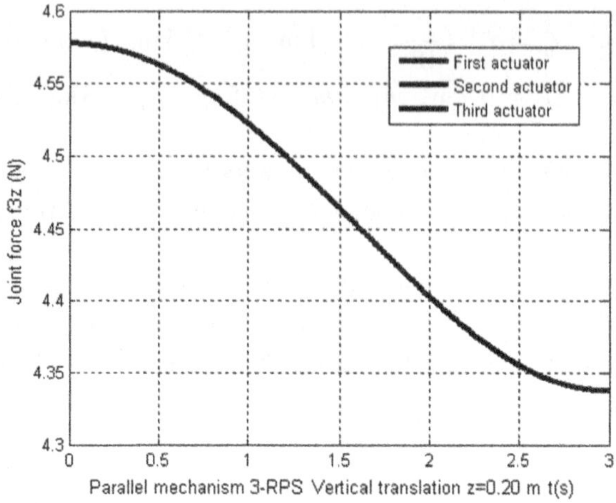

Fig. 9.28 Internal joint forces f_{3z}^j from legs (translation)

If the platform rotates about the x_G axis, the powers and the internal joint forces are also computed by the program and plotted against time as follows P_{10}^j (Fig. 9.29), f_{3x}^j (Fig. 9.30), f_{3y}^j (Fig. 9.31) and f_{3z}^j (Fig. 9.32).

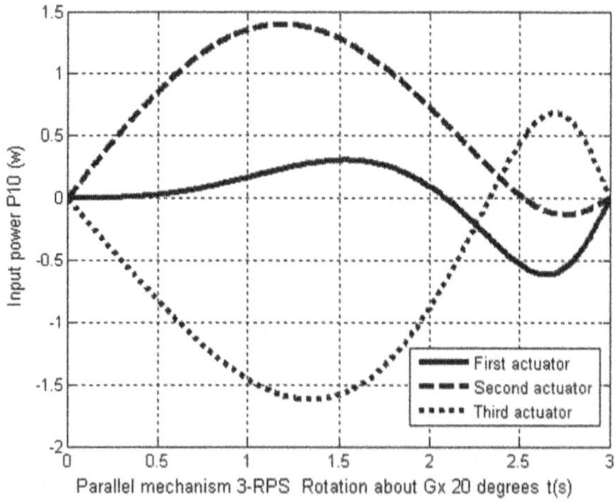

Fig. 9.29 Powers P_{10}^j of three actuators (rotation)

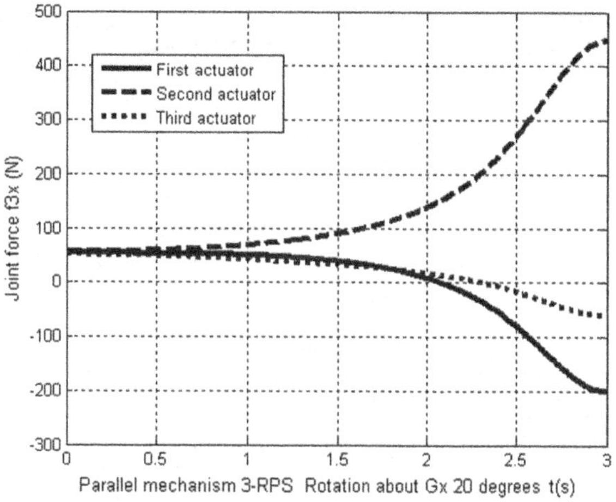

Fig. 9.30 Internal joint forces f_{3x}^{j} from legs (rotation)

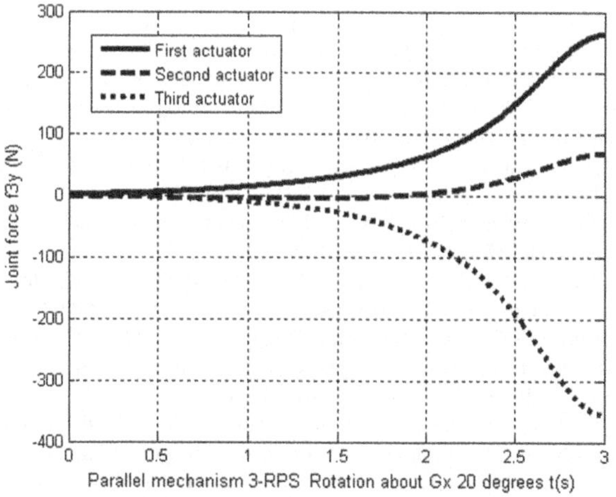

Fig. 9.31 Internal joint forces f_{3y}^{j} from legs (rotation)

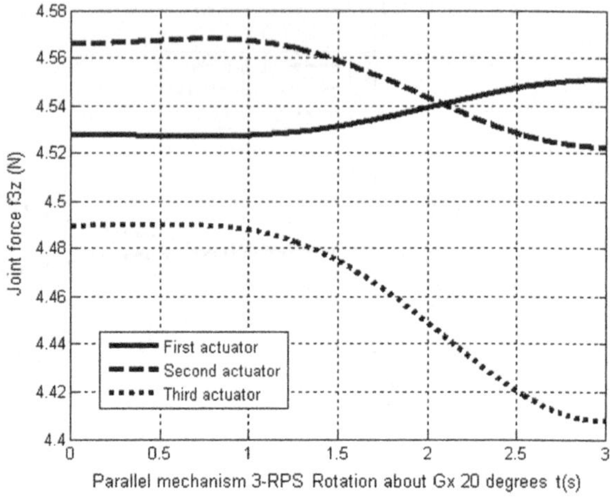

Fig. 9.32 Internal joint forces f_{3z}^{j} from legs (rotation)

References

1. Tsai, L.-W.: Robot Analysis: The Mechanics of Serial and Parallel Manipulators. Wiley, New York (1999)
2. Clavel, R.: Delta: a fast robot with parallel geometry. In: Proceedings of 18th International Symposium on Industrial Robots, Sydney, pp. 91–100 (1988)
3. Tsai, L.-W., Stamper, R.: A parallel manipulator with only translational degrees of freedom. In: Proceedings of ASME Design Engineering Technical Conferences, Irvine, CA (1996)
4. Chablat, D., Wenger, P., Staicu, S.: Dynamics of the orthoglide parallel robot. UPB Sci. Bull. Ser. D: Mech. Eng. **71**(3), 3–16 (2009)
5. Staicu, S.: Dynamics analysis of the Star parallel manipulator. Robot. Auton. Syst. **57**(11), 1057–1064 (2009)
6. Staicu, S.: Matrix modeling of inverse dynamics of spatial and planar parallel robots. Multibody Syst. Dyn. **27**(2), 239–265 (2012)
7. Li, Y.-W., Wang, J., Wang, L.-P., Liu, X.-J.: Inverse dynamics and simulation of a 3-DOF spatial parallel manipulator. In: Proceedings of the IEEE International Conference on Robotics & Automation, ICRA'2003, Taipei, vol. 3, pp. 4092–4097 (2003)
8. Carricato, M., Parenti-Castelli, V.: Singularity-free fully-isotropic translational parallel mechanisms. Int. J. Robot. Res. **21**(2), 161–164 (2002)
9. Chablat, D., Wenger, P.: Architecture optimization of a 3-DOF parallel mechanism for machining applications: the Orthoglide. IEEE Trans. Robot. Autom. **19**(3), 403–410 (2003)
10. Ibrahim, O., Khalil, W.: Inverse and direct dynamic models of hybrid robots. Mech. Mach. Theory **45**(4), 627–640 (2010)
11. Chen, G., Yu, W., Li, Q., Wang, H.: Dynamic modeling and performance analysis of the 3-PRRU 1T2R parallel manipulator without parasitic motion. Nonlinear Dyn. **90**(1), 339–353 (2017)
12. Dasgupta, B., Mruthyunjaya, T.S.: A Newton-Euler formulation for the inverse dynamics of the Stewart platform manipulator. Mech. Mach. Theory **33**(8), 1135–1152 (1998)
13. Li, Y., Xu, Q.: Kinematics and inverse dynamics analysis for a general 3-PRS spatial parallel mechanism. Robotica **23**(2), 219–229 (2005)
14. Miller, K., Clavel, R.: The Lagrange-based model of Delta-4 robot dynamics. Robotersysteme **8**, 49–54 (1992)

15. Xin, G., Deng, H., Zhong, G.: Closed-form dynamics of a 3-DOF spatial parallel manipulator by combining the lagrangian formulation with the virtual work principle. Nonlinear Dyn. **86** (2), 1329–1347 (2016)
16. Staicu, S., Zhang, D.: A novel dynamic modelling approach for parallel mechanisms analysis. Robot. Comput.-Integr. Manuf. **24**(1), 167–172 (2008)
17. Tsai, L.-W.: Solving the inverse dynamics of Stewart-Gough manipulator by the principle of virtual work. J. Mech. Des. **122**(1), 3–9 (2000)
18. Zhang, C.-D., Song, S.-M.: An efficient method for inverse dynamics of manipulators based on virtual work principle. J. Robot. Syst. **10**(5), 605–627 (1993)
19. Angeles, J.: Fundamentals of Robotic Mechanical Systems: Theory, Methods and Algorithms. Springer, New York (2002)
20. Gosselin, C., Angeles, J.: The optimum kinematic design of a spherical three-degree-of-freedom parallel manipulator. ASME J. Mech. Transm. Automat. Des. **111**(2), 202–207 (1989)
21. Staicu, S., Liu, X.-J., Wang, J.: Inverse dynamics of the HALF parallel manipulator with revolute actuators. Nonlinear Dyn. **50**(1–2), 1–12 (2007)
22. Staicu, S.: Dynamics of the spherical 3-U$\underline{P}$S/S parallel mechanism with prismatic actuators. Multibody Syst. Dyn. **22**(2), 115–132 (2009)
23. Wang, J., Gosselin, C.: A new approach for the dynamic analysis of parallel manipulators. Multibody Syst. Dyn. **2**(3), 317–334 (1998)
24. Khalil, W., Ibrahim, O.: General solution for the dynamic modeling of parallel robots. J. Intell. Robot. Sys. **49**(1), 19–37 (2007)
25. Baron, L., Angeles, J.: The direct kinematics of parallel manipulators under joint-sensor redundancy. IEEE Trans. Robot. Autom. **16**(1), 12–19 (2000)
26. Stewart, D.: A platform with six degrees of freedom. Proc. Inst. Mech. Eng., Part. I **180**(15), 371–386 (1965)
27. Liu, X.-J., Jeong, J., Kim, J.: A three translational DOFs parallel cube-manipulator. Robotica **21**(6), 645–653 (2003)
28. Xi, F., Angelico, O., Sinatra, R.: Tripod dynamics and its inertia effects. J. Mech. Des. **127**(1), 144–149 (2005)
29. Li, Y., Staicu, S.: Inverse dynamics of a 3-$\underline{P}$RC parallel kinematic machine. Nonlinear Dyn. **67**(2), 1031–1041 (2012)
30. Cheng, G., Shan, X.: Dynamics analysis of a parallel hip-joint simulator with four degree of freedoms (3R1T). Nonlinear Dyn. **70**(4), 2475–2486 (2012)
31. Wang, Z., Zhang, N., Chai, X., Li, Q.: Kinematic/dynamic analysis and optimization of a 2-URR-RRU parallel manipulator. Nonlinear Dyn. **88**(1), 503–519 (2017)
32. Kalani, H., Rezaei, A., Akbarzadeh, A.: Improved general solution for the dynamic modeling of Gough-Stewart platform based on principle of virtual work. Nonlinear Dyn. **83**(4), 2393–2418 (2016)
33. Geng, Z., Haynes, L.S., Lee, J.D., Carroll, R.L.: On the dynamic model and kinematic analysis of a class of Stewart platforms. Robot. Auton. Syst. **9**(4), 237–254 (1992)
34. Didrit, O., Petitot, M., Walter, E.: Guaranteed solution of direct kinematic problems for general configurations of parallel manipulator. IEEE Trans. Robot. Autom. **14**(2), 259–266 (1998)
35. Yuan, W., Tsai, M.: A novel approach for forward dynamic analysis of 3-PRS parallel manipulator with consideration of friction effect. Robot. Comput.-Integr. Manuf. **30**(3), 315–325 (2014)
36. Salinic, S.: Determination of joint reaction forces in a symbolic form in rigid multibody systems. Mech. Mach. Theory **46**(11), 1796–1810 (2011)

Chapter 10
Geared Parallel Mechanisms

Gears are widely used in machines for the purpose of transmission of rotary or rectilinear motion from one shaft to another. They are employed in a variety of industrial and technological applications such as automobiles, tractors, helicopter motor systems, machine tools, aircrafts, robotic mechanisms and others. In geared systems, the power can be transmitted from one body to another if the components have rolling contact. Rotation motion or translation motion is transmitted by the successive engagement of teeth, so that the two surfaces in contact roll theoretically without slipping and the velocities of components at the common point of contact are equal.

Spur gears are formed if the teeth are cut in a direction parallel to the axis of rotation. If the teeth-cutting is formed along a helix generated around the axis of the gear, we have *helical gear*. These two geared systems are used to transmit power between two parallel axes. Another type of gears is *bevel gears*, which are cut from cones and used in the case of intersecting shafts. In a *gear pair*, the relationship between the angular velocity of the driven gear and that of the driver gear depends on the diameters of the *base circles* of two components.

Industrial robots with orienting gear trains can perform several operations such as welding, flame cutting, spray painting, milling or assembling. Being comparatively simple and compact in size, bevel-gear wrist mechanisms can be sealed in a metallic box that keeps the device free of contamination. Furthermore, by using bevel-gear trains for power transmission, the actuators can be mounted remotely on the forearm, thereby reducing the weight and inertia of a robot manipulator.

S. Staicu, *Dynamics of Parallel Robots*, Parallel Robots: Theory and Applications,
https://doi.org/10.1007/978-3-319-99522-9_10

10.1 Kinematics and Dynamics Analysis of the Minuteman Cover Drive

Epicyclical gear trains are incorporated into the structure of industrial robots and have one, two or three output rotations. Including in their architecture both conical and cylindrical-toothed elements, the input axes of these mechanisms are parallel while the output axes can be nonparallel.

Gear drives are commonly used for speed reduction and torque amplification in mechanical systems. Numerous methods for kinematics analysis of epicyclical gear trains have been proposed by several researchers. Planetary gear trains with three degrees of freedom are adopted as the design concept for robotic wrist (Hsieh and Sheu [1], Ma and Gupta [2], Paul and Stevenson [3], White [4], Willis [5]).

The mechanism topology of the Minuteman cover drive consists of four moving links, four turning pairs and three gear pairs. Knowing the rotation motion of the output link, the inverse dynamic problem is solved using an approach based on the principle of virtual work, but it has been verified the results in the framework of the Lagrange equations. Finally, some simulation graphs for the input and output angles of rotation, the torque and the power of the actuator are obtained.

10.1.1 Inverse Kinematics Model

Recursive relations for kinematics of a one-degree-of-freedom orienting gear train will be developed in the following. First, we wish to find the overall speed reduction ratio of this mechanism. A matrix methodology for the kinematics analysis based on the concept of fundamental circuit of an open-loop chain is presented. This method involves the identification of all open-loop chains and the derivation of the geometric relationships between the orientation of the output link and the joint angles of the kinematical chains, including the input actuator displacements [6–8].

Let $O_0x_0y_0z_0(T_0)$ be a fixed Cartesian orthogonal frame, about which the mechanism moves. In the Minuteman cover drive, the first ring gear 0 is fixed to the ground, the sun gear $1a$ of radius r_1, mass m_1 and tensor of inertia $\hat{J}_1$ is the input link connected to link 0, while the moving ring gear $1b$ of mass m_4 and tensor of inertia $\hat{J}_4$ serves as output member. The compound planet gear $3a = 3b = 2c$ meshes with sun gear $1a$ as do two ring gears and is supported with a revolute joint by the carrier $2a = 2b = 1c$ as a connected coupling shaft of h in height, mass m_2 and tensor of inertia $\hat{J}_2$. The central body $3a$, of mass m_3 and tensor of inertia $\hat{J}_3$ is adjacent to carrier $2a$ and consists of two cylindrical gears of radius r_2, r_3, respectively. Otherwise, the reduced bi-coupled transmission becomes a simply one-DOF compounding planetary gear train (Fig. 10.1).

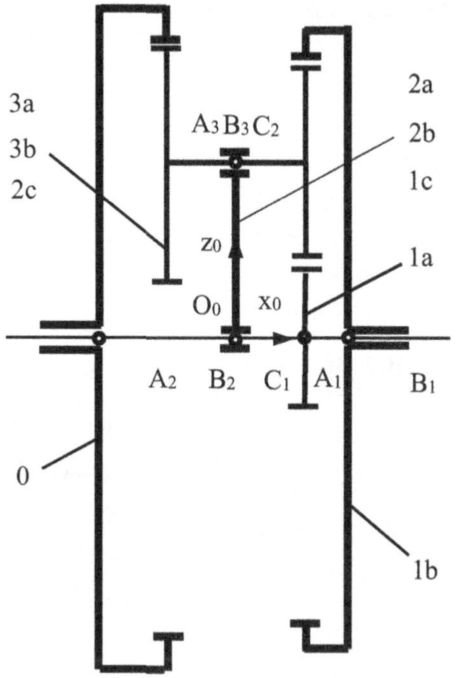

Fig. 10.1 Minuteman cover drive

Applying the method of successive displacements to geometric analysis of closed-loop chains, we call $a^{\varphi}_{k,k-1}$ the orthogonal transformation 3×3 matrix of relative rotation with the angle $\varphi^{A}_{k,k-1}$ of link T^{A}_{k} around z^{A}_{k} axis, for example. Pursuing the matrix approach which utilizes the *theory of fundamental circuits* developed by Tsai [9], there is a *real* or *fictitious* carrier for every gear pair in a *planetary spur gear train* and consequently a fundamental matrix equation for each loop can be written:

$$q_{k+1,k-1} = q^{\varphi}_{k+1,k} q^{\varphi}_{k,k-1}, \quad \varphi_{k,k-1} = N_{k+1,k-1}\varphi_{k+1,k} \; (k = 1, 2, 3), \qquad (10.1.1)$$

where $\varphi_{k,k-1}$ and $\varphi_{k+1,k}$ denote two successive relative angles of rotation of the carrier T_k and the planet gear T_{k+1}, respectively. The gear ratios of a gear pair is defined as

$$N_{k+1,k-1} = r_{k+1}/r_{k-1} = Z_{k+1}/Z_{k-1}, \qquad (10.1.2)$$

where r_{k-1}, r_{k+1} and Z_{k-1}, Z_{k+1} are the radius and the number of teeth of two gears, respectively (Fig. 10.2).

We consider the rotation angle φ^{A}_{10} of the actuator A_1 as a single variable which can give the instantaneous position of the mechanism (Fig. 10.3). Pursuing three kinematical chains $0 - 1a - 2a - 3a$, $0 - 1b - 2b - 3b$, $0 - 1c - 2c$, we obtain the following successive matrices of transformation [10]:

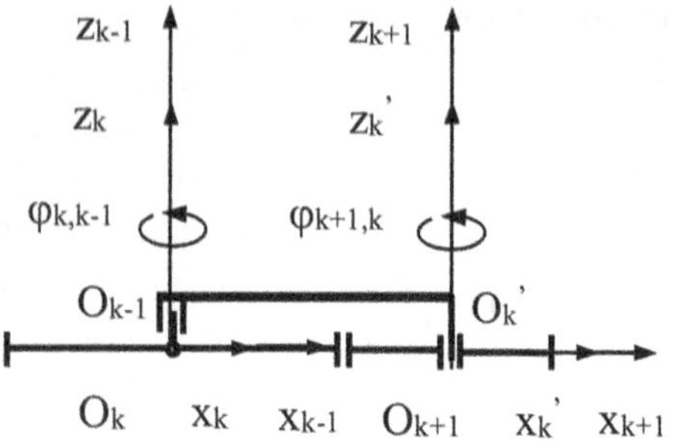

Fig. 10.2 Gear fundamental circuit

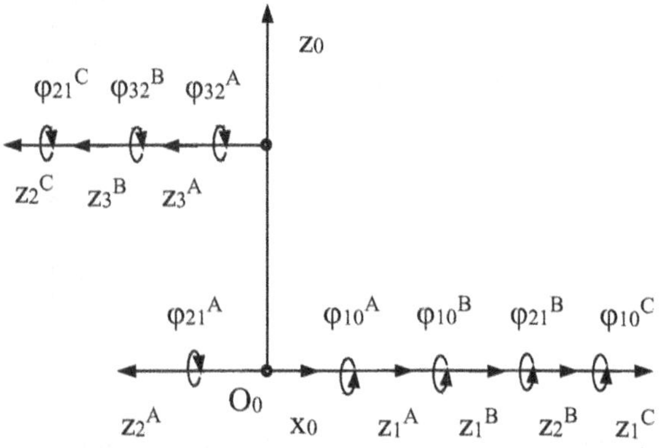

Fig. 10.3 Kinematical scheme of the mechanism

$$
\begin{aligned}
&a_{10} = a_{10}^{\varphi}\theta_1, \quad a_{21} = a_{21}^{\varphi}\theta_2, \quad a_{32} = a_{32}^{\varphi} \\
&b_{10} = b_{10}^{\varphi}\theta_1, \quad b_{21} = b_{21}^{\varphi}, \quad b_{32} = b_{32}^{\varphi}\theta_2 \\
&c_{10} = c_{10}^{\varphi}\theta_1, \quad c_{21} = c_{21}^{\varphi}\theta_2,
\end{aligned}
\tag{10.1.3}
$$

with the notations

$$
\begin{aligned}
&q_{k,k-1}^{\varphi} = rot(z, \varphi_{k,k-1}^{i}), \quad q_{k0} = \prod_{s=1}^{k} q_{k-s+1,k-s}, \quad (q = a, b, c), \quad (i = A, B, C) \\
&\theta_1 = rot\left(y, \frac{\pi}{2}\right), \quad \theta_2 = rot(y, \pi).
\end{aligned}
\tag{10.1.4}
$$

Let us suppose that the absolute motion of the end-effector attached at the output link $1b$ is a rotation of angle expressed by the analytical function

$$
\varphi_{10}^{B} = \phi = \phi_0\left[1 - \cos\left(\frac{\pi}{6}t\right)\right], \tag{10.1.5}
$$

where $2\phi_0$ is a parameter characterizing the final position of the end-effector.

The rotation conditions for the central planet gear $3a = 3b = 2c$ and some constraint geometric conditions established along the above three kinematical chains will be expressed by matrix equations

$$a_{30} = b_{30} = c_{20}, \quad a_{20}^T \vec{r}_{32}^{\,A} = b_{20}^T \vec{r}_{32}^{\,B} = c_{10}^T \vec{r}_{21}^{\,C}, \tag{10.1.6}$$

where

$$\begin{gathered}\vec{r}_{32}^{\,A} = -\vec{r}_{32}^{\,B} = -\vec{r}_{21}^{\,C} = r\vec{u}_1\, r = r_1 + r_2 \\ \vec{u}_1 = \begin{bmatrix}1\\0\\0\end{bmatrix}, \quad \vec{u}_2 = \begin{bmatrix}0\\1\\0\end{bmatrix}, \quad \vec{u}_3 = \begin{bmatrix}0\\0\\1\end{bmatrix}, \quad \tilde{u}_3 = \begin{bmatrix}0 & -1 & 0\\1 & 0 & 0\\0 & 0 & 0\end{bmatrix}.\end{gathered} \tag{10.1.7}$$

From Eq. (10.1.1) and the conditions (10.1.6), real-time evolution of all characteristic joint angles is easily obtained, as follows:

$$\begin{gathered}\varphi_{10}^B = \phi, \quad \varphi_{32}^A = \varphi_{32}^B = \varphi_{21}^C = \frac{\phi}{n_3 - n_2}, \quad \varphi_{10}^A = (n_1 + n_3)\varphi_{21}^C \\ \varphi_{21}^A = n_1\varphi_{21}^C, \quad \varphi_{21}^B = n_2\varphi_{21}^C, \quad \varphi_{10}^C = n_3\varphi_{21}^C \\ n_1 = \frac{r_2}{r_1}, \quad n_2 = \frac{r_2}{r_1 + 2r_2}, \quad n_3 = \frac{r_3}{r_1 + r_2 + r_3}.\end{gathered} \tag{10.1.8}$$

In the design of power transmission mechanisms such as speed reducers or automotive transmissions, it is necessary to analyze the speed ratios between their input and output members and, sometimes, angular velocities or angular accelerations of all intermediate members.

The conventional tabular or analytical method, which concentrates on planar epicyclical gear trains, is no longer applicable. To overcome this difficulty, Freudenstein et al. [11] applied the dual relative velocity and dual matrix of transformation for the analysis of epicyclical bevel-gear trains. The most straightforward approaches make apply of the theory of fundamental circuits used by Chang and Tsai [12], Hedman [13], Tsai et al. [8] and show that the kinematical analysis of geared robotic mechanisms can be accomplished by pursuing this systematic method.

Since a kinematical chain is an assemblage of links and joints, these can be symbolized in a more abstract form known as equivalent graph representation (Fig. 10.4). So, we use the equivalent *associated graph* to represent the topology of the mechanism and identify all fundamental circuits.

In the kinematical graph representation, we denote the links by vertices and the joints by edges (Yan and Hsieh [14, 15]). Two small concentric circles label the vertex denoting the fixed link 0. To distinguish between the pairs of connections, the gear pairs $1a$–$3a$, $1b$–$3b$, 0–$2c$ are designated by thick edges and the revolute joints 0 $1a$, $1a$–$2a$, $2a$–$3a$, $1a$–$1b$ by thin edges. Four-edged paths, which start from the base link 0 and end at the central link $3a = 3b = 2c$ consist of the vertices

Fig. 10.4 Associated graph of the mechanism

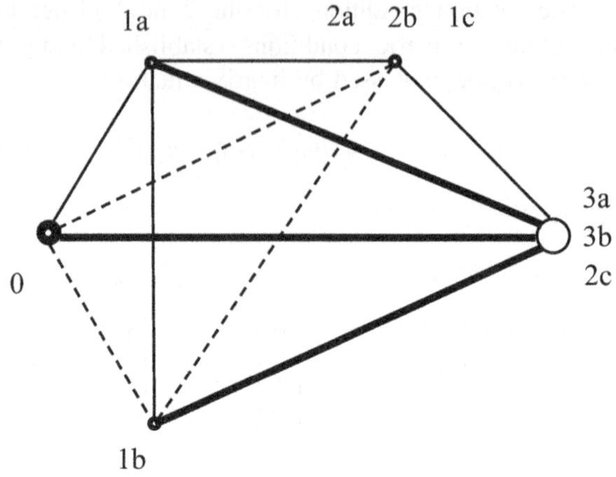

$1a, 2a, 3a$, $1b, 2b, 3b$ and $1c, 2c$. There are three independent loops, three fundamental circuits $(1a, 2a, 3a)$, $(1b, 2b, 3b)$, $(0, 1c, 2c)$ and we identify two *fictitious* carriers. The kinematics of an arbitrary element for each circuit is characterized by the angular velocity, and its associated skew-symmetric matrix given by recursive relations [16]:

$$\begin{aligned}&\vec{\omega}_{k0}^{i} = q_{k,k-1}\vec{\omega}_{k-1,0}^{i} + \vec{\omega}_{k,k-1}^{i}, \vec{\omega}_{k,k-1}^{i} = \dot{\varphi}_{k,k-1}^{i}\vec{u}_3\\&\tilde{\omega}_{k0}^{i} = q_{k,k-1}\tilde{\omega}_{k-1,0}^{i}q_{k,k-1}^{T} + \tilde{\omega}_{k,k-1}^{i}, \quad \tilde{\omega}_{k,k-1}^{i} = \dot{\varphi}_{k,k-1}^{i}\tilde{u}_3.\end{aligned} \tag{10.1.9}$$

Starting from the known rotation motion of the output link $1b$ by the relation (10.1.5), we develop the inverse kinematics problem and determine the velocities $\vec{v}_{k0}^{i}, \vec{\omega}_{k0}^{i}$ and the accelerations $\vec{\gamma}_{k0}^{i}, \vec{\varepsilon}_{k0}^{i}$ of each of the moving links. Based on the important remark $\omega_{k,k-1} = N_{k+1,k-1}\omega_{k+1,k}$, the derivatives with respect to time of the relations (10.1.8) lead to the relative angular velocities of all links as function of the angular velocity $\dot{\varphi}_{10}^{B} = \dot{\phi}$ of the output gear:

$$\begin{aligned}&\omega_{10}^{B} = \dot{\phi}, \quad \omega_{32}^{A} = \omega_{32}^{B} = \omega_{21}^{C} = \frac{\dot{\phi}}{n_3 - n_2}, \quad \omega_{10}^{A} = (n_1 + n_3)\omega_{21}^{C}\\&\omega_{21}^{A} = n_1\omega_{21}^{C}, \quad \omega_{21}^{B} = n_2\omega_{21}^{C}, \quad \omega_{10}^{C} = n_3\omega_{21}^{C}.\end{aligned} \tag{10.1.10}$$

Let us assume now that the mechanism has one *virtual motion*. Characteristic virtual velocities are given by the above relations (10.1.10):

$$\begin{aligned}&\omega_{10v}^{A} = 1, \quad \omega_{32v}^{A} = \omega_{32v}^{B} = \omega_{21v}^{C} = \frac{1}{n_1 + n_3}\\&\omega_{21v}^{A} = n_1\omega_{21v}^{C}, \quad \omega_{10v}^{C} = n_3\omega_{21v}^{C}, \quad \omega_{10v}^{B} = (n_3 - n_2)\omega_{21v}^{C}, \quad \omega_{21v}^{B} = n_2\omega_{21v}^{C}.\end{aligned} \tag{10.1.11}$$

These virtual velocities are required into the computation of the virtual work developed by the forces applied to the component elements of the gear train.

Concerning the relative angular accelerations $\varepsilon^i_{k,k-1} = \dot{\omega}^i_{k,k-1}$ of the compounding elements of the mechanism, these are immediately given by differentiating the above relations (10.1.10). Moreover, knowing the rotation motion of the output link $1b$ through the relations (10.1.5), angular acceleration $\vec{\varepsilon}^{\,i}_{k0}$, useful square matrix $\tilde{\omega}^i_{k0}\tilde{\omega}^i_{k0} + \tilde{\varepsilon}^i_{k0}$, velocities $\vec{v}^{\,i}_{k0}$ and accelerations $\vec{\gamma}^{\,i}_{k0}$ of each of the moving links are computed with recursive formulae:

$$
\begin{aligned}
\vec{\varepsilon}^{\,i}_{k0} &= q_{k,k-1}\vec{\varepsilon}^{\,i}_{k-1,0} + \varepsilon^i_{k,k-1}\vec{u}_3 + \omega^i_{k,k-1}q_{k,k-1}\tilde{\omega}^i_{k-1,0}q^T_{k,k-1}\vec{u}_3 \\
\tilde{\omega}^i_{k0}\tilde{\omega}^i_{k0} + \tilde{\varepsilon}^i_{k0} &= q_{k,k-1}\left(\tilde{\omega}^i_{k-1,0}\tilde{\omega}^i_{k-1,0} + \tilde{\varepsilon}^i_{k-1,0}\right)q^T_{k,k-1} \\
&\quad + \omega^i_{k,k-1}\omega^i_{k,k-1}\tilde{u}_3\tilde{u}_3 + \varepsilon^i_{k,k-1}\tilde{u}_3 + 2\omega^i_{k,k-1}q_{k,k-1}\tilde{\omega}^i_{k-1,0}q^T_{k,k-1}\tilde{u}_3 \\
\vec{v}^{\,i}_{k0} &= q_{k,k-1}\vec{v}^{\,i}_{k-1,0} + q_{k,k-1}\tilde{\omega}^i_{k-1,0}\vec{r}^{\,i}_{k,k-1} \\
\vec{\gamma}^{\,i}_{k0} &= q_{k,k-1}\vec{\gamma}^{\,i}_{k-1,0} + q_{k,k-1}\{\tilde{\omega}^i_{k-1,0}\tilde{\omega}^i_{k-1.0} + \tilde{\varepsilon}^i_{k-1,0}\}\vec{r}^{\,i}_{k,k-1}
\end{aligned}
\tag{10.1.12}
$$

Finally, the velocity $\vec{v}^{\,Ci}_k$ and the acceleration $\vec{\gamma}^{\,Ci}_k$ of mass centre of T^i_k rigid body are obtained from two known relations $\vec{v}^{\,Ci}_k = \vec{v}^{\,i}_{k0} + \tilde{\omega}^i_{k0}\vec{r}^{\,Ci}_k$, $\vec{\gamma}^{\,Ci}_k = \vec{\gamma}^{\,i}_{k0} + \{\tilde{\omega}^i_{k0}\tilde{\omega}^i_{k0} + \tilde{\varepsilon}^i_{k0}\}\vec{r}^{\,Ci}_k$. The above matrix relations (10.1.9) and (10.1.12) will further be used for the calculus of the wrench of inertia forces for every rigid body of the geared mechanism.

For simulation purposes let us consider a mechanism which has the following characteristics

$$
\begin{aligned}
&r_1 = 0.03\,\mathrm{m}, \quad r_2 = 0.02\,\mathrm{m}, \quad r_3 = 0.04\,\mathrm{m} \\
&\phi_0 = \pi, \quad r = h = 0.05\,\mathrm{m}, \quad \Delta t = 6\,\mathrm{s} \\
&m_1 = 0.3\,\mathrm{kg}, \quad m_2 = 0.25\,\mathrm{kg}, \quad m_3 = 0.6\,\mathrm{kg} \\
&m_4 = 0.85\,\mathrm{kg}, \quad M_r = -1.75\,\mathrm{Nm}.
\end{aligned}
$$

A program which implements the suggested algorithm is developed in MATLAB to solve first the inverse kinematics of the gear mechanism, assuming that for a period of six seconds the end-effector starts at rest from its initial position and is moving in a known rotation motion. A numerical study of the kinematics is carried out by computation of the input and output angles of rotation $\varphi^A_{10}, \varphi^B_{10}$, for example (Fig. 10.5).

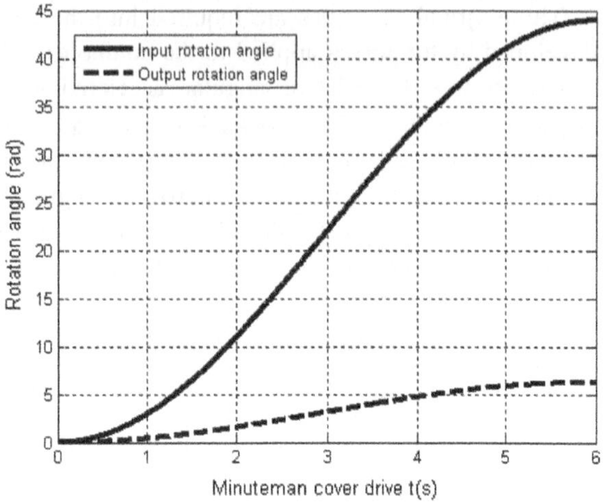

Fig. 10.5 Input and output rotation angles $\varphi_{10}^{A}, \varphi_{10}^{B}$

10.1.2 Equations of Motion

The objective of the dynamics here is to determine the input torques and powers which must be exerted by the actuators in order to produce a given trajectory of the end-effector. The derivation of a dynamics model has a very important effect in the determination of the actuator torque [7].

10.1.2.1 Principle of Virtual Work

The torque of moment $\vec{m}_{10}^{A} = m_{10}^{A}\vec{u}_3$ can control, by the intermediate of an electric motor, the motion of the geared mechanism.

In the inverse dynamic problem, one applies first the principle of virtual work in order to establish recursive matrix relations for the torque and the power of the active system. The mechanism of the gear train can artificially be transformed into a set of *two independent trees*, by *cutting* for the central link 3*a* in the equivalent graph its contacts 0–3*a*, 1*a*–3*a*, 1*b*–3*a* designated by thick edges and taking account of constraint conditions (Fig. 10.4).

It should be remarked that the vectors $\vec{f}_k^{*i} = -gm_k^i q_{k0}\vec{u}_3$, $\vec{m}_k^{*i} = -gm_k^i \tilde{r}_k^{Ci} q_{k0}\vec{u}_3$ and $\vec{f}_r^{*B} = \vec{0}$, $\vec{m}_r^{*B} = -M_r\vec{u}_3$ evaluate the influence due to the weight $m_k^i\vec{g}$ and the resistant torque M_r, respectively, and that the wrench of inertia forces of an arbitrary rigid body T_k^A, for example, is computed with respect to the centre of its joint A_k as follows:

$$
\begin{aligned}
\vec{f}_k^{\,in\,A} &= -m_k^A\left[\vec{\gamma}_{k0}^A + \left(\tilde{\omega}_{k0}^A\tilde{\omega}_{k0}^A + \tilde{\varepsilon}_{k0}^A\right)\vec{r}_k^{\,CA}\right] \\
\vec{m}_k^{\,in\,A} &= -\left[m_k^A\tilde{r}_k^{CA}\vec{\gamma}_{k0}^A + \hat{J}_k^A\vec{\varepsilon}_{k0}^A + \tilde{\omega}_{k0}^A\hat{J}_k^A\vec{\omega}_{k0}^A\right]. \\
&(k = 1, 2, 3)
\end{aligned}
\tag{10.1.13}
$$

Finally, using the notations $\vec{F}_{k0}^i = -\vec{f}_k^{\,in\,i} - \vec{f}_k^{\,*i}$, $\vec{M}_{k0}^i = -\vec{m}_k^{\,in\,i} - \vec{m}_k^{\,*i}(i = A, B)$, two recursive relations (7.19) will generate the set of vectors

$$
\begin{aligned}
&\vec{F}_3^A = \vec{F}_{30}^A, \quad \vec{M}_3^A = \vec{M}_{30}^A, \quad \vec{F}_2^A = \vec{F}_{20}^A + a_{32}^T\vec{F}_3^A, \quad \vec{F}_1^A = \vec{F}_{10}^A + a_{21}^T\vec{F}_2^A, \quad \vec{F}_1^B = \vec{F}_{10}^B \\
&\vec{M}_2^A = \vec{M}_{20}^A + a_{32}^T\vec{M}_3^A + \tilde{r}_{32}^A a_{32}^T\vec{F}_3^A, \quad \vec{M}_1^A = \vec{M}_{10}^A + a_{21}^T\vec{M}_2^A + \tilde{r}_{21}^A a_{21}^T\vec{F}_2^A, \quad \vec{M}_1^B = \vec{M}_{10}^B.
\end{aligned}
\tag{10.1.14}
$$

Applying *the equations of parallel robots dynamics* [17] based on the principle of virtual work, following matrix relation results for the torque of the actuator in a given displacement of end-effector

$$
m_{10}^A = \vec{u}_3^T\{\omega_{10v}^A\vec{M}_1^A + \omega_{21v}^A\vec{M}_2^A + \omega_{32v}^A\vec{M}_3^A + \omega_{10a}^{Bv}\vec{M}_1^B\}. \tag{10.1.15}
$$

The above relations (10.1.14) and (10.1.15) represent the *inverse dynamic model* of a Minuteman cover drive, provided that the inertial properties of the gears are known with sufficient accuracy and that friction is not significant. This dynamics approach developed here can be extended to any gyroscopic bevel-gear train with revolute actuators.

10.1.2.2 Lagrange Equations

A solution of the dynamics problem of the Minuteman cover drive can be developed based on Lagrange's equations of the second kind. A single generalized coordinate of the robot is represented by the rotation angle $\varphi_{10}^A = \alpha$ of the actuator. Lagrange's equation will be expressed by one differential relation

$$
\frac{d}{dt}\left\{\frac{\partial E}{\partial \dot{\alpha}}\right\} - \frac{\partial E}{\partial \alpha} = \Phi_\alpha, \tag{10.1.16}
$$

which contain the following generalized force

$$
\Phi_\alpha = m_{10}^A - \frac{n_3 - n_2}{n_1 + n_3}M_r + \left(\frac{1}{2}m_2 + m_3\right)gr\frac{n_3}{n_1 + n_3}\sin\left(\frac{n_3}{n_1 + n_3}\alpha\right). \tag{10.1.17}
$$

The components of the general expression of the kinetic energy $E = \sum_{\upsilon=1}^{3} E_{\upsilon}^{A} + E_{1}^{B}$ are expressed as an analytical function of the first derivative with respect to time $\dot{\alpha}$ of the generalized coordinate:

$$\begin{aligned} E_1^A &= \frac{1}{2}\vec{\omega}_{10}^{AT}\hat{J}_1\vec{\omega}_{10}^{A}, E_2^A = \frac{1}{2}\vec{\omega}_{20}^{AT}\hat{J}_3^A\vec{\omega}_{20}^{A} \\ E_3^A &= \frac{1}{2}m_3^A\vec{v}_{30}^{AT}\vec{v}_{30}^{A} + \frac{1}{2}\vec{\omega}_{30}^{AT}\hat{J}_3^A\vec{\omega}_{30}^{A}, E_1^B = \frac{1}{2}\vec{\omega}_{10}^{BT}\hat{J}_4\vec{\omega}_{10}^{B}, \end{aligned} \qquad (10.1.18)$$

where the absolute angular velocities have the expressions:

$$\begin{aligned} \vec{\omega}_{10}^{A} &= \dot{\alpha}\,\vec{u}_3, \quad \vec{\omega}_{20}^{A} = \dot{\alpha}\frac{n_3}{n_1+n_3}\vec{u}_3, \quad \vec{\omega}_{30}^{A} = \dot{\alpha}\frac{n_3-1}{n_1+n_3}\vec{u}_3 \\ \vec{v}_{30}^{A} &= -\dot{\alpha}\,r\frac{n_3}{n_1+n_3}\vec{u}_2, \quad \vec{\omega}_{10}^{B} = \dot{\alpha}\frac{n_3-n_2}{n_1+n_3}\vec{u}_3. \end{aligned} \qquad (10.1.19)$$

In the inverse dynamics problem, a calculus of the derivatives with respect to time $\frac{d}{dt}\left\{\frac{\partial E}{\partial \dot{\alpha}}\right\}$ of all above functions leads quickly to the same expression (10.1.15) for the input torque m_{10}^{A} required by the actuator, now given as an *analytical solution*:

$$\begin{aligned} m_{10}^{A} = {} & \frac{1}{(n_1+n_3)^2}\{(n_1+n_3)^2 J_{1z} + n_3^2(J_{2z}+m_3 r^2) + (n_3-1)^2 J_{3z} \\ & + (n_1-n_3)^2 J_{4z}\}\ddot{\alpha} + \frac{n_3-n_2}{n_1+n_3}M_r - \left(\frac{1}{2}m_2+m_3\right)gr\frac{n_3}{n_1+n_3}\sin\left(\frac{n_3}{n_1+n_3}\alpha\right). \end{aligned} \qquad (10.1.20)$$

A computer program solves the inverse dynamics modelling of the robot, using MATLAB software. Assuming that the *resistant torque* of constant moment $M_r = 1.75$ Nm applied at end-effector and the weights $m_k^A\vec{g}$ of compounding rigid bodies constitute the external forces acting on the mechanism during its evolution, a numerical computation in dynamics is developed, based on the determination of the input torque m_{10}^{A} (Fig. 10.6) and its active power $P_{10}^{A} = \omega_{10}^{A}m_{10}^{A}$ (Fig. 10.7), using both (10.1.15) and (10.1.20) formulae. The time-history evolution of the torque and power required by the active system is shown for a period of six seconds of motion.

Based on the principle of virtual work, which has been verified in the conceptual framework of Lagrange's equations, the dynamics approach is very efficient and establishes a direct recursive determination of the variation in real time of the torque and power of the actuator.

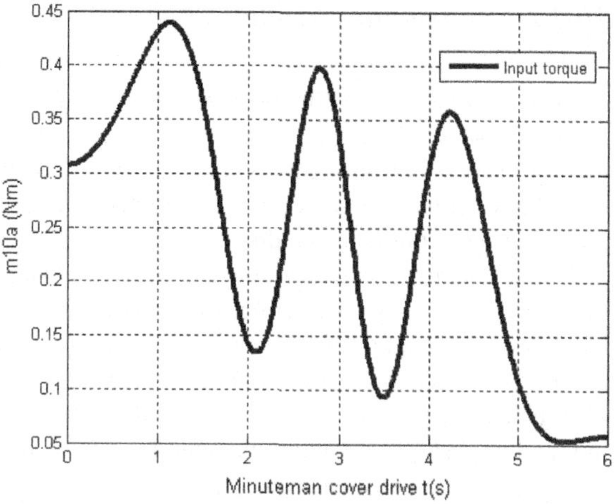

Fig. 10.6 Input torque m_{10}^{A} of the actuator

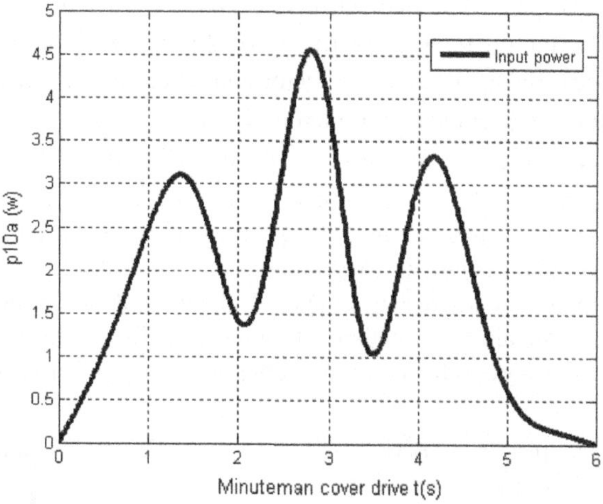

Fig. 10.7 Input power P_{10}^{A} of the actuator

10.2 Inverse Dynamics of a 2-DOF Orienting Gear Train

Generally, these mechanical systems have conical and cylindrical-toothed elements in their structure, where the input axes are all parallel while the output axes are orthogonal. The rotary orientation movements are usually performed around the axes of a Cartesian orthogonal frame, having its axes linked to the last arm of the robot's positioning mechanism. Planetary gear trains with three degrees of freedom are adopted as the design concept for the robotic wrist [1, 4, 5].

Recursive matrix relations for the kinematics and dynamics of a 2-DOF orienting gear train which has a non-symmetrical kinematical scheme are developed in this study. The robot wrist must rotate around two orthogonal axes and consequently the mechanism has two degrees of freedom.

10.2.1 Kinematics Model

A matrix methodology for the kinematics analysis based on the concept of fundamental circuit of an open-loop chain is now presented. This method involves the identification of all open-loop chains and the derivation of the geometric relationships between the orientation of the end-effector and the joint angles of the chains, including the input actuator displacements [7, 12, 14].

The mechanism topology consists of five moving links, five revolute pairs and three bevel gear pairs. Therefore, the wrist is a 2-DOF *spherical mechanism* which has a limited rotation range about the vertical joint axis.

Let $Oxyz$ be a fixed frame, about which the mechanism moves. There are two kinematical chains $0 - 1a - 2a - 3a$ and $0 - 1b - 2b - 3b$ starting from the fixed link and ending to a common element designated by the end-effector $c = 3a = 3b$.

In the wrist mechanism, the gear $1a$, of radius r_1^A, mass m_1^A and tensor of inertia $\hat{J}_1^A$ adjoins the link $2a$ of l_1 in length, which can serve as carrier for the $2b$–$3b$ bevel gear pair. This element includes a bevel gear of radius r_3^A, mass m_3^A and tensor of inertia $\hat{J}_3^A$. Four gears $1a, 2a, 1b, 2b$ are sun gears while $c = 3a$ is a bevel planet gear linked at the end-effector and adjacent to carrier $2a$. The gear $1b$ of radius r_1^B, mass m_1^B and tensor of inertia $\hat{J}_1^B$ is connected to a second gear $2b$ of radius r_3^B, mass m_3^B and tensor of inertia $\hat{J}_3^B$. By including the end-effectors of length l_2 at the third gear $c = 3b = 3a$ of radius r_5^B, mass m_5^B, tensor of inertia $\hat{J}_5^B$, we obtain an assembly which is free to undergo arbitrary two concurrent rotations with respect to the common centre O. We remarks that the links $1a$ and $1b$ share one fixed common joint axis x and that $2a$ and $2b$ share another common joint axis z(Fig. 10.8).

Pursuing the method of successive displacements to the geometric analysis of closed-loop chains, the variable angles $\varphi_{k,k-1}$ of rotation about the joint axis z_k is

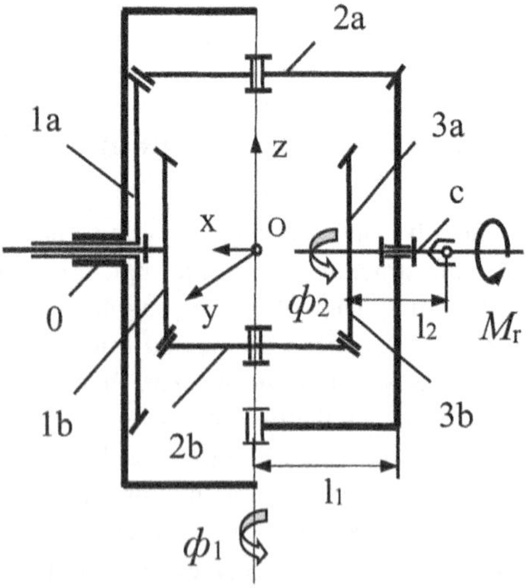

Fig. 10.8 The 2-DOF orienting gear train

the parameter needed to bring the next link from a reference configuration to the next configuration. Let $a^{\varphi}_{k,k-1}$ be the orthogonal transformation 3×3 matrix of relative rotation with the angle $\varphi^{A}_{k,k-1}$ of link T^{A}_{k} around z^{A}_{k} axis, for example. In the kinematics analysis of constrained systems we are interested in deriving a matrix equation relating the location of an arbitrary T_k body to the joint variables.

Based on the matrix *theory of fundamental circuits* aroused by Tsai [9], there is a *real* or *fictitious* carrier for each gear pair in a *planetary bevel gear train* and a fundamental matrix equation for each loop can be written as follows

$$a_{k+1,k-1} = a^{\varphi}_{k+1,k}\theta^{\delta_k}_{y}a_{k,k-1}, \quad b_{k+1,k-1} = b^{\varphi}_{k+1,k}\theta^{\delta_k}_{y}b_{k,k-1}, \quad \varphi^{i}_{k,k-1} = N^{i}_{k+1,k-1}\varphi^{i}_{k+1,k}, \quad (i = A, B)$$

$$\delta_k = \alpha_{k-1} + \alpha_{k+1}, \quad \theta^{\delta_k}_{y} = rot(y, \delta_k) = \begin{bmatrix} \cos\delta_k & 0 & -\sin\delta_k \\ 0 & 1 & 0 \\ \sin\delta_k & 0 & \cos\delta_k \end{bmatrix}, \tag{10.2.1}$$

where $\varphi^{i}_{k,k-1}$ and $\varphi^{i}_{k+1,k}$ denote the relative angles of rotation of the carrier T^{i}_{k} and the planet gear T^{i}_{k+1}, respectively, while $\alpha_{k-1}, \alpha_{k+1}$ are the angles that characterize the geometry of two connected gears T_{k-1} and T_{k+1} (Fig. 10.9).

Denoting by r_{k-1}, r_{k+1} and Z_{k-1}, Z_{k+1} the radius and the number of teeth of two gears, respectively, the gear *ratios* of a gear pair is defined as $N_{k+1,k-1} = r_{k+1}/r_{k-1} = Z_{k+1}/Z_{k-1}$.

The motions of the five parts of the gear train are several *concurrent rotations* around the fixed point O. To simplify the graphical image of the kinematical scheme of the gear mechanism, in what follows we will represent the intermediate

Fig. 10.9 Gear fundamental wrist circuit

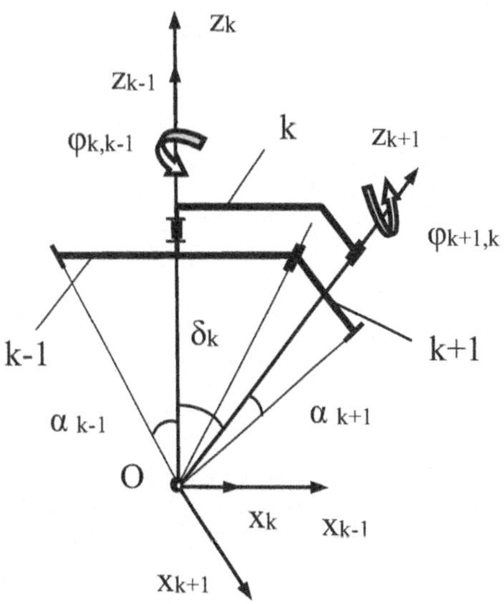

reference systems by only two axes, whit the z_k axis represented for every component element T_k. It is noted that the relative rotation with angle $\varphi_{k,k-1}$ of the body T_k must always be pointed about the direction of the z_k axis.

Four appropriate frames for the first kinematical chain and five frames for the second circuit could be fixed in the same origin O. However, to distinguish the two loops, the kinematical circuits A and B are sketched separately in the form of two independent cycles (Figs. 10.10 and 10.11).

We consider the rotation angles $\varphi_{10}^{A}, \varphi_{10}^{B}$ of the actuators as parameters giving the instantaneous position of the mechanism. Seven relative angles of rotation $\varphi_{21}^{A}, \varphi_{32}^{A}, \varphi_{43}^{A}, \varphi_{21}^{B}, \varphi_{32}^{B}, \varphi_{43}^{B}, \varphi_{54}^{B}$ are the variables in the inverse kinematics problem.

Starting from the reference origin O and pursuing two independent input-output circuits A and B, we obtain the following successive matrices of transformation [18]:

$$\begin{aligned} &a_{10} = a_{10}^{\varphi}\theta_1, \quad a_{21} = a_{21}^{\varphi}\theta_2, \quad a_{32} = a_{32}^{\varphi}\theta_1^{T}, \quad a_{43} = a_{43}^{\varphi}\theta_1 \\ &b_{10} = b_{10}^{\varphi}\theta_1, \quad b_{21} = b_{21}^{\varphi}\theta_2, \quad b_{32} = b_{32}^{\varphi}\theta_1, \quad b_{43} = b_{43}^{\varphi}\theta_2, \quad b_{54} = b_{54}^{\varphi}\theta_1, \end{aligned} \tag{10.2.2}$$

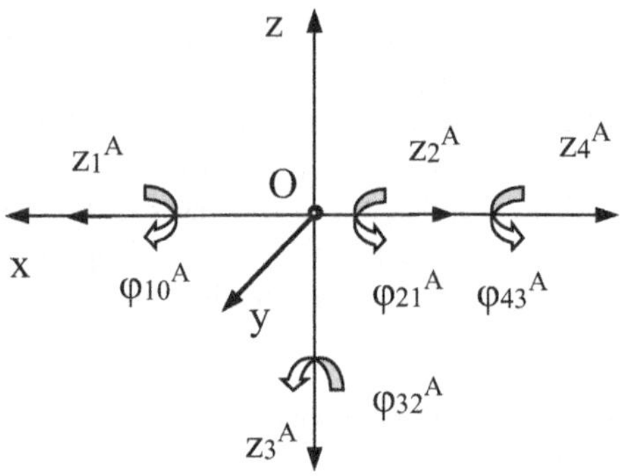

Fig. 10.10 Moving frames sequence of circuit A

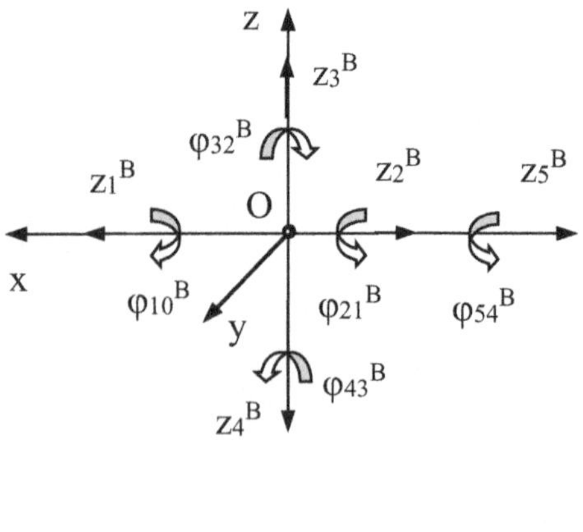

Fig. 10.11 Moving frames sequence of circuit B

with the same notations as above (10.1.4)

$$q^{\varphi}_{k,k-1} = rot(z, \varphi^{i}_{k,k-1}), \quad q_{k0} = \prod_{s=1}^{k} q_{k-s+1,k-s}, \quad \theta_1 = rot\left(y, \frac{\pi}{2}\right), \quad \theta_2 = rot(y, \pi)$$
$$(i = A, B) \ (q = a, b) \ (k = 1, 2, 3, 4, 5). \tag{10.2.3}$$

Let us suppose that the absolute motion of the platform attached at the planet gear $c = 3a = 3b$ is a rotation about the centre O. In the inverse geometric problem however, the orientation of the end-effector is ascertained from the intermediate of the two Euler angles ϕ_1, ϕ_2, which are expressed by two given analytical functions

$$\phi_l = \phi_l^{\circ}\left[1 - \cos\left(\frac{\pi}{6}t\right)\right], \quad (l = 1, 2), \tag{10.2.4}$$

where $2\phi_l^{\circ}$ represents the maximum of the angle of rotation ϕ_l. The resistant torque M_r and two independent output parameters $\varphi^{A}_{32} = \phi_1$, $\varphi^{B}_{54} = \phi_2$, giving the orientation of the end-effector, are all represented in Fig. 10.8. Representing the orientation of the end-effector $c = 3a = 3b$ (Fig. 10.12) in the fixed frame $Oxyz$, the product of the successive known matrices of rotation

$$c_{10} = c_1^{\phi}\theta_2, \quad c_{21} = c_2^{\phi}\theta_1 \tag{10.2.5}$$

leads to the transposed rotation matrix

$$c_{20}^{T} = c_{10}^{T}c_{21}^{T} = \theta_2^{T}c_1^{\phi T}\theta_1^{T}c_2^{\phi T}, \tag{10.2.6}$$

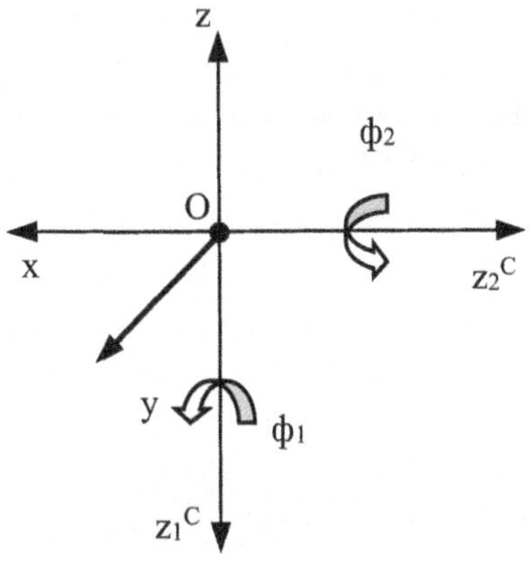

Fig. 10.12 Moving frames sequence of end-effector

where

$$c_l^{\phi T} = \begin{bmatrix} \cos\phi_l & -\sin\phi_l & 0 \\ \sin\phi_l & \cos\phi_l & 0 \\ 0 & 0 & 1 \end{bmatrix}, \quad (l = 1, 2). \tag{10.2.7}$$

The constraining geometric conditions for the rotation of the end-effector are given by the equations $a_{40} = b_{50} = c_{20}$. From these constraints and Eq. (10.2.1), we obtain the real-time evolution of all characteristic joint angles, as follows:

$$\begin{aligned} &\varphi_{10}^A = \varphi_{21}^A = \frac{\phi_1}{n_{31}^A}, \quad \varphi_{32}^A = \phi_1, \quad \varphi_{43}^A = \phi_2 \\ &\varphi_{10}^B = \varphi_{21}^B = \frac{1}{n_{31}^B}\left(-\phi_1 + \frac{\phi_2}{n_{53}^B}\right), \quad \varphi_{32}^B = n_{31}^B \varphi_{10}^B, \quad \varphi_{43}^B = \frac{\phi_2}{n_{53}^B}, \quad \varphi_{54}^B = \phi_2 \\ &n_{31}^A = \frac{r_1^A}{r_3^A}, \quad n_{31}^B = \frac{r_1^B}{r_3^B}, \quad n_{53}^B = \frac{r_3^B}{r_5^B}. \end{aligned} \tag{10.2.8}$$

The analysis of the kinematics of bevel-gear wrist mechanisms of gyroscopic structure is very complex, due to the fact that the carriers and planet gears may possess simultaneous angular velocities about nonparallel axes. As assembly of links and joints, the topology of the kinematical chain can be symbolized in the form of an *associated graph* (Fig. 10.13), where the links are represented by vertices and the joints by edges, and two small concentric circles label the vertex denoting the fixed link 0. The gear pairs $1a$–$2a$, $1b$–$2b$, $2b$–$3b$ are designated by thick edges and revolute joints 0–$1a$, 0–$1b$, $2a$–$3a$, 0–$1b$, 0–$2b$ by thin edges. Starting from the base link 0 and ending at the end-effector link $3a = 3b = c$, four edged paths consist of vertices $1a, 2a, 3a$ and $1b, 2b, 3b$. There are three independent loops, two fundamental circuits and we identify three *fictitious* carriers.

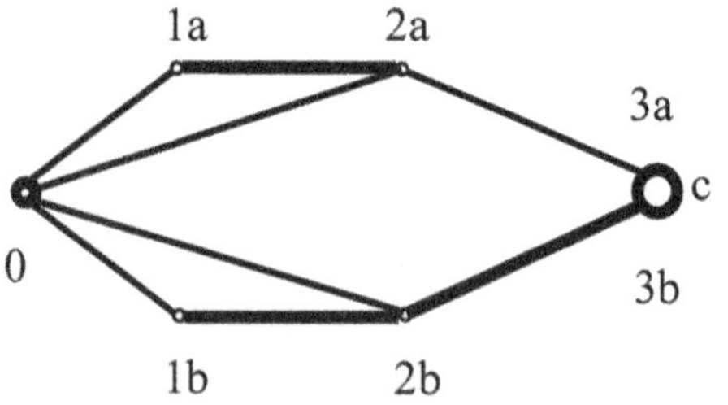

Fig. 10.13 Associated graph of the mechanism

Kinematics of any element for each of the circuits is characterized by skew-symmetric matrices given by the recursive relations:

$$\tilde{\omega}^i_{k0} = q_{k,k-1}\tilde{\omega}^i_{k-1,0}q^T_{k,k-1} + \tilde{\omega}^i_{k,k-1}, \quad \tilde{\omega}^i_{k,k-1} = \dot{\varphi}^i_{k,k-1}\tilde{u}_3 \ (q = a, b) \ (i = A, B), \tag{10.2.9}$$

where $\tilde{u}_3$ is a skew-symmetric matrix associated to $\vec{u}_3$. These matrices are *associated* with the angular velocities

$$\vec{\omega}^i_{k0} = q_{k,k-1}\vec{\omega}^i_{k-1,0} + \vec{\omega}^i_{k,k-1}, \quad \vec{\omega}^i_{k,k-1} = \dot{\varphi}^i_{k,k-1}\vec{u}_3. \tag{10.2.10}$$

Knowing the rotation motion of the end-effector through the relations (10.2.4), will allow us to determine the velocities $\vec{v}^i_{k0}, \vec{\omega}^i_{k0}$ of each of the moving links. Using the relation $\omega_{k,k-1} = N_{k+1,k-1}\omega_{k+1,k}$, the derivatives with respect to time of the relations (10.2.8) lead to the expressions of relative angular velocities of all links as a function of two angular velocities $\dot{\phi}_1, \dot{\phi}_2$ of the end-effector:

$$\begin{aligned} &\omega^A_{10} = \omega^A_{21} = \frac{\dot{\phi}_1}{n^A_{31}}, \quad \omega^A_{32} = \omega^C_1 = \dot{\phi}_1, \quad \omega^A_{43} = \omega^C_2 = \dot{\phi}_2 \\ &\omega^B_{10} = \omega^B_{21} = \frac{1}{n^B_{31}}\left(-\dot{\phi}_1 + \frac{\dot{\phi}_2}{n^B_{53}}\right), \quad \omega^B_{32} = n^B_{31}\omega^B_{10}, \quad \omega^B_{43} = \frac{\dot{\phi}_2}{n^B_{53}}, \quad \omega^B_{54} = \omega^C_2 = \dot{\phi}_2. \end{aligned} \tag{10.2.11}$$

Assuming that the mechanism has successively two independent *virtual motions*, the characteristic velocities are expressed by the relations (10.2.11) as function of robot's position. First, we consider the following input angular velocities $\omega^{Av}_{10a} = 1$, $\omega^{Bv}_{10a} = 0$ and we obtain the set of virtual velocities:

$$\begin{aligned} &\omega^{Av}_{21a} = 1, \quad \omega^{Av}_{32a} = n^A_{31}, \quad \omega^{Av}_{43a} = n^A_{31}n^B_{53} \\ &\omega^{Bv}_{21a} = 0, \quad \omega^{Bv}_{32a} = 0, \quad \omega^{Bv}_{43a} = n^A_{31}, \quad \omega^{Bv}_{54a} = n^A_{31}n^B_{53} \\ &\omega^{Cv}_{1a} = n^A_{31}, \quad \omega^{Cv}_{2a} = n^A_{31}n^B_{53}. \end{aligned} \tag{10.2.12}$$

A second virtual motion is defined by the input velocities $\omega^{Bv}_{10b} = 1$, $\omega^{Av}_{10b} = 0$, and the following set results:

$$\begin{aligned} &\omega^{Bv}_{21b} = 1, \quad \omega^{Bv}_{32b} = n^B_{31}, \quad \omega^{Bv}_{43b} = n^B_{31}, \quad \omega^{Bv}_{54b} = n^B_{31}n^B_{53} \\ &\omega^{Av}_{21b} = 0, \quad \omega^{Av}_{32b} = 0, \quad \omega^{Av}_{43b} = n^B_{31}n^B_{53}, \quad \omega^{Cv}_{1b} = 0, \quad \omega^{Cv}_{2b} = n^B_{31}n^B_{53}. \end{aligned} \tag{10.2.13}$$

Concerning the relative angular accelerations of the compounding elements of the mechanism, these are immediately given by $\varepsilon^i_{k,k-1} = \dot{\omega}^i_{k,k-1}$, by deriving the relations to the velocities (10.2.11). The accelerations $\vec{\gamma}^i_{k0}, \vec{\varepsilon}^i_{k0}$ of each of the moving links and the useful square matrices $\tilde{\omega}^i_{k0}\tilde{\omega}^i_{k0} + \tilde{\varepsilon}^i_{k0}$ are given by the general formulae (10.1.12), while the velocity $\vec{v}^{Ci}_k$ and the acceleration $\vec{\gamma}^{Ci}_k$ of the mass center of T^i_k rigid body are calculated from two basic relations $\vec{v}^{Ci}_k = \tilde{\omega}^i_{k0}\vec{r}^{Ci}_k$, $\vec{\gamma}^{Ci}_k = \{\tilde{\omega}^i_{k0}\tilde{\omega}^i_{k0} + \tilde{\varepsilon}^i_{k0}\}\vec{r}^{Ci}_k$.

The following characteristics of a geared mechanism are considered for simulation:

$$l_1 = 0.03\,\text{m},\quad l_2 = 0.055\,\text{m},\quad \phi^*_1 = \frac{\pi}{4}, \phi^*_2 = \pi,\quad M_r = 0.01\,\text{Nm}, \Delta t = 6\,\text{s}$$
$$r^A_1 = 0.025\,\text{m},\quad r^A_3 = 0.04\,\text{m},\quad r^B_1 = 0.02\,\text{m},\quad r^B_3 = 0.035\,\text{m},\quad r^B_5 = 0.015\,\text{m}$$
$$m^A_1 = 0.25\,\text{kg},\quad m^A_3 = 0.4\,\text{kg},\quad m^A_4 = 1\,\text{kg},\quad m^B_1 = 0.2\,\text{kg},\quad m^B_3 = 0.35\,\text{kg},\quad m^B_5 = 1\,\text{kg}.$$

A software program and a numerical study has been developed in MATLAB to solve the kinematics of the orienting gear train. For illustration, it is assumed that for a period of six second the end-effector starts at rest from its initial position and is moving in a known rotation motion. The input angles of rotation $\varphi^A_{10}, \varphi^B_{10}$ (Fig. 10.14) of two revolute actuators and the angles $\varphi^A_{32}, \varphi^B_{32}$ of two second gears (Fig. 10.15), for example, are computed.

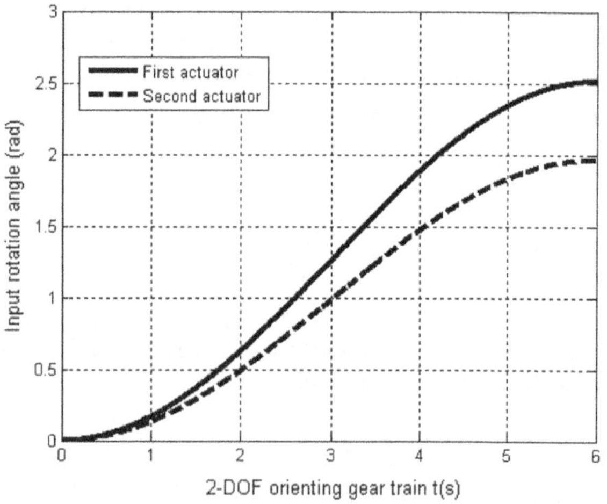

Fig. 10.14 Input rotation angles $\varphi^A_{10}, \varphi^B_{10}$ of two actuators

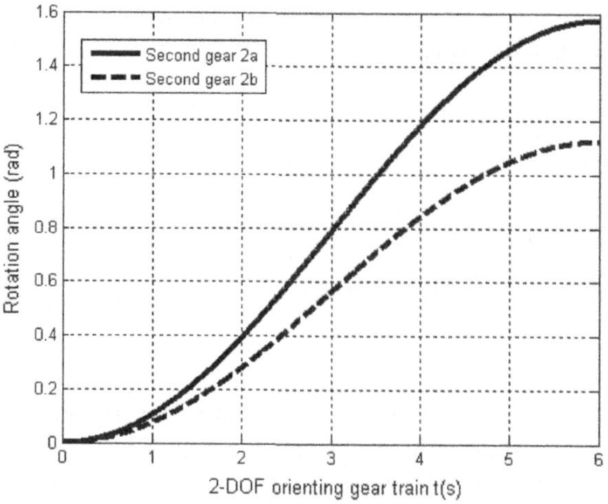

Fig. 10.15 Rotation angles $\varphi_{32}^{A}, \varphi_{32}^{B}$ of two second gears

10.2.2 *Dynamics Modelling*

10.2.2.1 Principle of Virtual Work

Two torques of moment $\vec{m}_{10}^{A} = m_{10}^{A}\vec{u}_3$, $\vec{m}_{10}^{B} = m_{10}^{B}\vec{u}_3$ control by the intermediary of electric motors, the motion of the orienting gear train.

The fundamental principle of virtual work is used, in order to derive an inverse dynamics model and establish some recursive matrix relations for the torques and the powers of the active systems.

The parallel mechanism is artificially transformed into a set of *three independent trees*, by *cutting* for the end-effector c in the equivalent graph (Fig. 10.13) its contacts $2a$–c, $2b$–c and taking account of the constraining conditions.

Considering that the end-effector's motion is given, the position, the velocity and the acceleration of the centre of mass are computed for each element. The force of inertia $\vec{f}_k^{\,in\,A}$ and the moment of inertia forces $\vec{m}_k^{\,in\,A}$ of an arbitrary rigid body T_k^A, for example, are determined with respect to the common centre of rotation O:

$$\vec{f}_k^{\,in\,A} = -m_k^A\left[\vec{\gamma}_{k0}^{A} + \left(\tilde{\omega}_{k0}^{A}\tilde{\omega}_{k0}^{A} + \tilde{\varepsilon}_{k0}^{A}\right)\vec{r}_k^{\,CA}\right], \quad \vec{m}_k^{\,in\,A} = -\left[m_k^A\tilde{r}_k^{\,CA}\vec{\gamma}_{k0}^{A} + \hat{J}_k^A\vec{\varepsilon}_{k0}^{A} + \tilde{\omega}_{k0}^{A}\hat{J}_k^A\vec{\omega}_{k0}^{A}\right]. \tag{10.2.14}$$

On the other hand, the wrench of two vectors $\vec{f}_k^{\,*A} = -gm_k^A a_{k0}\vec{u}_3$, $\vec{m}_k^{\,*A} = -gm_k^A\tilde{r}_k^{\,CA}a_{k0}\vec{u}_3$ evaluates the influence of the action of the weight $m_k^A\vec{g}$ on the same element T_k^A.

In the interests of real-time control, considering the gravitational effect and the action of a resistant torque M_r through the vectors $\vec{f}_{2r}^{\,*C} = \vec{0}$, $\vec{m}_{2r}^{\,*C} = -M_r\vec{u}_3$, the relevant objective of a dynamic model is to determine the input torques which must be exerted by the actuators in order to produce a given trajectory of the end-effector.

Starting from *the equations of parallel robots dynamics* [16], following compact matrix relations results

$$m_{10}^{A} = \vec{u}_3^T \left[\vec{M}_1^A + \omega_{21a}^{Av}\vec{M}_2^A + \omega_{32a}^{Av}\vec{M}_3^A + \omega_{21a}^{Bv}\vec{M}_2^B + \omega_{32a}^{Bv}\vec{M}_3^B + \omega_{1a}^{Cv}\vec{M}_1^C + \omega_{2a}^{Cv}\vec{M}_2^C\right] \tag{10.2.15}$$

for the torque of first actuator, and

$$m_{10}^{B} = \vec{u}_3^T \left[\vec{M}_1^B + \omega_{21b}^{Bv}\vec{M}_2^B + \omega_{32b}^{Bv}\vec{M}_3^B + \omega_{21b}^{Av}\vec{M}_2^A + \omega_{32b}^{Av}\vec{M}_3^A + \omega_{1b}^{Cv}\vec{M}_1^C + \omega_{2b}^{Cv}\vec{M}_2^C\right] \tag{10.2.16}$$

for the torque of second actuator, where the vectors $\vec{F}_k^i, \vec{M}_k^i (i = A, B, C)$ are obtained using above recursive relations (10.1.14).

The relations (10.2.15) and (10.2.16) represent the *inverse dynamic model* of the 2-DOF orienting gear train.

10.2.2.2 Lagrange Equations

A solution of the dynamics problem of the orienting gear train has been developed using Lagrange formalism. The generalized coordinates of the robot are represented by the rotation angles of two actuators:$s_1 = \varphi_{10}^A, s_2 = \varphi_{10}^B$.

Lagrange's equations will be expressed by two differential relations

$$\frac{d}{dt}\left\{\frac{\partial E}{\partial \dot{s}_l}\right\} - \frac{\partial E}{\partial s_l} = \Phi_l \ \ (l = 1, 2), \tag{10.2.17}$$

which contain the following generalized forces

$$\Phi_1 = m_{10}^A - M_r n_{31}^A n_{53}^B, \quad \Phi_2 = m_{10}^B - M_r n_{31}^B n_{53}^B. \tag{10.2.18}$$

The components of the general expression of the kinetic energy $E = \sum_{v=1}^{2} (E_v^A + E_v^B) + E_3^B$ are expressed as analytical functions of the first derivatives with respect to time of the generalized coordinates:

$$\begin{aligned} E_1^A &= \frac{1}{2}\vec{\omega}_{10}^{AT}\hat{J}_1^A\vec{\omega}_{10}^A, \quad E_2^A = \frac{1}{2}\vec{\omega}_{20}^{AT}\hat{J}_3^A\vec{\omega}_{20}^A \\ E_1^B &= \frac{1}{2}\vec{\omega}_{10}^{BT}\hat{J}_1^B\vec{\omega}_{10}^B, \quad E_2^B = \frac{1}{2}\vec{\omega}_{20}^{BT}\hat{J}_3^B\vec{\omega}_{20}^B, \quad E_3^B = \frac{1}{2}\vec{\omega}_{30}^{BT}\hat{J}_5^B\vec{\omega}_{30}^B, \end{aligned} \tag{10.2.19}$$

where the absolute angular velocities have the expressions:

$$\begin{aligned}&\vec{\omega}_{10}^{A} = \dot{s}_1\vec{u}_3, \quad \vec{\omega}_{20}^{A} = \dot{s}_1 n_{31}^{A}\vec{u}_3, \quad \vec{\omega}_{10}^{B} = \dot{s}_2\vec{u}_3\\ &\vec{\omega}_{20}^{B} = \dot{s}_2 n_{31}^{B}\vec{u}_3, \quad \vec{\omega}_{30}^{B} = -\dot{s}_1 n_{31}^{A}\vec{u}_1 + (n_{31}^{A}\dot{s}_1 + n_{31}^{B}\dot{s}_2)n_{53}^{B}\vec{u}_3.\end{aligned} \tag{10.2.20}$$

A calculus of the derivatives with respect to time $\frac{d}{dt}\left\{\frac{\partial E}{\partial \dot{s}_l}\right\}(l = 1, 2)$ of the above functions leads to same expressions (10.2.15), (10.2.16) for the input torques m_{10}^{A}, m_{10}^{B} required by the actuators, here given as *analytical solutions*:

$$\begin{aligned}m_{10}^{A} &= n_{31}^{A}n_{53}^{B}M_r + \left\{\frac{J_{1z}^{A}}{n_{31}^{A}} + J_{3z}^{A}n_{31}^{A} + J_{5x}^{B}n_{31}^{A}\right\}\ddot{\phi}_1 + J_{5z}^{B}n_{31}^{A}n_{53}^{B}\ddot{\phi}_2\\ m_{10}^{B} &= n_{31}^{B}n_{53}^{B}M_r - \left\{\frac{J_{1z}^{B}}{n_{31}^{B}} + J_{3z}^{B}n_{31}^{B}\right\}\ddot{\phi}_1 + \left\{\frac{J_{1z}^{B}}{n_{31}^{B}n_{53}^{B}} + \frac{J_{3z}^{B}n_{31}^{B}}{n_{53}^{B}} + J_{5z}^{B}n_{31}^{B}n_{53}^{B}\right\}\ddot{\phi}_2.\end{aligned} \tag{10.2.21}$$

Based on the algorithm derived from the above relations (10.2.15), (10.2.16) or (10.2.21), a computer program solves the inverse dynamics modelling of the robot, using MATLAB software. Assuming that the resistant torque of constant moment $M_r = 0.01$ Nm applied at the end-effector and the weights $m_k^{A}\vec{g}$ of compounding rigid bodies constitute the external forces acting on the mechanism during its evolution, a numerical computation in the dynamics is developed, based on the determination of the two input torques m_{10}^{A}, m_{10}^{B} (Fig. 10.16) and their active powers $P_{10}^{A} = \omega_{10}^{A}m_{10}^{A}$ and $P_{10}^{B} = \omega_{10}^{B}m_{10}^{B}$ (Fig. 10.17), using both the formulae (10.2.15), (10.2.16) and (10.2.21). The time-history evolution of the torques and the powers required by the two active systems are shown for a period of six seconds of motion. We note that the movement of the spherical wrist mechanism is generated, in fact, by two fixed motors, operating by some complex intermediate cylindrical or conical

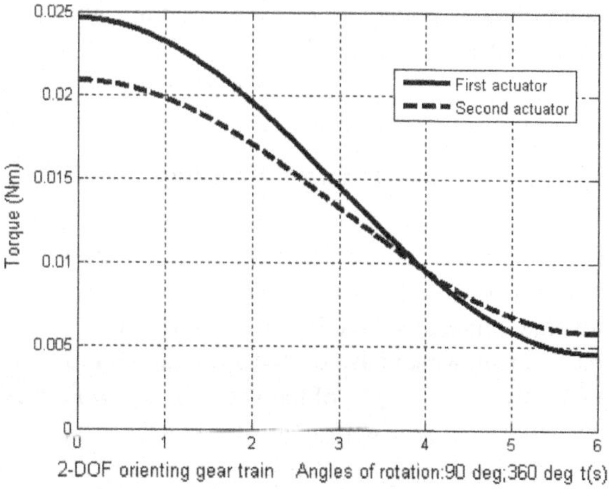

Fig. 10.16 Input torques m_{10}^{A}, m_{10}^{B} of two actuators

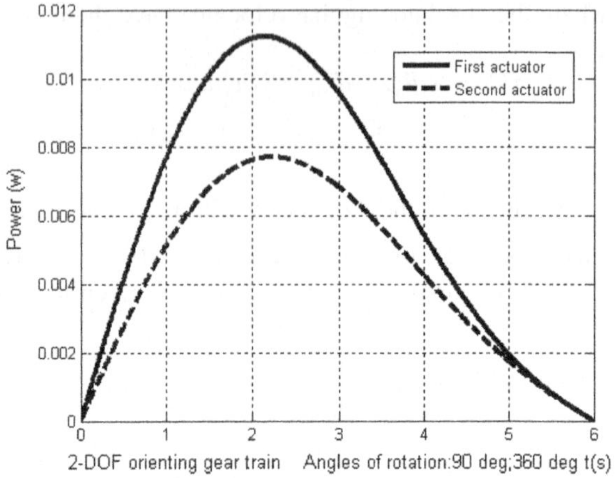

Fig. 10.17 Input powers P_{10}^{A}, P_{10}^{B} of two actuators

planetary reducers. Hence, the computed powers P_{10}^{A}, P_{10}^{B} are only the power requirements effectively transmitted by the two active links and not the powers consumed and delivered by the motors.

Using the principle of virtual powers, the new approach described above is very efficient and establishes a direct recursive determination of the variation in real-time of torques and powers of the actuators.

10.3 Dynamics Analysis of the Cincinnati-Milacron Wrist Robot

Recursive matrix relations for kinematics and dynamics of a Cincinnati-Milacron planetary bevel-gear mechanism, which has a non-symmetrical kinematical schema, are developed in the following. The end-effectors must rotate about three non-orthogonal axes, so that the orienting wrist mechanism has three degrees of freedom [9].

10.3.1 Geometric and Kinematics Model

Let $Ox_0y_0z_0$ be a fixed Cartesian frame, about which the *oblique wrist* mechanism moves (Fig. 10.18). The mechanism topology consists of six moving links, six revolute joints and three bevel gear pairs. Therefore, the wrist is a 3-DOF spherical mechanism, which has non-limited rotational ranges about the joint axes. There are two active gears $1b$, $1c$ of radii $r_1^B = r_1, r_1^C = r_5$, masses m_1^B, m_1^C and inertia tensors $\hat{J}_1^B, \hat{J}_1^C$. Three kinematical chains $0 - 1a - 2a - 3a - 4a$, $0 - 1b - 2b$ and $0 - 1c - 2c$ are identified, starting from the forearm and ending with a common element

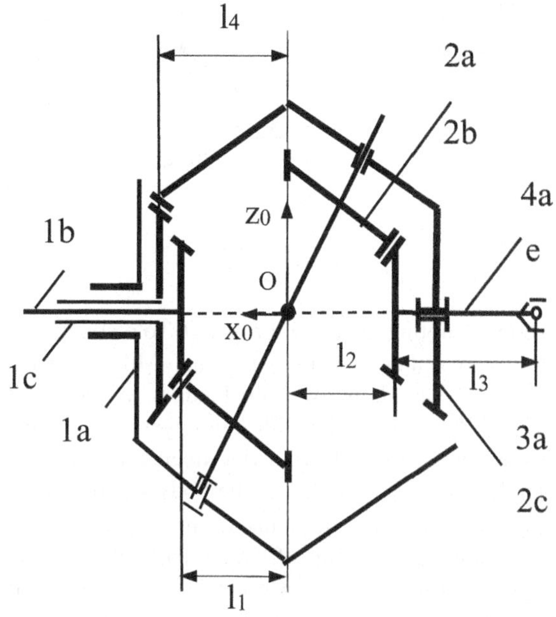

Fig. 10.18 Cincinnati-Milacron wrist robot

as end-effector $e = 4a$. A significant feature of this wrist device is that there is no physical interference between any of the links.

In the Cincinnati-Milacron wrist, the link $1a$ as a box of characteristic length l_0, mass m_1^A and tensor of inertia $\hat{J}_1^A$, one of the three driving parts of the spherical mechanism, serves as carrier for the $1b$–$2b$ and $1c$–$2c$ bevel gear pairs, while the link $3a$, of characteristic length r, mass m_3^A and tensor of inertia $\hat{J}_3^A$, serves as carrier for the $2a$–$4a$ bevel gear pair. The rotation of gear $1c$ is transmitted to link $3a$ via a special gear pair, while the rotation of second active gear $1b$ is transmitted to gear $4a$ through two coaxial gears. The two gears of radii r_2, r_3, total mass m_2^A and total inertia tensor $\hat{J}_2^A$ are planet gears rigidly connected together by a single shaft of known length that is housed in both carriers $1a$ and $3a$. The axes of rotation of three motors are supported by bearings housed in the forearm.

Including the end-effector of length l_3 at the last gear $4a = e$ of radius $r_4^A = r_4$, we obtain an assembly of total mass m_4^A and total tensor of inertia $\hat{J}_4^A$ that is free to arbitrary undergo three concurrent rotations with respect to the common centre O. Concerning the output motion of the end-effector, we remark that the wrist *roll motion* is achieved by rotating link $1a$ about z_1^A-axis while the *pitch motion* is accomplished by relative rotating link $3a$ with respect to link $1a$ about the z_3^A-axis. The assemblage of links and joints in the topology of the kinematical chains can be symbolized in the known form of an equivalent graph (Fig. 10.19), where we denote the links by vertices and the joints by edges [14, 15]. Two small concentric

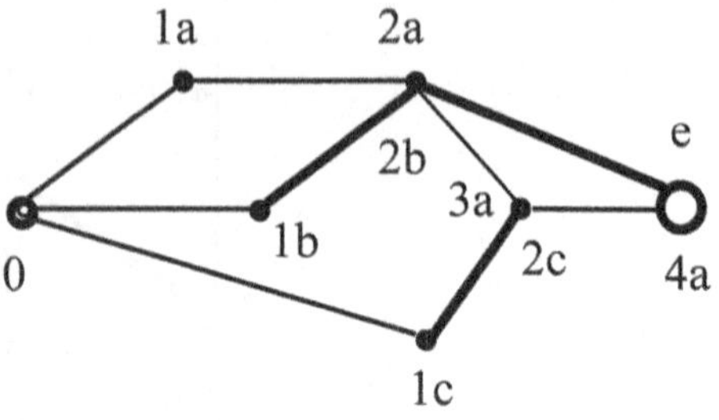

Fig. 10.19 Associated graph of the mechanism

circles label the vertex denoting the fixed forearm 0. To distinguish between the pair types, in the graph representation of the Cincinnati-Milacron wrist the three gear pairs 1*b*–2*b*, 1*c*–2*c*, 2*a*–4*a* are drawn with thick edges and the revolute joints 0–1*a*, 1*a*–2*a*, 2*a*–3*a*, 3*a*–4*a*, 0–1*b* and 0–1*c* with thin edges. There are three significant independent loops and we identify three serial kinematical chains.

The variable angles $\varphi_{k,k-1}$ of rotation of the body T_k about the joint axis z_k are the parameters needed to bring the next link from a reference configuration to the subsequent configuration. The motions of the six parts of the wrist mechanism are all concurrent rotations around the fixed point *O*. A matrix methodology for the kinematics analysis based on the concept of *fundamental circuit* of an open-loop chain involves the identification of open-loop chains and the derivation of the geometric relationships (10.2.1) between the orientation of the end-effector and the joint angles of compounding links [1, 5].

To simplify the graphical image of the kinematical scheme of the spherical mechanism, in the following we will represent the intermediate reference systems by only two axes. Consequently, four appropriate frames for the first circuit *A*, three frames for the second kinematical chain *B* and three frames for the last circuit *C* are fixed in a the same origin *O* (Fig. 10.20).

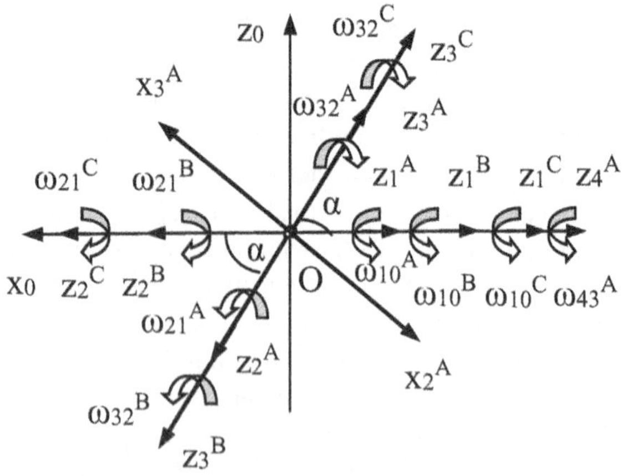

Fig. 10.20 Moving frames sequence

Let us consider the rotation angles $\varphi_{10}^{A}, \varphi_{10}^{B}, \varphi_{10}^{C}$ of the three actuators A_1, B_1, C_1 as variables that give the instantaneous position of the mechanism. Starting from the origin O and pursuing three independent serial circuits A, B, C: $0-1a-2a-3a-4a$, $0-1b-2b$ and $0-1c-2c$, we obtain the following successive matrices of transformation [10]

$$a_{10} = a_{10}^{\varphi}\theta_1^T, \quad a_{21} = a_{21}^{\varphi}a_\alpha\theta_2, \quad a_{32} = a_{32}^{\varphi}\theta_2, \quad a_{43} = a_{43}^{\varphi}a_\alpha^T$$
$$b_{10} = b_{10}^{\varphi}\theta_1^T, \quad b_{21} = b_{21}^{\varphi}\theta_2, \quad b_{32} = b_{32}^{\varphi}a_\alpha, \quad c_{10} = c_{10}^{\varphi}\theta_1^T, \quad c_{21} = c_{21}^{\varphi}\theta_2, \quad c_{32} = c_{32}^{\varphi}a_\alpha\theta_2, \tag{10.3.1}$$

where are denoted

$$a_\alpha = rot(y, \alpha), \quad \theta_1 = rot\left(y, \frac{\pi}{2}\right), \quad \theta_2 = rot(y, \pi)$$
$$q_{k,k-1}^{\varphi} = rot(z, \varphi_{k,k-1}^{i}), \quad q_{k0} = \prod_{s=1}^{k} q_{k-s+1,k-s} \quad (i = A, B, C) \quad (q = a, b) \tag{10.3.2}$$

In the inverse geometric problem, however, a complete description of the orientation of the end-effector in the fixed frame is known fro the intermediate of three Euler angles ϕ_1, ϕ_2, ϕ_3 associated with three successive rotations, which can be expressed by the analytical functions

$$\phi_l = \phi_l^*\left[1 - \cos\left(\frac{\pi}{6}t\right)\right], \quad (l = 1, 2, 3), \tag{10.3.3}$$

where $2\phi_l^*$ represents the maximum of the angle of rotation ϕ_l.

Since all rotations take place successively about the moving coordinate axes, the general rotation matrix $d_{30} = d_{32}d_{21}d_{10}$ of the end-effector e from $Ox_0y_0z_0$ to $Ox_4^Ay_4^Az_4^A$ reference system is obtained by multiplying the three transformation matrices $d_{10} = d_1^{\phi}\theta_1, d_{21} = d_2^{\phi}a_\alpha, d_{32} = d_3^{\phi}a_\alpha^T$, where $d_l^{\phi} = rot(z, \phi_l)$, $(l = 1, 2, 3)$.

The *constraining geometric conditions* for the rotation of the end-effector are given by the identities

$$a_{20} = b_{30}, \quad a_{30} = c_{30}, \quad a_{40} = d_{30}. \tag{10.3.4}$$

From these conditions and the general equations (10.2.1), we obtain the real-time evolution of all characteristic joint angles, as follows:

$$\varphi_{10}^{A} = \phi_1, \quad \varphi_{21}^{A} = -\phi_2 + \frac{\phi_3}{n_2}, \quad \varphi_{32}^{A} = \frac{\phi_3}{n_2}, \quad \varphi_{43}^{A} = \phi_3$$
$$\varphi_{10}^{B} = \phi_1 + \frac{1}{n_1}\left(-\phi_2 + \frac{\phi_3}{n_2}\right), \quad \varphi_{21}^{B} = \frac{1}{n_1}\left(-\phi_2 + \frac{\phi_3}{n_2}\right), \quad \varphi_{32}^{B} = -\phi_2 + \frac{\phi_3}{n_2}$$
$$\varphi_{10}^{C} = \phi_1 + \frac{\phi_2}{n_3}, \quad \varphi_{21}^{C} = \frac{\phi_2}{n_3}, \quad \varphi_{32}^{C} = \phi_2$$
$$n_1 = \frac{r_1}{r_2}, \quad n_2 = \frac{r_3}{r_4}, \quad n_3 = \frac{r_5}{r}. \tag{10.3.5}$$

Based on the remark $\omega_{k,k-1} = N_{k+1,k-1}\omega_{k+1,k}$, the derivatives with respect to time of these relations lead to the relative angular velocities of all links as a function of the angular velocities $\omega_1 = \dot{\phi}_1, \omega_2 = \dot{\phi}_2$, $\omega_3 = \dot{\phi}_3$ of the end-effector

$$\omega_{10}^{A} = \dot{\phi}_1, \quad \omega_{21}^{A} = -\dot{\phi}_2 + \frac{\dot{\phi}_3}{n_2}, \quad \omega_{32}^{A} = \frac{\dot{\phi}_3}{n_2}, \quad \omega_{43}^{A} = \dot{\phi}_3$$
$$\omega_{10}^{B} = \dot{\phi}_1 + \frac{1}{n_1}\left(-\dot{\phi}_2 + \frac{\dot{\phi}_3}{n_2}\right), \quad \omega_{21}^{B} = \frac{1}{n_1}\left(-\dot{\phi}_2 + \frac{\dot{\phi}_3}{n_2}\right), \quad \omega_{32}^{B} = -\dot{\phi}_2 + \frac{\dot{\phi}_3}{n_2}$$
$$\omega_{10}^{C} = \dot{\phi}_1 + \frac{\dot{\phi}_2}{n_3}, \quad \omega_{21}^{C} = \frac{\dot{\phi}_2}{n_3}, \quad \omega_{32}^{C} = \dot{\phi}_2. \tag{10.3.6}$$

Starting from the relationship $\vec{\omega}_{10} = J\vec{\omega}$, the expression of the Jacobian matrix is easily written in an invariant form

$$J = \begin{bmatrix} 1 & 0 & 0 \\ 1 & -1/n_1 & 1/(n_1 n_2) \\ 1 & 1/n_3 & 0 \end{bmatrix}, \quad J^{-1} = \begin{bmatrix} 1 & 0 & 0 \\ -n_3 & 0 & n_3 \\ -n_2(n_1 + n_3) & n_1 n_2 & n_2 n_3 \end{bmatrix}, \tag{10.3.7}$$

where $\vec{\omega}_{10} = \begin{bmatrix} \omega_{10}^{A} & \omega_{10}^{B} & \omega_{10}^{C} \end{bmatrix}^T$ denotes the input vector of three angular velocities and $\vec{\omega} = \begin{bmatrix} \dot{\phi}_1 & \dot{\phi}_2 & \dot{\phi}_3 \end{bmatrix}^T$ is the vector of joint angular velocities of the end-effector. This square invertible matrix is an essential element for the analysis of the loci into robot workspace.

Knowing the general rotation motion of the end-effector $e = 4a$ by the relations (10.3.3), the kinematics of an element T_k^i for each circuit is characterized by absolute angular velocities $\vec{\omega}_{k0}^{i} = q_{k,k-1}\vec{\omega}_{k-1,0}^{i} + \vec{\omega}_{k,k-1}^{i}$ and their *associated* skew-symmetric matrices $\tilde{\omega}_{k0}^{i} = q_{k,k-1}\tilde{\omega}_{k-1,0}^{i} q_{k,k-1}^{T} + \tilde{\omega}_{k,k-1}^{i}$.

The relative angular accelerations $\varepsilon_{k,k-1}^{i} = \dot{\omega}_{k,k-1}^{i}$ are given by deriving the relations of the velocities (10.3.6), while the accelerations $\vec{\gamma}_{k0}^{i}, \vec{\varepsilon}_{k0}^{i}$ of each of the moving links and the square matrices $\tilde{\omega}^i_{k0}\tilde{\omega}^i_{k0}+\tilde{\varepsilon}^i_{k0}$ are computed using general formulae (10.1.12). The velocity $\vec{v}^{Ci}_k$ and the acceleration $\vec{\gamma}^{Ci}_k$ of mass centre of T^i_k rigid body are calculated from two known matrix relations

$$\vec{v}^{Ci}_k = \tilde{\omega}^i_{k0}\vec{r}^{Ci}_k, \quad \vec{\gamma}^{Ci}_k = \{\tilde{\omega}^i_{k0}\tilde{\omega}^i_{k0}+\tilde{\varepsilon}^i_{k0}\}\vec{r}^{Ci}_k \tag{10.3.8}$$

where

$$\begin{aligned} &z^{CA}_1 = 0, \quad z^{CA}_2 = 0.5(l_1 - l_2), \quad z^{CA}_3 = r/3, \quad z^{CA}_4 = l_2 + 0.8l_3 \\ &z^{CB}_1 = -l_1, \quad z^{CC}_1 = -l_4, \quad \vec{r}^{Ci}_k = \begin{bmatrix} 0 & 0 & z^{Ci}_k \end{bmatrix}^T, \quad (i = A, B, C). \end{aligned} \tag{10.3.9}$$

10.3.2 Dynamics Modelling of the Robot

Three torques of moment $\vec{m}^A_{10} = m^A_{10}\vec{u}_3$, $\vec{m}^B_{10} = m^B_{10}\vec{u}_3$, $\vec{m}^C_{10} = m^C_{10}\vec{u}_3$ can control by the intermediary of electric motors, the motion of the wrist mechanism.

10.3.2.1 Principle of Virtual Work

Neglecting the frictional forces and considering the action of a resistant torque $\vec{m}^{*E}_{2r} = -M_r\vec{u}_3$, the objective of the dynamic model is to determine the input torques which must be exerted by the actuators in order to produce a given trajectory of the end-effector.

To apply the principle of virtual work, in the inverse dynamic problem, the parallel mechanism can artificially be transformed into a set of three open serial chains subjected to the constraints by cutting open at two bevel gear pairs $1b-2b$ and $1c-2c$. The wrench of inertia force of an arbitrary rigid body T^A_k, for example,

$$\vec{f}^{\,inA}_k = -m^A_k(\tilde{\omega}^A_{k0}\tilde{\omega}^A_{k0}+\tilde{\varepsilon}^A_{k0})\vec{r}^{CA}_k, \quad \vec{m}^{inA}_k = -\hat{J}^A_k\vec{\varepsilon}^A_{k0} - \tilde{\omega}^A_{k0}\hat{J}^A_k\vec{\omega}^A_{k0} \tag{10.3.10}$$

and the wrench of its weight

$$\vec{f}^{*A}_k = -gm^A_k a_{k0}\vec{u}_3, \quad \vec{m}^{*A}_k = -gm^A_k\tilde{r}^{CA}_k a_{k0}\vec{u}_3 \ \ (k = 1, 2, \ldots, 6) \tag{10.3.11}$$

are all computed with respect to the common centre of rotation O.

Let us assume now that the robot successively has three independent virtual motions. First, we consider the following input angular velocities $\omega^{Av}_{10a} = 1$, $\omega^{Bv}_{10a} = 0$, $\omega^{Cv}_{10a} = 0$. A set of requisite virtual velocities is obtained from (10.3.6):

$$\omega^{Av}_{21a} = -n_1, \quad \omega^{Av}_{32a} = -(n_1+n_3), \quad \omega^{Av}_{43a} = -n_2(n_1+n_3). \tag{10.3.12}$$

A second virtual motion is defined by the input velocities $\omega_{10b}^{Bv} = 1$, $\omega_{10b}^{Cv} = 0$, $\omega_{10b}^{Av} = 0$, and a new set results:

$$\omega_{21b}^{Av} = n_1, \quad \omega_{32b}^{Av} = n_1, \quad \omega_{43b}^{Av} = n_1 n_2. \tag{10.3.13}$$

Finally, from the third virtual motion $\omega_{10c}^{Cv} = 1$, $\omega_{10c}^{Av} = 0$, $\omega_{10c}^{Bv} = 0$ we obtain

$$\omega_{21c}^{Av} = 0, \quad \omega_{32c}^{Av} = n_3, \quad \omega_{43c}^{Av} = n_2 n_3. \tag{10.3.14}$$

Applying *the equations of parallel robots dynamics* [19], derived from the principle of virtual work, compact matrix relations results for the torques of three actuators:

$$\begin{aligned}
m_{10}^{A} &= \vec{u}_3^{T}\{\omega_{10a}^{Av}\vec{M}_1^{A} + \omega_{21a}^{Av}\vec{M}_2^{A} + \omega_{32a}^{Av}\vec{M}_3^{A} + \omega_{43a}^{Av}\vec{M}_4^{A}\}\\
m_{10}^{B} &= \vec{u}_3^{T}\{\omega_{10b}^{Bv}\vec{M}_1^{B} + \omega_{21b}^{Av}\vec{M}_2^{A} + \omega_{32b}^{Av}\vec{M}_3^{A} + \omega_{43b}^{Av}\vec{M}_4^{A}\}\\
m_{10}^{C} &= \vec{u}_3^{T}\{\omega_{10c}^{Cv}\vec{M}_1^{C} + \omega_{21c}^{Av}\vec{M}_2^{A} + \omega_{32c}^{Av}\vec{M}_3^{A} + \omega_{43c}^{Av}\vec{M}_4^{A}\},
\end{aligned} \tag{10.3.15}$$

where the significant vectors $\vec{F}_k^{i}, \vec{M}_k^{i} (i = A, B, C)$ are obtained using above recursive relations (10.1.14). These relations represent the *inverse dynamic model* of the Cincinnati-Milacron wrist mechanism.

10.3.2.2 Lagrange Equations

The generalized coordinates of the Cincinnati-Milacron mechanism are represented by the three rotation angles of the actuators: $s_1 = \varphi_{10}^{A}$, $s_2 = \varphi_{10}^{B}$, $s_3 = \varphi_{10}^{C}$. Lagrange's equations will be expressed by three differential relations

$$\frac{d}{dt}\left\{\frac{\partial L}{\partial \dot{s}_j}\right\} - \frac{\partial L}{\partial s_j} = \Phi_j \quad (j = 1, 2, 3), \tag{10.3.16}$$

containing the generalized forces

$$\Phi_1 = m_{10}^{A} + M_r n_2 (n_1 + n_3), \quad \Phi_2 = m_{10}^{B} - M_r n_1 n_2, \quad \Phi_3 = m_{10}^{C} - M_r n_2 n_3. \tag{10.3.17}$$

The components of the general expression $L = \sum_{k=1}^{4} L_k^{A} + L_1^{B} + L_1^{C}$ of the *Lagrange function* are given as analytical functions of the generalized coordinates and their first derivatives with respect to time:

$$L_k^{A} = \frac{1}{2}\vec{\omega}_{k0}^{AT}\hat{J}_k^{A}\vec{\omega}_{k0}^{A} - m_k^{A} g \vec{u}_3^{T} a_{k0}^{T}\vec{r}_k^{CA}, \quad L_1^{B} = \frac{1}{2}\vec{\omega}_{10}^{BT}\hat{J}_1^{B}\vec{\omega}_{10}^{B}, \quad L_1^{C} = \frac{1}{2}\vec{\omega}_{10}^{CT}\hat{J}_1^{C}\vec{\omega}_{10}^{C}, \tag{10.3.18}$$

where the requisite angular velocities have the expressions:

$$\begin{aligned}&\vec{\omega}_{10}^{A} = \dot{s}_1\vec{u}_3, \quad \vec{\omega}_{21}^{A} = -n_1(\dot{s}_1 - \dot{s}_2)\vec{u}_3, \quad \vec{\omega}_{32}^{A} = -(n_1 + n_3)(\dot{s}_1 - \dot{s}_3)\vec{u}_3\\&\vec{\omega}_{43}^{A} = -n_2[(n_1 + n_3)\dot{s}_1 - n_1\dot{s}_2 - n_3\dot{q}_3]\vec{u}_3, \quad \vec{\omega}_{10}^{B} = \dot{s}_2\vec{u}_3, \quad \vec{\omega}_{10}^{C} = \dot{s}_3\vec{u}_3.\\&\vec{\omega}_{20}^{A} = a_{21}\vec{\omega}_{10}^{A} + \vec{\omega}_{21}^{A}, \quad \vec{\omega}_{30}^{A} = a_{32}\vec{\omega}_{20}^{A} + \vec{\omega}_{32}^{A}, \quad \vec{\omega}_{40}^{A} = a_{43}\vec{\omega}_{30}^{A} + \vec{\omega}_{43}^{A}.\end{aligned} \tag{10.3.19}$$

The first derivatives of orthogonal matrices $q_{k,k-1}$ are expressed as follows:

$$\begin{aligned}&\dot{q}_{k,k-1} = \dot{\varphi}_{k,k-1}^{i}\tilde{u}_3^T q_{k,k-1}, \quad \dot{q}_{k,k-1}^{T} = \dot{\varphi}_{k,k-1}^{i} q_{k,k-1}^{T}\tilde{u}_3\\&\frac{\partial q_{k,k-1}}{\partial \varphi_{k,k-1}^{i}} = \tilde{u}_3^T q_{k,k-1}, \quad \frac{\partial q_{k,k-1}^{T}}{\partial \varphi_{k,k-1}^{i}} = q_{k,k-1}^{T}\tilde{u}_3, \quad (q = a, b, c).\end{aligned} \tag{10.3.20}$$

In the inverse dynamics problem, the derivatives with respect to time $\frac{d}{dt}\left\{\frac{\partial L}{\partial \dot{q}_j}\right\}(j = 1, 2, 3)$ of the above functions leads ultimately to the same expressions (10.3.15) for the input torques $m_{10}^{A}, m_{10}^{C}, m_{10}^{C}$ required by actuators.

For simulation purposes let us consider a mechanism which has the following characteristics

$$\begin{aligned}&l_0 = 0.075\,\text{m}, \quad l_1 = 0.045\,\text{m}, \quad l_2 = 0.035\,\text{m}, \quad l_3 = 0.035\,\text{m}, \quad l_4 = 0.055\,\text{m}\\&r = 0.065\,\text{m}, \quad r_1 = 0.025\,\text{m}, \quad r_2 = 0.025\,\text{m}, \quad r_3 = 0.02\,\text{m}, \quad r_4 = 0.02\,\text{m}, \quad r_5 = 0.03\,\text{m}\\&m_1^A = 0.75\,\text{kg}, \quad m_2^A = 0.6\,\text{kg}, \quad m_3^A = 0.6\,\text{kg}, \quad m_4^A = 1.05\,\text{kg}, \quad m_1^B = 0.35\,\text{kg}, \quad m_1^C = 0.4\,\text{kg}\\&\phi_1^* = \frac{\pi}{2}, \quad \phi_2^* = \frac{\pi}{3}, \quad \phi_3^* = \pi, \quad \alpha = \frac{\pi}{3}, \quad M_r = 0.05\,\text{Nm}, \quad \Delta t = 6\,\text{s}.\end{aligned}$$

Assuming that for a period of six seconds the end-effector is moving in a given rotation motion, a first numerical study of the kinematics has been developed in MATLAB by computation of the input angles of rotation $\varphi_{10}^{A}, \varphi_{10}^{B}, \varphi_{10}^{C}$ of the three revolute actuators (Fig. 10.21). Based on the algorithm derived from the relations

Fig. 10.21 Input rotation angles $\varphi_{10}^{A}, \varphi_{10}^{B}, \varphi_{10}^{C}$

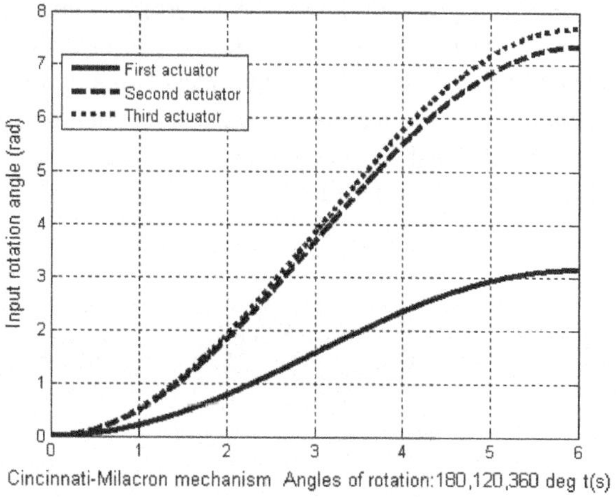

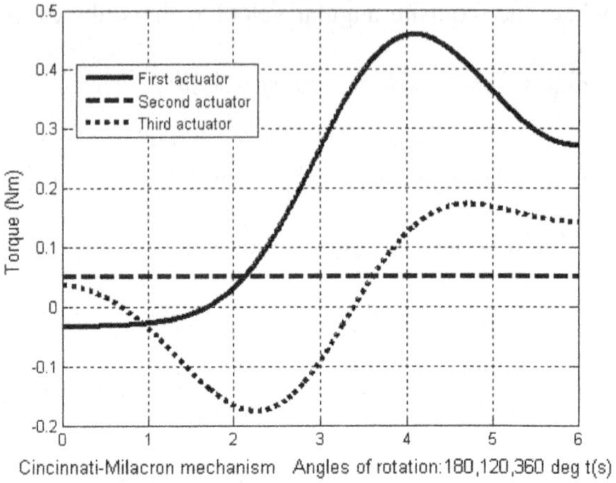

Fig. 10.22 Input torques $m_{10}^{A}, m_{10}^{B}, m_{10}^{C}$ of three actuators

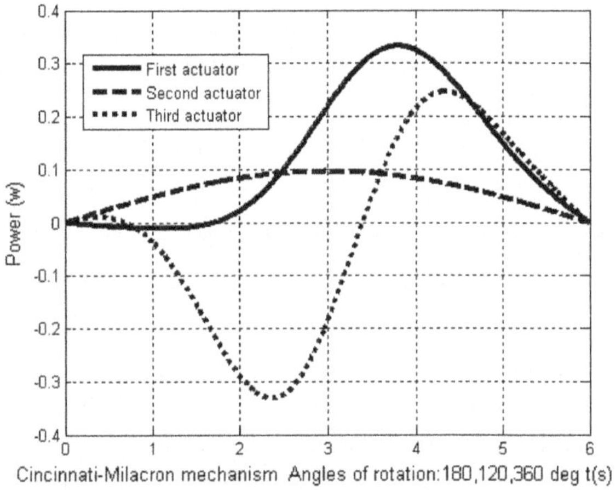

Fig. 10.23 Input powers $P_{10}^{A}, P_{10}^{B}, P_{10}^{C}$ of three actuators

(10.3.15), a program solves the inverse dynamics modelling of the mechanism. Considering that a resistant torque of constant moment $M_r = 0.05\,\text{Nm}$ applied at the end-effectors and the weights $m_k^i \vec{g}$ of rigid bodies constitute the external forces acting on the mechanism, time-history evolution of input torques $m_{10}^{A}, m_{10}^{B}, m_{10}^{C}$ (Fig. 10.22) and active powers $P_{10}^{A} = \omega_{10}^{A} m_{10}^{A}$, $P_{10}^{B} = \omega_{10}^{B} m_{10}^{B}$, $P_{10}^{C} = \omega_{10}^{C} m_{10}^{C}$ (Fig. 10.23) are sketched.

References

1. Hsieh, L-C., Sheu, K-B.: Conceptual design of planetary bevel-gear trains for robotic wrist. In: Proceedings of 9th World Congress on the Theory of Machines and Mechanisms, Milan, pp. 1861–1865 (1995)
2. Ma, R., Gupta, K.C.: On the motion of oblique geared robot wrists. J. Robot. Syst. **5**, 509–520 (1988)
3. Paul, R.P., Stevenson, C.N.: Kinematics of robot wrists. Int. J. Robot. Res. **2**(1), 31–38 (1983)
4. White, G.: Epicyclical gears applied to early steam engines. Mech. Mach. Theory **23**(1), 25–37 (1988)
5. Willis, R.J.: On the kinematics of the closed epicyclical differential gears. J. Mech. Des. **104**, 712–723 (1982)
6. Tsai, L.-W.: An application of the linkage characteristic polynomial to the topological synthesis of planetary gear trains. ASME J. Mech. Transm. Automat. Des. **109**(3), 329–336 (1987)
7. Tsai, L.-W.: The kinematics of spatial robotic bevel-gear trains. IEEE J. Robot. Autom. **4**(2), 150–156 (1988)
8. Tsai, L.-W., Chen, D.-Z., Lin, T.-W.: Dynamic analysis of geared robotic mechanisms using graph theory. J. Mech. Des. **120**(2), 240–244 (1998)
9. Tsai, L.-W.: Robot Analysis: The Mechanics of Serial and Parallel Manipulators. Wiley, New York (1999)
10. Staicu, S.: Inverse dynamics of a planetary gear train for robotics. Mech. Mach. Theory **43**(7), 918–927 (2008)
11. Freudenstein, F., Longman, R.W., Chen, C.K.: Kinematic analysis of robotic bevel-gear train. ASME J. Mech. Transm. Automat. Des. **106**(3), 371–375 (1984)
12. Chang, S.-L., Tsai, L.-W.: Topological synthesis of articulated gear mechanisms. IEEE J. Robot. Automat. **6**(1), 97–103 (1989)
13. Hedman, A.: Transmission analysis: automatic derivation of relationships. J. Mech. Des. **115** (4), 1031–1037 (1993)
14. Yan, H-S., Hsieh, L-C.: Kinematic analysis of general planetary gear trains. In: Proceedings of 8th World Congress on the Theory of Machines and Mechanisms, Prague, pp. 153–157 (1991)
15. Yan, H.-S., Hsieh, L.-C.: Conceptual design of gear differentials for automatic vehicles. J. Mech. Des. **116**, 565–570 (1994)
16. Staicu, S., Liu, X.-J., Wang, J.: Inverse dynamics of the HALF parallel manipulator with revolute actuators. Nonlinear Dyn. **50**(1–2), 1–12 (2007)
17. Staicu, S., Zhang, D.: A novel dynamic modelling approach for parallel mechanisms analysis. Robot. Comput.-Integr. Manuf. **24**(1), 167–172 (2008)
18. Chablat, D., Wenger, P., Staicu, S.: Dynamics of the orthoglide parallel robot. UPB Sci. Bull., Ser. D: Mech. Eng. **71**(3), 3–16 (2009)
19. Staicu, S.: Matrix modeling of inverse dynamics of spatial and planar parallel robots. Multibody Syst. Dyn. **27**(2), 239–265 (2012)

Chapter 11
Mobile Wheeled Robots

While the classical holonomic constraints limit the freedom of motion through the position variables, and thus simultaneously through the velocity variables, the non-holonomic constraints, realized purely mechanically via (for example) rolling wheels, lead to a restriction of the velocities only, not of the positions. The non-holonomic constraints pertain to the kinematical constraints and can be either scleronomic or rheonomic. A transport car with two independent wheels is characterized by the fact that the mid-point cannot move in either of the two body-fixed directions, owing to the friction forces on the wheels.

11.1 Kinematics and Dynamics of a Mobile Robot Provided with Caster Wheel

Wheeled mobile robots are pre-programmable multi-functional systems with autonomous control designed to move material, parts, tools or specialized devices through variable programmed motions for the performance of a variety of tasks. They consist of a mobile platform and a number of cylindrical wheels, which undertake a rolling with friction motion on a fixed or mobile surface. Such vehicles are used in automated industrial processes mainly for the transport of material and tools over distances much greater than their own dimensions.

In order to achieve robot locomotion, wheeled mobile robots are widely used in many applications. The mobile robot is capable of autonomous motion because it is equipped with motors that are driven by an on-board computer. The property of autonomy is understood as the ability to independently make intelligent decisions and circumstances change. These machineries are used in inaccessible environments that tend to be cluttered with lots of moving and fixed unknown obstacles, in agricultural settings or for special medical procedures. This is why the study of mobile robots dynamics has acquired more and more importance [1, 2].

S. Staicu, *Dynamics of Parallel Robots*, Parallel Robots: Theory and Applications,
https://doi.org/10.1007/978-3-319-99522-9_11

Wheeled mobile robots are a class of mechanical systems characterized by kinematical constraints that cannot be integrated. The condition of rolling motion without slipping and side-slipping between the wheel and the contact surface necessitates the presence of non-holonomic constraints, which represent the kinematics particularity of this kind of robot. The common characteristic of mobile robots is that they cannot autonomously produce a velocity which is transversal to the axis of their wheels. The non-holonomic constraints reduce the mobile robot's instantaneous velocity degrees of freedom; hence, robots have only two actuated joints. On the other hand, mobile robots are more complex to control than serial and parallel robots, due to non-holonomic constraints. However, at the instantaneous velocities level, mobile robots can be treated mathematically as a *special type of parallel robot*, having different connections in parallel to the ground.

Angeles [1] studied some interesting aspects of the mobile robots dynamics using the formalism of Lagrange equations. Other authors [3, 4] gave a characterization of the mechanical non-holonomic systems. Volterra, Appel and Ceaplighin used also Lagrange's equations and multiplier formalism in the dynamics of motion with non-holonomic links. Recently, neural networks have emerged as powerful tools for learning dynamics of highly non-linear systems [5].

The analysis of problems of two-wheeled mobile robot dynamics is mainly undertaken to solve successfully the control of the motion of such systems. Simple models are very often accepted for a system description, even though they do not take into account the mass of the many mobile elements. Authors describing the dynamics of such systems use classical equations taken from Newtonian mechanics, and most typically, the approaches describe the motion of these systems using second-order Lagrange equations [6–8].

The kinematics and dynamics of a rolling robot, consisting of a platform, two conventional wheels and a crank that controls the motion of a free rolling caster wheel, are what is analyzed in the present study. Based on several matrix relations of connectivity, the characteristic velocities and accelerations of this non-holonomic mechanical system are derived. Using the principle of virtual work, expressions and graphs for the time-history evolution of torques and powers required by the two driving wheels are established. The results have been verified using the conceptual framework of second-order with their multipliers.

11.1.1 Kinematics Model of the Robot

Let us consider a mobile robot, sketched by Angeles [1], with three conventional wheels that can roll on a horizontal surface without slipping (Fig. 11.1). This kind of differentially-driven robot needs three non-collinear support points in order not to topple over. In practice, the robot can turn on the spot by giving opposite speeds to its two actuated wheels. The mobile robots are made up of a rigid frame with non-deformable wheels, and sometimes they move on fixed horizontal ground. Sometimes, the mobile platform is used to carry and support a modular manipulator

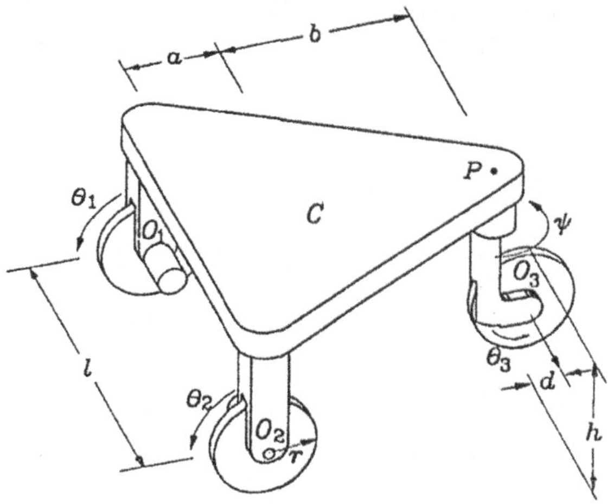

Fig. 11.1 General layout of wheeled mobile robot

and usually it has two driving wheels and one supporting wheel. Two driving wheels can be acted upon independently in order to achieve a differential drive in case of turning. The supporting passive wheel is usually introduced to balance the mobile platform.

In the kinematically simplified scheme of the robot, we will represent the intermediate reference systems by only two axes. It is noted that the relative rotation with the angle θ_i must be always about the direction of the z_k axis. An arbitrary inertial base frame $Ox_0y_0z_0(T_0)$ is fixed on the horizontal plane, while a moving central reference frame $Gx_1y_1z_1$ is attached to the moving platform, such that y_1 axis is parallel to the coaxial line of the two driving wheels, while the x_1 axis passes through the midpoint of the line segment connecting the two wheel centres. The platform is an isosceles triangle with the dimensions l for the base and $a+b$ for its height, having mass m_1 and the tensor of inertia $\hat{J}_1$. Two cylindrical coaxial driving wheels of the same radius r are fixed to the frames $A_2x_2^Ay_2^Az_2^A$, $B_2x_2^By_2^Bz_2^B$ and connect to the chassis by means of revolute joints in two points $A_2 = O_1$ and $B_2 = O_2$. They have masses $m_2^A = m_2^B$ and the central tensors of inertia $\hat{J}_2^A = \hat{J}_2^B$. A crank $C_2C_3 = PO_3$ is jointed to the moving platform in the point $C_2 = P$ of the triangle. Its mass and tensor of inertia with respect C_2 are, respectively, m_2^C and $\hat{J}_2^C$. This rigid element can permanently orientate the motion of a passive rolling castor wheel of small radius r_0, mass m_3^C and tensor of inertia $\hat{J}_3^C$ (Fig. 11.2). Mounted in a swivel frame, the caster wheel is a non-actuated wheel and has no kinematic function; its only purpose is to keep the robot in balance.

Let us now analyze the motion of the robot on a curved trajectory in the turning period between two permanent rectilinear motions. The non-holonomic constraints reduce the mobile robot's velocity degrees of freedom and hence the robot has only two actuated joints.

In the forward velocity kinematics (FVK), we will consider that two independent input rotation angles θ_1, θ_2 of the driven wheels can determine completely the

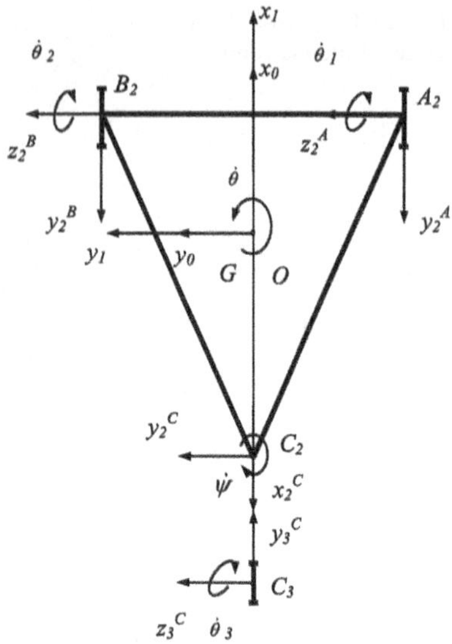

Fig. 11.2 Kinematical scheme of the mobile robot

instantaneous position and orientation of the robot. Thus, since the platform has a planar motion, its position with respect to a fixed reference frame $Ox_0y_0z_0$ with origin O on the horizontal ground is given by the coordinates x_{10}, y_{10}, H and by the angle of rotation θ, which form the following matrices:

$$\vec{r}_{10} = \begin{bmatrix} x_{10} \\ y_{10} \\ H \end{bmatrix}, \quad a_{10} = rot(z, \theta) = \begin{bmatrix} \cos\theta & \sin\theta & 0 \\ -\sin\theta & \cos\theta & 0 \\ 0 & 0 & 1 \end{bmatrix}. \qquad (11.1.1)$$

Let us denote by $\theta_i (i = 1, 2, 3)$ the rotation angles of the three wheels and by $\theta_4 = \psi$ the relative angle of rotation of the crank about the apex C_2 of the platform. The mobile robot moves over a horizontal plane, so that the rotation axis is permanently perpendicular to the plane of translation.

In the study of the kinematics of mobile robots, we are interested in deriving a matrix equation relating the location of an arbitrary body T_k to the joint variables. If every link is connected to least two other links, the chain forms one or more independent closed loops. We call the matrix $a^{\varphi}_{k,k-1}$, for example, the orthogonal transformation 3×3 matrix of relative rotation with the angle $\varphi^{A}_{k,k-1}$ of link T^A_k around the z^A_k axis. We obtain the following orthogonal transformation matrices in the reference frames [9]:

$$a^A_{21} = a^{\theta_1}_z a_1 a_2, \quad a^B_{21} = a^{\theta_2}_z a_1 a_2, \quad a^C_{21} = a^{\theta_4}_z a_3, \quad a^C_{32} = a^{\theta_3}_z a_1 a_2, \qquad (11.1.2)$$

where

$$a_1 = rot\left(y, \frac{\pi}{2}\right), \quad a_2 = rot\left(z, \frac{\pi}{2}\right), \quad a_3 = rot(y, \pi), \quad a_z^{\theta_1} = rot(z, \theta_i) \quad (i = 1, 2, 3, 4). \tag{11.1.3}$$

If the distance $A_2B_2 = l$ between the two actuated wheels is known, as well as the characteristic dimensions d, h of the crank PO_3, then the following vectors give the invariable positions of the revolute joints A_2, B_2, C_2:

$$\vec{r}_{21}^{\,A} = \begin{bmatrix} a \\ -l/2 \\ -h_0 \end{bmatrix}, \quad \vec{r}_{21}^{\,B} = \begin{bmatrix} a \\ l/2 \\ -h_0 \end{bmatrix}, \quad \vec{r}_{21}^{\,C} = \begin{bmatrix} -b \\ 0 \\ 0 \end{bmatrix}, \quad \vec{r}_{32}^{\,C} = \begin{bmatrix} d \\ 0 \\ h \end{bmatrix}. \tag{11.1.4}$$

Thus, the kinematics of the elements of the robot is completely characterized by five relative angular velocities

$$\begin{aligned} &\omega_{10} = \dot{\theta}\vec{u}_3, \quad \vec{\omega}_{21}^{\,A} = \dot{\theta}_1\vec{u}_3, \quad \vec{\omega}_{21}^{\,B} = \dot{\theta}_2\vec{u}_3 \\ &\vec{\omega}_{21}^{\,C} = \dot{\theta}_4\vec{u}_3, \quad \vec{\omega}_{32}^{\,C} = \dot{\theta}_3\vec{u}_3, \quad \vec{u}_3 = \begin{bmatrix} 0 & 0 & 1 \end{bmatrix}^T, \end{aligned} \tag{11.1.5}$$

which are *associated* with following the skew-symmetric matrices

$$\begin{aligned} &\tilde{\omega}_{10} = \dot{\theta}\tilde{u}_3, \quad \tilde{\omega}_{21}^{A} = \dot{\theta}_1\tilde{u}_3, \quad \tilde{\omega}_{21}^{B} = \dot{\theta}_2\tilde{u}_3 \\ &\tilde{\omega}_{21}^{C} = \dot{\theta}_4\tilde{u}_3, \quad \tilde{\omega}_{32}^{C} = \dot{\theta}_3\tilde{u}_3, \quad \tilde{u}_3 = \begin{bmatrix} 0 & -1 & 0 \\ 1 & 0 & 0 \\ 0 & 0 & 0 \end{bmatrix}. \end{aligned} \tag{11.1.6}$$

Since the analyzed system of three rolling wheels are characterized by non-holonomic constraints, the following *matrix conditions of connectivity* are capable of establishing five analytical relations between the characteristic velocities of a two-degree-of-freedom mobile robot:

$$\begin{aligned} &\vec{v}_{10} + \tilde{\omega}_{10}\vec{r}_{21}^{\,A} = \begin{bmatrix} r\dot{\theta}_1 & 0 & 0 \end{bmatrix}^T, \quad \vec{v}_{10} + \tilde{\omega}_{10}\vec{r}_{21}^{\,B} = \begin{bmatrix} r\dot{\theta}_2 & 0 & 0 \end{bmatrix}^T \\ &a_{21}^C(\vec{v}_{10} + \tilde{\omega}_{10}\vec{r}_{21}^{\,C}) + (a_{21}^C\tilde{\omega}_{10}a_{21}^{CT} + \tilde{\omega}_{21}^C)\vec{r}_{32}^{\,C} = \begin{bmatrix} -r_0\dot{\theta}_3 & 0 & 0 \end{bmatrix}^T. \end{aligned} \tag{11.1.7}$$

These constraint conditions are satisfied if all wheels do not slip transversally and do not slip longitudinally, so that the distance over which the outer wheel surface rotates equals the distance traveled by the point on the rigid body to which the wheel axle is attached.

Indeed, we assume in the *FVK* problem that the position and orientation of the mechanism at a given instant will be completely determined by the input rotation angles of the two actuated wheels, namely:

$$\theta_1 = \theta_1^* \left[1 - \cos\left(\frac{\pi}{3}t\right)\right], \theta_2 = \theta_2^* \left[1 - \cos\left(\frac{\pi}{3}t\right)\right]. \tag{11.1.8}$$

Therefore, the relations (11.1.7) can provide first the Jacobian matrix and secondly the expressions of the characteristic velocities of the moving platform:

$$\begin{gathered}\omega_{10} = \dot{\theta} = \frac{r}{l}(\dot{\theta}_1 - \dot{\theta}_2), \quad v_{10}^x = r\dot{\theta}_1 - \frac{1}{2}l\dot{\theta}, \quad v_{10}^y = -a\dot{\theta} \\ \begin{bmatrix}\dot{x}_{10} & \dot{y}_{10} & 0\end{bmatrix}^T = a_{10}^T \begin{bmatrix}v_{10}^x & v_{10}^y & 0\end{bmatrix}^T.\end{gathered} \tag{11.1.9}$$

Concerning the kinematics of the crank PO_3 and the passive caster wheel jointed in the point $O_3 = C_3$, from the same matrix conditions (11.1.7) we will derive a significant differential equation and a relation containing the angular velocities $\omega_{21}^C = \dot{\theta}_4 = \dot{\psi}, \omega_{32}^C = \dot{\theta}_3$, as follows:

$$\begin{gathered}d\dot{\psi} + r\dot{\theta}_1 \sin\psi - [d + 0.5l\sin\psi + (a+b)\cos\psi]\dot{\theta} = 0 \\ r_0\dot{\theta}_3 = [(a+b)\sin\psi - 0.5l\cos\psi]\dot{\theta} + r\dot{\theta}_1\cos\psi.\end{gathered} \tag{11.1.10}$$

In the forward position kinematics, the estimating of the relative angle of rotation ψ and the absolute pose of the moving platform must be performed by integration of the above velocity equation (11.1.10).

In order to determine new *conditions of connectivity* at level of the accelerations, we can proceed to derive the matrix relations (11.1.9). Thus, the characteristic accelerations of the moving platform are immediately obtained:

$$\begin{gathered}\varepsilon_{10} = \dot{\omega}_{10} = \ddot{\theta} = \frac{r}{l}(\ddot{\theta}_1 - \ddot{\theta}_2), \quad \gamma_{10}^x = r\ddot{\theta}_1 + a\dot{\theta}^2 - \frac{1}{2}l\ddot{\theta}, \quad \gamma_{10}^y = r\dot{\theta}\dot{\theta}_1 - a\ddot{\theta} - \frac{1}{2}l\dot{\theta}^2 \\ \begin{bmatrix}\ddot{x}_{10} & \ddot{y}_{10} & 0\end{bmatrix}^T = a_{10}^T \begin{bmatrix}\gamma_{10}^x & \gamma_{10}^y & 0\end{bmatrix}^T.\end{gathered} \tag{11.1.11}$$

Note that the absolute velocities $\vec{v}_{k0}^C, \vec{\omega}_{k0}^C$, the accelerations $\vec{\gamma}_{k0}^C, \vec{\varepsilon}_{k0}^C$ and the useful square matrices $\tilde{\omega}_{k0}^C\tilde{\omega}_{k0}^C + \tilde{\varepsilon}_{k0}^C$ of the third leg OC_2C_3 of the robot, for example, can be computed with some recursive matrix formulae:

$$\begin{gathered}\vec{v}_{20}^C = a_{21}^C\{\vec{v}_{10} + \tilde{\omega}_{10}\vec{r}_{21}^C\}, \quad \vec{v}_{30}^C = a_{32}^C\{\vec{v}_{20}^C + \tilde{\omega}_{20}^C\vec{r}_{32}^C\} \\ \vec{\omega}_{20}^C = a_{21}^C\vec{\omega}_{10} + \vec{\omega}_{21}^C, \quad \vec{\omega}_{30}^C = a_{32}^C\vec{\omega}_{20}^C + \vec{\omega}_{32}^C \\ \vec{\gamma}_{20}^C = a_{21}^C\{\vec{\gamma}_{10} + (\tilde{\omega}_{10}\tilde{\omega}_{10} + \tilde{\varepsilon}_{10})\vec{r}_{21}^C\}, \quad \vec{\gamma}_{30}^C = a_{32}^C\{\vec{\gamma}_{20}^C + (\tilde{\omega}_{20}^C\tilde{\omega}_{20}^C + \tilde{\varepsilon}_{20}^C)\vec{r}_{32}^C\} \\ \vec{\varepsilon}_{20}^C = a_{21}^C\vec{\varepsilon}_{10} + \vec{\varepsilon}_{21}^C + a_{21}^C\tilde{\omega}_{10}a_{21}^{CT}\vec{\omega}_{21}^C, \quad \vec{\varepsilon}_{30}^C = a_{32}^C\vec{\varepsilon}_{20}^C + \vec{\varepsilon}_{32}^C + a_{32}^C\tilde{\omega}_{20}^C a_{32}^{CT}\vec{\omega}_{32}^C \\ \tilde{\omega}_{20}^C\tilde{\omega}_{20}^C + \tilde{\varepsilon}_{20}^C = a_{21}^C\{\tilde{\omega}_{10}\tilde{\omega}_{10} + \tilde{\varepsilon}_{10}\}a_{21}^{CT} + \tilde{\omega}_{21}^C\tilde{\omega}_{21}^C + \tilde{\varepsilon}_{21}^C + 2a_{21}^C\tilde{\omega}_{10}a_{21}^{CT}\tilde{\omega}_{21}^C \\ \tilde{\omega}_{30}^C\tilde{\omega}_{30}^C + \tilde{\varepsilon}_{30}^C = a_{32}^C\{\tilde{\omega}_{20}^C\tilde{\omega}_{20}^C + \tilde{\varepsilon}_{20}^C\}a_{32}^{CT} + \tilde{\omega}_{32}^C\tilde{\omega}_{32}^C + \tilde{\varepsilon}_{32}^C + 2a_{32}^C\tilde{\omega}_{20}^C a_{32}^{CT}\tilde{\omega}_{32}^C.\end{gathered} \tag{11.1.12}$$

11.1.2 Equations of Motion

Two electric motors that generate the torques $\vec{M}_1 = M_1\vec{u}_3$ and $\vec{M}_2 = M_2\vec{u}_3$, which have the direction of the common axis $A_2B_2 = O_1O_2$, control the evolution of the driving wheels A_2, B_2 and transmit the motion at the passive caster wheel C_2.

11.1.2.1 Principle of Virtual Work

Based on the principle of virtual work, we develop an *inverse dynamic problem* in order to establish the variation of the torques M_1, M_2 and the powers P_1, P_2 developed by the two active wheels, during the evolution of the robot between the initial position and another, which corresponds to the stationary motion.

A first set of *virtual characteristic velocities* of the robot bodies result easily from the above constraining conditions (11.1.7), (11.1.9), namely:

$$\begin{aligned}
&\omega_{1a}^{v} = 1, \quad \omega_{2a}^{v} = 0, \quad \omega_{10a}^{v} = \frac{r}{l}, \quad v_{10a}^{xv} = \frac{r}{2}, \quad v_{10a}^{yv} = -a\frac{r}{l} \\
&\omega_{3a}^{v} = \frac{r}{r_0}\left(\frac{1}{2}\cos\psi + \frac{a+b}{l}\sin\psi\right), \quad \omega_{4a}^{v} = \frac{r}{d}\left(-\frac{1}{2}\sin\psi + \frac{a+b}{l}\cos\psi\right) + \frac{r}{l} \\
&\omega_{\varphi a}^{v} = \omega_{4a}^{v} - \omega_{10a}^{v}.
\end{aligned} \tag{11.1.13}$$

It is noteworthy that the simulation runs do not account for either external dissipation, such as rolling friction between the wheels and ground, or for internal dissipation such as friction in the bearings. Consequently, the virtual work produced by the forces of constraint at the joints is zero. Hence, the following compact expression of the torque applied to the right driving wheel A_2

$$\begin{aligned}
M_1 = {} & \vec{v}_{10a}^{\,vT}\vec{F}_{10} + \vec{\omega}_{10a}^{\,vT}\vec{M}_{10} + \vec{v}_{20a}^{\,AvT}\vec{F}_{20}^{A} + \vec{\omega}_{20a}^{\,AvT}\vec{M}_{20}^{A} + \vec{v}_{20a}^{\,BvT}\vec{F}_{20}^{B} + \vec{\omega}_{20a}^{\,BvT}\vec{M}_{20}^{B} \\
& + \vec{v}_{20a}^{\,CvT}\vec{F}_{20}^{C} + \vec{\omega}_{20a}^{\,CvT}\vec{M}_{20}^{C} + \vec{v}_{30a}^{\,CvT}\vec{F}_{30}^{C} + \vec{\omega}_{30a}^{\,CvT}\vec{M}_{30}^{C},
\end{aligned} \tag{11.1.14}$$

with its analytical form result

$$\begin{aligned}
M_1 = {} & m_1\omega_{10a}^{v}\left(\frac{1}{4}l^2\ddot{\theta} - \frac{1}{2}lr\ddot{\theta}_1 - ar\dot{\theta}\dot{\theta}_1 + a^2\ddot{\theta}\right) + m_1\omega_{1a}^{v}\left(r^2\ddot{\theta}_1 - \frac{1}{2}lr\ddot{\theta} + ar\dot{\theta}^2\right) + J_1^{z}\ddot{\theta}\omega_{10a}^{v} \\
& + \frac{1}{2}m_2^{A}r^2\left(3\omega_{1a}^{v}\ddot{\theta}_1 + \frac{1}{2}\omega_{10a}^{v}\ddot{\theta}\right) + \frac{1}{2}m_2^{B}r^2\left(3\omega_{2a}^{v}\ddot{\theta}_2 + \frac{1}{2}\omega_{10a}^{v}\ddot{\theta}\right) \\
& + m_2^{C}\omega_{3a}^{v}r_0\left[r_0\ddot{\theta}_3 - (d - x_2^{C})\omega_{\varphi}^2\right] + m_2^{C}\omega_{\varphi a}^{v}\left[r_0(d + x_2^{C})\dot{\theta}_3\omega_{\varphi} + \frac{d^3(3h+d)}{3(h+d)}\varepsilon_{\varphi}\right] \\
& + \frac{1}{2}m_3^{C}r_0^2\left(3\omega_{3a}^{v}\ddot{\theta}_3 + \frac{1}{2}\omega_{\varphi a}^{v}\varepsilon_{\varphi}\right), \quad \omega_{\varphi} = \dot{\psi} - \dot{\theta}, \; \varepsilon_{\varphi} = \ddot{\psi} - \ddot{\theta}.
\end{aligned} \tag{11.1.15}$$

The force of inertia $\vec{f}_{k0}^{C}$, the torque $\vec{m}_{k0}^{C}$ of inertia forces and the vectors $\vec{F}_{k0}^{C}$, $\vec{M}_{k0}^{C}$ of the third leg OC_2C_3, for example, have the following general form:

$$\vec{f}_{k0}^{C} = -m_k^C\{\vec{\gamma}_{k0}^{C} + (\tilde{\omega}_{k0}^{C}\tilde{\omega}_{k0}^{C} + \tilde{\varepsilon}_{k0}^{C})\vec{r}_k^{C}\}, \quad \vec{m}_{k0}^{C} = -\{m_k^C\tilde{r}_k^C\vec{\gamma}_{k0}^{C} + \hat{J}_k^C\vec{\varepsilon}_{k0}^{C} + \tilde{\omega}_{k0}^{C}\hat{J}_k^C\vec{\omega}_{k0}^{C}\}$$
$$\vec{F}_{k0}^{C} = -\vec{f}_{k0}^{C}, \quad \vec{M}_{k0}^{C} = -\vec{m}_{k0}^{C}, \quad \vec{M}_{k0}^{C} = -\vec{m}_{k0}^{C}, \quad (k = 2, 3), \tag{11.1.16}$$

where the accelerations $\vec{\gamma}_{k0}^{C}$, $\vec{\varepsilon}_{k0}^{C}$ and the square matrices $\tilde{\omega}_{k0}^{C}\tilde{\omega}_{k0}^{C} + \tilde{\varepsilon}_{k0}^{C}$ can be computed with the relations (11.1.12).

For the torque M_2 of the couple applied to the left driving wheel B_2, an analogous expression to (11.1.15) results:

$$\begin{aligned} M_2 = {} & m_1\omega_{10b}^{v}\left(\frac{1}{4}l^2\ddot{\theta} + \frac{1}{2}lr\ddot{\theta}_2 - ar\dot{\theta}\dot{\theta}_2 + a^2\ddot{\theta}\right) + m_1\omega_{2b}^{v}\left(r^2\ddot{\theta}_2 + \frac{1}{2}lr\ddot{\theta} + ar\dot{\theta}^2\right) + J_1^z\ddot{\theta}\omega_{10b}^{v} \\ & + \frac{1}{2}m_2^B r^2\left(3\omega_{2b}^{v}\ddot{\theta}_2 + \frac{1}{2}\omega_{10b}^{v}\ddot{\theta}\right) + \frac{1}{2}m_2^A r^2\left(3\omega_{1b}^{v}\ddot{\theta}_1 + \frac{1}{2}\omega_{10b}^{v}\ddot{\theta}\right) \\ & + m_2^C\omega_{3b}^{v}r_0[r_0\ddot{\theta}_3 - (d - x_2^C)\omega_{\varphi}^2] + m_2^C\omega_{\varphi b}^{v}\left[r_0(d + x_2^C)\dot{\theta}_3\omega_{\varphi} + \frac{d^3(3h+d)}{3(h+d)}\varepsilon_{\varphi}\right] \\ & + \frac{1}{2}m_3^C r_0^2\left(3\omega_{3b}^{v}\ddot{\theta}_3 + \frac{1}{2}\omega_{\varphi b}^{v}\varepsilon_{\varphi}\right), \end{aligned} \tag{11.1.17}$$

where new virtual velocities must be introduced:

$$\begin{aligned} & \omega_{1b}^{v} = 0, \quad \omega_{2b}^{v} = 1, \quad \omega_{10b}^{v} = -\frac{r}{l}, \quad v_{10b}^{xv} = \frac{r}{2}, \quad v_{10b}^{yv} = a\frac{r}{l} \\ & \omega_{3b}^{v} = \frac{r}{r_0}\left(\frac{1}{2}\cos\psi - \frac{a+b}{l}\sin\psi\right), \quad \omega_{4b}^{v} = -\frac{r}{d}\left(\frac{1}{2}\sin\psi + \frac{a+b}{l}\cos\psi\right) - \frac{r}{l} \\ & \omega_{\varphi b}^{v} = \omega_{4b}^{v} - \omega_{10b}^{v}. \end{aligned} \tag{11.1.18}$$

The relations (11.1.14)–(11.1.18) constitute the *inverse dynamics model* of the mobile robot fitted with a caster wheel.

11.1.2.2 Lagrange Equations

The dynamics problem of mobile robots can be solved using Lagrange's equations of the second kind. With non-holonomic constraints, the moving mechanism is a mechanical system with two degrees of freedom. The generalized coordinates of the robot are represented by seven parameters: $q_1 = x_{10}$, $q_2 = y_{10}$, $q_3 = \theta$, $q_4 = \theta_1$, $q_5 = \theta_2$, $q_6 = \theta_3$, $q_7 = \psi$.

Lagrange's equations, with their multipliers $\lambda_1, \lambda_2, \ldots, \lambda_5$, will be expressed by seven differential relations

$$\frac{d}{dt}\left\{\frac{\partial E}{\partial \dot{q}_k}\right\} - \frac{\partial E}{\partial q_k} = \Phi_k + \sum_{s=1}^{5} \lambda_s c_{sk} \quad (k = 1, 2, \ldots, 7), \tag{11.1.19}$$

which contain following seven generalized forces: $\Phi_1 = 0,\ \Phi_2 = 0,\ \Phi_3 = 0,$ $\Phi_4 = M_1,\ \Phi_5 = M_2,\ \Phi_6 = 0,\ \Phi_7 = 0.$

Five kinematical conditions of constraint $\sum_{k=1}^{7} c_{sk}\dot{q}_k = 0, (s = 1, 2, \ldots 5)$, given by (11.1.9), (11.1.10) can be concentred in a matrix form, as follows $C\dot{\vec{q}} = 0$, where

$$C = \begin{bmatrix} \cos\theta & \sin\theta & l/2 & -r & 0 & 0 & 0 \\ \cos\theta & \sin\theta & -l/2 & 0 & -r & 0 & 0 \\ -\sin\theta & \cos\theta & a & 0 & 0 & 0 & 0 \\ -\cos\varphi & \sin\varphi & -b\sin\psi & 0 & 0 & r_0 & 0 \\ \sin\varphi & \cos\varphi & -(d + b\cos\psi) & 0 & 0 & 0 & d \end{bmatrix} \tag{11.1.20}$$
$$\dot{\vec{q}} = \begin{bmatrix} \dot{x}_{10} & \dot{y}_{10} & \dot{\theta} & \dot{\theta}_1 & \dot{\theta}_2 & \dot{\theta}_3 & \dot{\psi} \end{bmatrix}^T, \quad \dot{\varphi} = \dot{\psi} - \dot{\theta}$$

The components of the general expression of kinetic energy $E = \sum_{j=1}^{5} E_j$ are expressed as analytical functions of the generalized coordinates and their first derivatives with respect to time:

$$\begin{aligned}
E_1 &= \frac{1}{2}m_1\left(\dot{x}_{10}^2 + \dot{y}_{10}^2\right) + \frac{1}{2}J_1^z\dot{\theta}^2 \\
E_2 &= \frac{1}{2}m_2^A\left[\left(\dot{x}_{10}\cos\theta + \dot{y}_{10}\sin\theta + \frac{l}{2}\dot{\theta}\right)^2 + \left(-\dot{x}_{10}\sin\theta + \dot{y}_{10}\cos\theta + a\dot{\theta}\right)^2\right] \\
&\quad + \frac{1}{8}m_2^A r^2\left(\dot{\theta}^2 + 2\dot{\theta}_1^2\right) \\
E_3 &= \frac{1}{2}m_2^B\left[\left(\dot{x}_{10}\cos\theta + \dot{y}_{10}\sin\theta - \frac{l}{2}\dot{\theta}\right)^2 + \left(-\dot{x}_{10}\sin\theta + \dot{y}_{10}\cos\theta + a\dot{\theta}\right)^2\right] \\
&\quad + \frac{1}{8}m_2^B r^2\left(\dot{\theta}^2 + 2\dot{\theta}_2^2\right) \\
E_4 &= \frac{1}{2}m_2^C\left[\left(\dot{x}_{10}\cos\theta + \dot{y}_{10}\sin\theta\right)^2 + \left(-\dot{x}_{10}\sin\theta + \dot{y}_{10}\cos\theta - b\dot{\theta}\right)^2\right] \\
&\quad + m_2^C\left(\dot{x}_{10}\sin\varphi + \dot{y}_{10}\cos\varphi - b\dot{\theta}\cos\psi\right)x_2^C\dot{\varphi} + \frac{1}{6}m_2^C\frac{d^3}{h+d}\dot{\varphi}^2 \\
E_5 &= \frac{1}{2}m_3^C\left[\left(-\dot{x}_{10}\cos\varphi + \dot{y}_{10}\sin\varphi - b\dot{\theta}\sin\psi\right)^2\right. \\
&\quad \left. + \left(\dot{x}_{10}\sin\varphi + \dot{y}_{10}\cos\varphi - b\dot{\theta}\cos\psi + d\dot{\varphi}\right)^2\right] + \frac{1}{8}m_3^C r_0^2\left(\dot{\varphi}^2 + 2\dot{\theta}_3^2\right).
\end{aligned} \tag{11.1.21}$$

A long and tedious calculus of the partial derivatives of all the above functions leads to an algebraic system of seven relations. In the inverse dynamics problem, after elimination of the five multipliers λ_s, finally we can obtain with certainty the same analytical expressions (11.1.15), (11.1.17) for the torques M_1, M_2 required by the two wheels.

If the above complete equations (11.1.15), (11.1.17) do not take into account the small masses of the caster wheel C_3 and of the crank C_2C_3, the expression of the torques M_1, M_2 for a symmetrical mobile robot can be written in the simplified form:

$$\begin{aligned} M_1 &= m_1 r^2 \left[\frac{1}{4}(\ddot{\theta}_1 + \ddot{\theta}_2) + \frac{a^2}{l^2}(\ddot{\theta}_1 - \ddot{\theta}_2) - \frac{ar}{l^2}\dot{\theta}_2(\dot{\theta}_1 - \dot{\theta}_2)\right] \\ &\quad + \frac{1}{2} m_2^A r^2 \left[3\ddot{\theta}_1 + \frac{r^2}{l^2}(\ddot{\theta}_1 - \ddot{\theta}_2)\right] + J_1^z \frac{r^2}{l^2}(\ddot{\theta}_1 - \ddot{\theta}_2) \\ M_2 &= m_1 r^2 \left[\frac{1}{4}(\ddot{\theta}_1 + \ddot{\theta}_2) + \frac{a^2}{l^2}(\ddot{\theta}_2 - \ddot{\theta}_1) - \frac{ar}{l^2}\dot{\theta}_1(\dot{\theta}_2 - \dot{\theta}_1)\right] \\ &\quad + \frac{1}{2} m_2^A r^2 \left[3\ddot{\theta}_2 + \frac{r^2}{l^2}(\ddot{\theta}_2 - \ddot{\theta}_1)\right] + J_1^z \frac{r^2}{l^2}(\ddot{\theta}_2 - \ddot{\theta}_1). \end{aligned} \tag{11.1.22}$$

Some results concerning the torques M_1 and M_2 are obtained in [10, 11], using the dynamical equations of motion based on Maggie's approach or Lagrange's formalism.

11.1.3 Simulations

As an application, we will analyze the motion simulation of a robot which has the following characteristics:

$$\begin{aligned} &x_2^C = \frac{1}{2}d^2/(h+d), \quad J_1^z = \frac{1}{3}m_1 l^2 \\ &m_1 = 10\,\text{kg}, \quad m_2^A = m_2^B = 2\,\text{kg}, \quad m_2^C = 1\,\text{kg}, \quad m_3^C = 0.25\,\text{kg} \\ &l = 0.5\,\text{m}, \quad a = l\sqrt{3}/6, \quad b = 2a, \quad d = 0.025\,\text{m}, \quad r = 0.1\,\text{m}, \\ &r_0 = 0.025\,\text{m}, \quad h_0 = 0.2\,\text{m}, \quad H = h_0 + r, \quad h = H - r_0. \end{aligned} \tag{11.1.23}$$

Three important manoeuvres can be implemented.

1. *Rectilinear translation*

Considering two equal kinematic inputs

$$\theta_1^* = \theta_2^* = \frac{\pi}{2}, \quad x_{10} = r\theta_1, \quad y_{10} = 0, \quad \theta = 0, \quad \psi = 0, \quad \theta_3 = \frac{r}{r_0}\theta_1, \tag{11.1.24}$$

the centre G of the platform moves along the direction of the fixed axis x_0 with a variable acceleration. The first and second derivatives of the input angles are equal, since the load conditions have a symmetrical representation. The two driving wheels are actuated by equal torques:

$$M_1 = M_2 = \frac{1}{2} r^2 \ddot{\theta}_1 \left(m_1 + 3m_2^A + m_2^C + \frac{3}{2} m_3^C \right). \tag{11.1.25}$$

As can be seen from Figs. 11.3 and 11.4 it is proved to be true that the torques and powers of the driving wheels are permanently equal to one another.

If the masses of the ensemble crank-caster wheel are supposed as negligible, we obtain the known expression

$$M_1 = M_2 = \frac{1}{2} r^2 \ddot{\theta}_1 (m_1 + 3m_2^A). \tag{11.1.26}$$

2. *Rotation motion* about the vertical axis passing through the midpoint of the actuated wheel axle A_2B_2, with

$$\theta_1^* = -\theta_2^* = \frac{\pi}{2}, \quad v_{10}^x = 0, \quad v_{10}^y = -2a\frac{r}{l}\dot{\theta}_1, \quad \dot{\theta} = 2\frac{r}{l}\dot{\theta}_1. \tag{11.1.27}$$

Starting from the initial position, the crank and the caster wheel have a transition motion first. The angular velocities $\dot{\psi}, \dot{\theta}_3$ follow two functions:

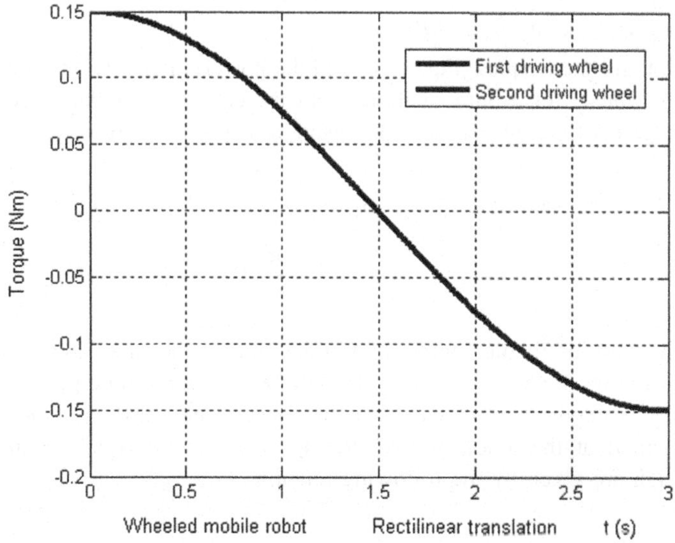

Fig. 11.3 Torques of driving wheels (translation)

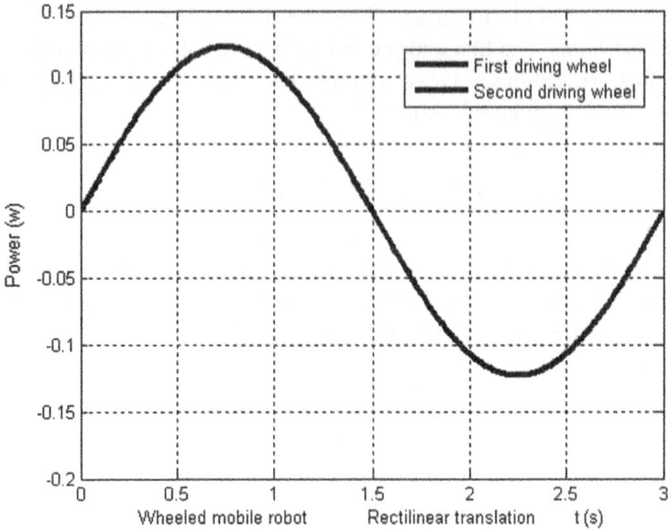

Fig. 11.4 Powers of driving wheels (translation)

$$\dot{\psi} = \left(1 + \frac{a+b}{d}\cos\psi\right)\dot{\theta}, \quad \dot{\theta}_3 = \frac{a+b}{r_0}\dot{\theta}\sin\psi. \tag{11.1.28}$$

We notice a rapid increase in the relative angle ψ (Fig. 11.5) and a diminishing of the angular velocity $\dot{\psi}$ at a characteristic angle ψ_0, as follows: $\cos\psi_0 = -\frac{d}{a+b}$. In a stabilized rotation, the centre C_3 describes a circle of radius $R = (a+b)\sin\psi_0$ with an angular velocity $\dot{\theta}_3 = \frac{R}{r_0}\dot{\theta}$.

Two plots for the torques (Fig. 11.6) and the powers (Fig. 11.7) of the right and left driving wheels for a pure rotation about the centerline of the wheel axle are displayed against time during $\Delta t = 3$ s, using a new set of virtual velocities:

$$\begin{aligned} &\omega_{1a}^{v} = 1, \quad \omega_{2a}^{v} = 0, \quad \omega_{10a}^{v} = \frac{r}{l}, \quad \omega_{3a}^{v} = \frac{R}{r_0}\omega_{10a}^{v}, \quad \omega_{\phi a}^{v} = -\omega_{10a}^{v} \\ &\omega_{1b}^{v} = 0, \quad \omega_{2b}^{v} = 1, \quad \omega_{10b}^{v} = -\frac{r}{l}, \quad \omega_{3b}^{v} = \frac{R}{r_0}\omega_{10b}^{v}, \quad \omega_{\phi b}^{v} = -\omega_{10b}^{v}. \end{aligned} \tag{11.1.29}$$

The irregularities in the graphs of torques during the first 0.85 s correspond precisely with the period of transition to stabilized rotation motion.

Ignoring as negligible the masses of the caster wheel and the crank, the two torques applied at the axles of two driving wheels with equal magnitude and opposite sign are given by the following relation

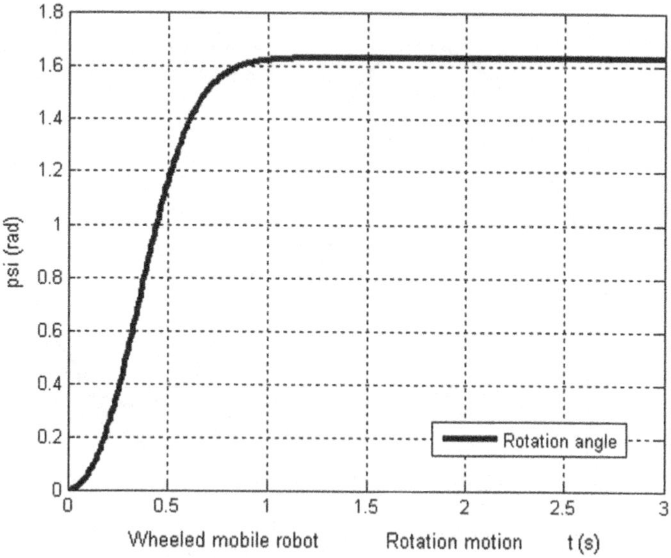

Fig. 11.5 Rotation angle ψ in transition period

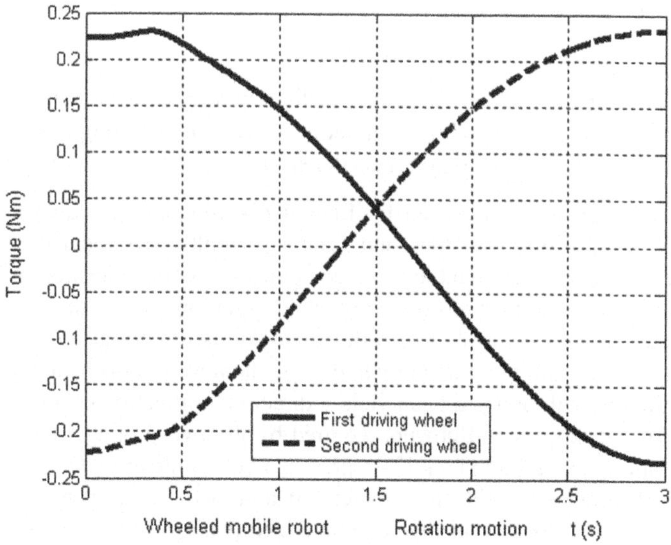

Fig. 11.6 Torques of driving wheels (rotation)

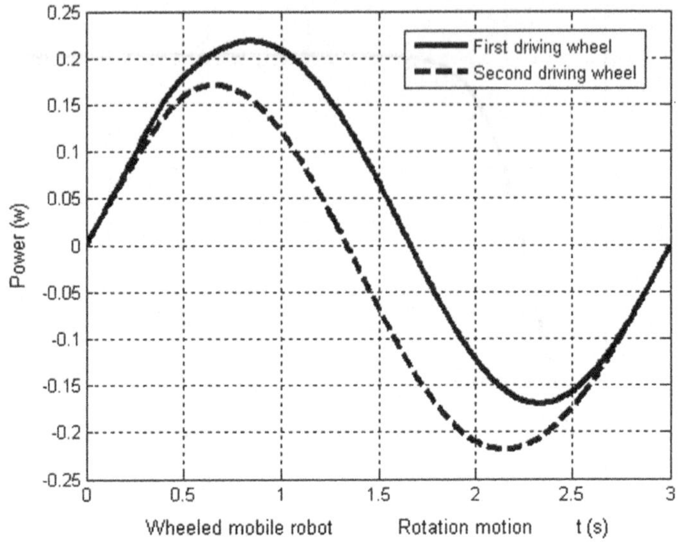

Fig. 11.7 Powers of driving wheels (rotation)

$$M_1 = |M_2| = \left[2\frac{r^2}{l^2}(J_1^z + m_1 a^2) + m_2^A r^2 \left(\frac{3}{2} + \frac{r^2}{l^2} \right) \right] \ddot{\theta}_1. \tag{11.1.30}$$

3. *Curvilinear motion*: $\theta_2^* = 2\theta_1^* = \frac{\pi}{2}$. The time-history of the torques (Fig. 11.8) and powers (Fig. 11.9) of two actuators A_2, B_2 will be computed if the platform's centre G moves along a curvilinear trajectory (Fig. 11.10).

The number of relations given by Lagrange's equations is equal to the total number of the position variables and Lagrange multipliers inclusive. Also, the analytical calculations involved in these equations present an elevated risk of error. So, applying Lagrange formalism we obtain 12 equations, while the Newton-Euler classic method brings us to a system of 36 equations. Ignoring the friction forces but taking into account the gravitational effects, the new approach can eliminate all forces of the internal joints and can yield a direct determination of the time-history evolution of the torques and powers required by the two driving wheels.

The relations (11.1.15), (11.1.17) represent two complete explicit equations of the dynamic simulation and, in a context of automatic command, can easily be transformed into a robust model for computerized control of the mobile robot fitted with a caster wheel.

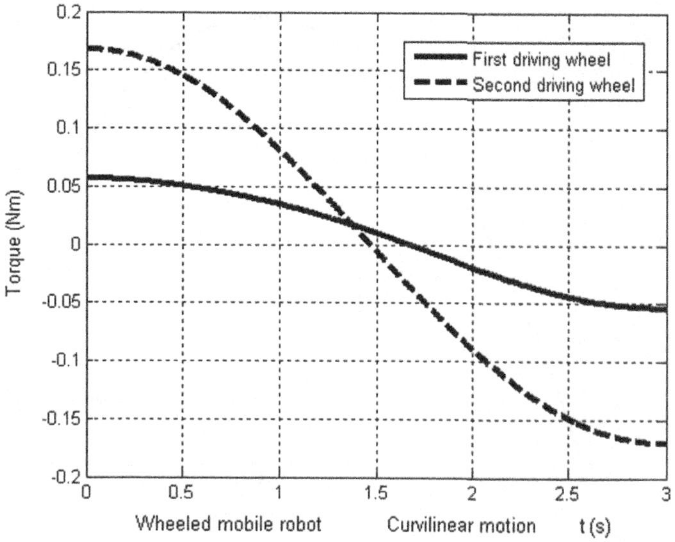

Fig. 11.8 Torques of driving wheels (curvilinear motion)

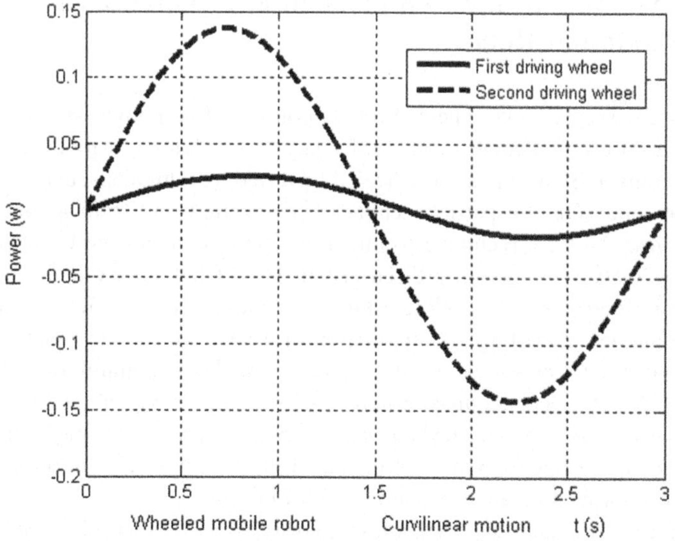

Fig. 11.9 Powers of driving wheels (curvilinear motion)

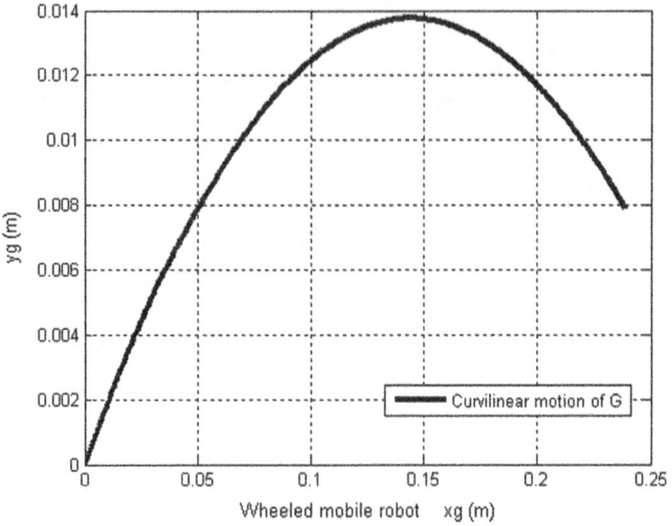

Fig. 11.10 Trajectory of the centre *G*

11.2 Dynamics of the Non-holonomic Two-Wheeled Pendulum Robot

A literature survey on two-wheeled robots confirms two known systems: SCOUT and Ginger-Segway. Pathak et al. [12] analyse the dynamics modelling and the position control of a series of wheeled inverted pendulums, such as Segway, Quasimoro and JOE, by partial feedback linearization and from a controllability point of view. Using a recursive formulation, the kinematics model with a global singularity analysis is carefully discussed in [13]. Chakraborty and Ghosal have presented in their works [14, 15] the kinematics and a set of differential equations for the dynamics modelling and simulation of a wheeled mobile robot.

The Quasimoro prototype of the mobile wheeled-pendulum of Salerno and Angeles [16] is a special quasi-holonomic mechanical system, comprising two driving wheels and an intermediate central body carrying the payload. Salerno, Ostrovskaya and Angeles studied in the paper [17] the dynamics of a rolling robot, using second-order Lagrange equations with multipliers.

The kinematics and the dynamics analysis of a two-wheeled mobile robot comprising an oscillating intermediate central body are established in the following. We consider that the angular displacements of the wheels and the rotation angle of the intermediate body determine the position and orientation of the robot. Using the principle of virtual powers or the Lagrange equations of the second kind, we formulate a matrix model of Quasimoro, which represents a two-input non-linear dynamical system with *three* outputs. This model is validated by means of simulations of dynamic responses of the system to different inputs and initial conditions.

The results of the analysis are highly significant for the robot design and control. It turns out that the major disturbance in accomplishing tasks at hand is caused by the oscillations of the intermediate central body. Hence, the introduction of design corrections and the derivation of a suitable control algorithm should be considered, in order to render robot performances the least sensitive to this disturbance [12, 21].

11.2.1 Kinematics Modelling

Quasimoro is a wheeled mobile robot (*WMR*) consisting of two wheels of conventional type of the same radius r, which are actuated by two independent motors, and of an intermediate central body that contains the control equipment, the actuation system, the power supply and the transmission mechanism. During motion, both robot wheels can roll without slipping on a *horizontal* planar surface and are assumed to be permanently in contact with this surface.

A *fictitious* horizontal platform, attached to the frame $C_2x_1y_1z_1$ of axis C_2x_1 linking the centres of the wheels, has a planar motion. Its position with respect to an inertial reference frame $Ox_0y_0z_0$ with the origin O fixed to the ground surface is given by the coordinates $x_{10}, y_{10}, z_{10} = r$ and by the orientation angle ψ, which form the following matrices

$$\vec{r}_{10} = \begin{bmatrix} x_{10} & y_{10} & z_{10} \end{bmatrix}^T, \quad a_{10} = rot(z, \psi). \tag{11.2.1}$$

Two cylindrical coaxial wheels linked to the frames $A_2x_2^Ay_2^Az_2^A$ and $B_2x_2^By_2^Bz_2^B$ are coupled to an intermediate body by means of revolute joints in points A_2 and B_2 (Fig. 11.11). They have known masses $m_2^A = m_2^B = m$ and the tensors of inertia

$$\hat{J}_2^A = \hat{J}_2^B = \hat{J} = diag\left[\frac{mr^2}{4}, \frac{mr^2}{4}, \frac{mr^2}{2}\right]. \tag{11.2.2}$$

The central cylinder of mass $m_2^C = M$ and tensor of inertia

$$\hat{J}_2^C = diag[J_1, J_2, J_3]. \tag{11.2.3}$$

is attached to the frame $C_2x_2^Cy_2^Cz_2^C$, which is jointed at the midpoint C_2 of the line linking the mass centres of two wheels and represents the intermediate body.

Let us denote by θ_1, θ_2 the rotation angles of the wheels about the $A_2z_2^A, B_2z_2^B$ axes, respectively, and by θ_3 the instantaneous inclination angle of the intermediate central body about the $C_2z_2^C$ axis.

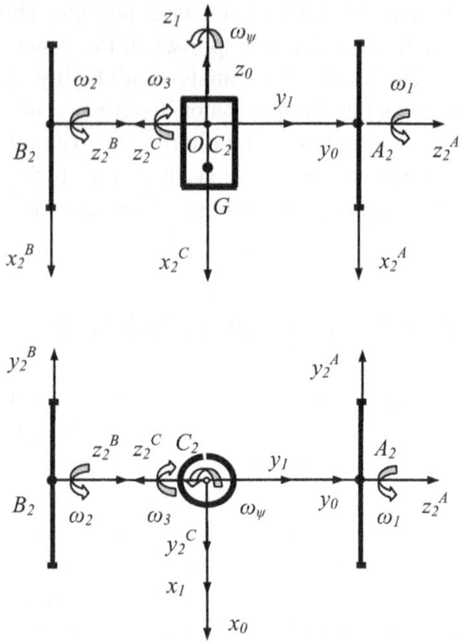

Fig. 11.11 Kinematical scheme of the rolling pendulum robot

Starting from the platform's frame $C_2x_1y_1z_1$, we obtain the following transformation matrices in the mobile reference frames [18]:

$$a_{21}^{A} = a_z^{\theta_1} a_1 a_2, \quad a_{21}^{B} = a_z^{\theta_2} a_1 a_2, \quad a_{21}^{C} = a_z^{\theta_3} a_1 a_2^T, \tag{11.2.4}$$

where

$$a_1 = rot\left(y, \frac{\pi}{2}\right), \quad a_2 = rot\left(z, \frac{\pi}{2}\right), \quad a_z^{\theta_i} = rot(z, \theta_i) \ (i = 1, 2, 3). \tag{11.2.5}$$

Defining as $A_2B_2 = l$ the distance between the centres of the wheels and $C_2G = d$ the displacement of the centre G, the following vectors give the invariable relative positions of the points A_2, B_2, C_2, G:

$$\vec{r}_{21}^{A_2} = \begin{bmatrix} 0 & \frac{l}{2} & 0 \end{bmatrix}^T, \quad \vec{r}_{21}^{B_2} = \begin{bmatrix} 0 & -\frac{l}{2} & 0 \end{bmatrix}^T, \vec{r}_{21}^{C_2} = \vec{0}, \quad \vec{r}_2^{A_2} = \vec{0}, \quad \vec{r}_2^{B_2} = \vec{0}, \quad \vec{r}_2^{G} = [d \quad 0 \quad 0]. \tag{11.2.6}$$

The kinematics of the three elements is completely characterized by the translation velocity

$$\dot{\vec{r}}_{10} = \begin{bmatrix} \dot{x}_{10} & \dot{y}_{10} & 0 \end{bmatrix}^T \tag{11.2.7}$$

and by the angular velocities given below

$$\vec{\omega}_{10} = \dot{\psi}\vec{u}_3, \quad \vec{\omega}_{21}^{A} = \dot{\theta}_1\vec{u}_3, \quad \vec{\omega}_{21}^{B} = \dot{\theta}_2\vec{u}_3, \quad \vec{\omega}_{21}^{C} = \dot{\theta}_3\vec{u}_3, \quad \vec{u}_3 = \begin{bmatrix} 0 & 0 & 1 \end{bmatrix}^T$$
$$\omega_1 = \dot{\theta}_1, \quad \omega_2 = \dot{\theta}_2, \quad \omega_3 = \dot{\theta}_3, \quad \omega_{\psi} = \dot{\psi}. \tag{11.2.8}$$

The speed difference between the wheels generates rotation of the vehicles. Moreover, differentially-driven robots can rotate on the spot.

Assuming that the wheels roll without slipping on the ground surface, three analytical relations between the characteristic velocities of the three-degrees-of-mobility robot express the non-holonomic constraints:

$$\vec{v}_{10} + \tilde{\omega}_{10}\vec{r}_{21}^{A_2} = \begin{bmatrix} r\dot{\theta}_1 & 0 & 0 \end{bmatrix}^T, \quad \vec{v}_{10} + \tilde{\omega}_{10}\vec{r}_{21}^{B_2} = \begin{bmatrix} r\dot{\theta}_2 & 0 & 0 \end{bmatrix}^T, \tag{11.2.9}$$

where

$$\vec{v}_{10} = a_{10}\dot{\vec{r}}_{10} = \begin{bmatrix} \dot{x}_{10}\cos\psi + \dot{y}_{10}\sin\psi \\ -\dot{x}_{10}\sin\psi + \dot{y}_{10}\cos\psi \\ 0 \end{bmatrix},$$
$$\tilde{\omega}_{10} = a_{10}\dot{a}_{10}^T = \dot{\psi}\tilde{u}_3 = \begin{bmatrix} 0 & -\dot{\psi} & 0 \\ \dot{\psi} & 0 & 0 \\ 0 & 0 & 0 \end{bmatrix}. \tag{11.2.10}$$

Deriving $\psi = \rho(\theta_2 - \theta_1)$, above *conditions of connectivity* (11.2.9) can provide the matrix of the Jacobian and express the characteristic velocities of the mechanism

$$\dot{\psi} = \rho(\dot{\theta}_2 - \dot{\theta}_1), \quad \dot{x}_{10} = \frac{r}{2}(\dot{\theta}_1 + \dot{\theta}_2)\cos\psi, \quad \dot{y}_{10} = \frac{r}{2}(\dot{\theta}_1 + \dot{\theta}_2)\sin\psi, \quad \rho = \frac{r}{l}, \tag{11.2.11}$$

In order to determine the conditions of connectivity of accelerations, we derive the matrix relations (11.2.11):

$$\vec{\gamma}_{10} + (\tilde{\omega}_{10}\tilde{\omega}_{10} + \tilde{\varepsilon}_{10})\vec{r}_{21}^{A_2} = \begin{bmatrix} r\ddot{\theta}_1 & 0 & 0 \end{bmatrix}^T + \tilde{\omega}_{10}\begin{bmatrix} r\dot{\theta}_1 & 0 & 0 \end{bmatrix}^T$$
$$\vec{\gamma}_{10} + (\tilde{\omega}_{10}\tilde{\omega}_{10} + \tilde{\varepsilon}_{10})\vec{r}_{21}^{B_2} = \begin{bmatrix} r_1\ddot{\theta}_2 & 0 & 0 \end{bmatrix}^T + \tilde{\omega}_{10}\begin{bmatrix} r\dot{\theta}_2 & 0 & 0 \end{bmatrix}^T, \tag{11.2.12}$$

where

$$\vec{\gamma}_{10} = a_{10}\ddot{\vec{r}}_{10} = \begin{bmatrix} \ddot{x}_{10}\cos\psi + \ddot{y}_{10}\sin\psi \\ -\ddot{x}_{10}\sin\psi + \ddot{y}_{10}\cos\psi \\ 0 \end{bmatrix}. \tag{11.2.13}$$

Thus, the characteristic accelerations $\ddot{\psi}$, $\ddot{x}_{10}, \ddot{y}_{10}$ are immediately obtained:

$$\begin{aligned} \ddot{\psi} &= \rho(\ddot{\theta}_2 - \ddot{\theta}_1) \\ \ddot{x}_{10} &= \frac{r}{2}(\ddot{\theta}_1 + \ddot{\theta}_2)\cos\psi - \frac{r\rho}{2}(\dot{\theta}_2^2 - \dot{\theta}_1^2)\sin\psi . \\ \ddot{y}_{10} &= \frac{r}{2}(\dot{\theta}_1 + \dot{\theta}_2)\sin\psi + \frac{r\rho}{2}(\dot{\theta}_2^2 - \dot{\theta}_1^2)\cos\psi \end{aligned} \quad (11.2.14)$$

To describe the kinematical state of each body T_2^A, T_2^B, T_2^C with respect to fixed frame $Ox_0y_0z_0$, we express the angular velocity $\vec{\omega}_{20}$ and the linear velocity $\vec{v}_{20}$ of the reference origin

$$\vec{v}_{20} = a_{21}(\vec{v}_{10} + \tilde{\omega}_{10}\vec{r}_{21}), \quad \vec{\omega}_{20} = a_{21}\vec{\omega}_{10} + \vec{\omega}_{21}. \quad (11.2.15)$$

Performing the derivatives with respect to time of Eq. (11.2.15), we obtain the accelerations $\vec{\gamma}_{20}$ and $\vec{\varepsilon}_{20}$:

$$\vec{\gamma}_{20} = a_{21}\{\vec{\gamma}_{10} + (\tilde{\omega}_{10}\tilde{\omega}_{10} + \tilde{\varepsilon}_{10})\vec{r}_{21}\}, \quad \vec{\varepsilon}_{20} = a_{21}\vec{\varepsilon}_{10} + \vec{\varepsilon}_{21} + a_{21}\tilde{\omega}_{10}a_{21}^T\vec{\omega}_{21} \quad (11.2.16)$$

and a useful characteristic square matrix [19]

$$\tilde{\omega}_{20}\tilde{\omega}_{20} + \tilde{\varepsilon}_{20} = a_{21}\{\tilde{\omega}_{10}\tilde{\omega}_{10} + \tilde{\varepsilon}_{10})a_{21}^T + \tilde{\omega}_{21}\tilde{\omega}_{21} + \tilde{\varepsilon}_{21} + 2a_{21}\tilde{\omega}_{10}a_{21}^T\tilde{\omega}_{21}. \quad (11.2.17)$$

11.2.2 Dynamics Modelling

11.2.2.1 Principle of Virtual Work

Internal active torques $\vec{M}_1 = M_1\vec{u}_3$ and $\vec{M}_2 = M_2\vec{u}_3$, which are generated by two electric motors mounted in the frame of intermediary rigid body, transmit the motion to the two wheels. Having the direction of the common axis A_2B_2, these moments can control the motion accomplished by the active wheels and the intermediate body. We will study the *direct dynamic problem*, in order to establish the evolution of the rotation angles $\theta_1, \theta_2, \theta_3$ and the variation of the angular velocities $\dot{\theta}_1, \dot{\theta}_2, \dot{\theta}_3$ during the transient motion of $\Delta t = 1$ s between the initial position and one which corresponds to another inactivated motion, knowing the variation of two applied torques M_1, M_2. Thus, we will use the principle of virtual work.

Indeed, we assume that the two torque pulses of amplitudes M_1° and M_2° are at an arbitrary moment known from the following sinusoidal waveform functions

$$M_i(t) = M_i^\circ \sin \pi t, \quad (i = 1, 2), \quad t \in [0, 1], \tag{11.2.18}$$

where, for example, $M_1^\circ = 0.1\,\mathrm{Nm}$.

The fundamental principle of virtual work states that a mechanical system is under dynamic equilibrium if and only if the virtual work developed by all external, internal and inertia forces vanish during any general virtual displacement which is compatible with the kinematic constraints [20].

The force of inertia and the resultant moment of the forces of inertia have the following general form:

$$\vec{F}_{i0} = -m_i\{\vec{\gamma}_{i0} + (\tilde{\omega}_{i0}^2 + \tilde{\varepsilon}_{i0})\vec{r}_i^{\,G}\}, \quad \vec{M}_{i0} = -\{m_i\tilde{r}_i^G\vec{\gamma}_{i0} + \hat{J}_i\vec{\varepsilon}_{i0} + \tilde{\omega}_{i0}\hat{J}_i\vec{\omega}_{i0}\}, \tag{11.2.19}$$

where m_i and $\hat{J}_i$ are the mass and the tensor of inertia of body T_i, respectively.

A set of *virtual velocities* of the robot bodies result from relations (11.2.9) and (11.2.15), namely:

$$\begin{aligned}
&\omega_{1a}^v = 1, \quad \omega_{2a}^v = 0, \quad \omega_{3a}^v = 0, \quad \omega_{10a}^v = -\rho \\
&\vec{v}_{20}^{AT} = [-r\sin\theta_1 \quad -r\cos\theta_1 \quad 0] \\
&\vec{v}_{20}^{BT} = [0 \quad 0 \quad 0] \\
&\vec{v}_{20}^{CT} = [\tfrac{r}{2}\sin\theta_3 \quad \tfrac{r}{2}\cos\theta_3 \quad 0] \\
&\vec{\omega}_{20}^{AT} = [\rho\cos\theta_1 \quad -\rho\sin\theta_1 \quad 1] \\
&\vec{\omega}_{20}^{BT} = [\rho\cos\theta_2 \quad -\rho\sin\theta_2 \quad 0] \\
&\vec{\omega}_{20}^{CT} = [\rho\cos\theta_3 \quad -\rho\sin\theta_3 \quad 0]
\end{aligned} \tag{11.2.20}$$

The following matrix expression of the torque applied to the wheel A_2 results

$$M_1 = \vec{v}_{20}^{AT}\vec{F}_{20}^{A} + \vec{\omega}_{20}^{AT}\vec{M}_{20}^{A} + \vec{v}_{20}^{BT}\vec{F}_{20}^{B} + \vec{\omega}_{20}^{BT}\vec{M}_{20}^{B} + \vec{v}_{20}^{CT}\vec{F}_{20}^{C} + \vec{\omega}_{20}^{CT}\vec{M}_{20}^{C}, \tag{11.2.21}$$

with its definitive analytical form

$$\begin{aligned}
M_1 = {} & A_{11}\ddot{\theta}_1 + A_{12}\ddot{\theta}_2 + A_{13}\ddot{\theta}_3 + A_{14}\dot{\theta}_1\dot{\theta}_2 + A_{15}\dot{\theta}_2\dot{\theta}_3 \\
& + A_{16}\dot{\theta}_3\dot{\theta}_1 + A_{17}\dot{\theta}_1^2 + A_{18}\dot{\theta}_2^2 + A_{19}\dot{\theta}_3^2,
\end{aligned} \tag{11.2.22}$$

where

$$\begin{aligned}
A_{11} &= 0.5mr^2(3+\rho^2)+\rho^2(J_1\cos^2\theta_3+J_2\sin^2\theta_3)+0.25Mr^2\\
A_{12} &= -0.5mr^2\rho^2-\rho^2(J_1\cos^2\theta_3+J_2\sin^2\theta_3)+0.25Mr^2\\
A_{13} &= 0.5Mrd\,\cos\theta_3\\
A_{14} &= Mrd\rho^2\sin\theta_3\\
A_{15} &= 2\rho^2(J_1-J_2)\sin\theta_3\cos\theta_3\\
A_{16} &= -A_{15},\quad A_{17}=0,\quad A_{18}=-A_{14}\\
A_{19} &= -0.5Mrd\sin\theta_3.
\end{aligned} \tag{11.2.23}$$

The relation (11.2.21) represent precisely the *dynamics model* of the *WMR*, where dynamical effects, including the Coriolis and centrifugal forces coupling and the gravitational actions, are considered in this explicit equation.

For the moment M_2 of the torque applied to the wheel B_2 we obtain an analogous final expression:

$$\begin{aligned}
M_2 = {}& A_{21}\ddot{\theta}_1+A_{22}\ddot{\theta}_2+A_{23}\ddot{\theta}_3+A_{24}\dot{\theta}_1\dot{\theta}_2+A_{25}\dot{\theta}_2\dot{\theta}_3\\
&+A_{26}\dot{\theta}_3\dot{\theta}_1+A_{27}\dot{\theta}_1^2+A_{28}\dot{\theta}_2^2+A_{29}\dot{\theta}_3^2
\end{aligned} \tag{11.2.24}$$

where

$$\begin{aligned}
&A_{21}=A_{12},\quad A_{22}=A_{11},\quad A_{23}=A_{13}\\
&A_{24}=A_{14},\quad A_{25}=A_{16},\quad A_{26}=A_{15}\\
&A_{27}=A_{18},\quad A_{28}=A_{17},\quad A_{29}=A_{19}.
\end{aligned} \tag{11.2.25}$$

A third virtual displacement of the intermediate body corresponds to the angular velocities

$$\omega_{3c}^{v}=1,\quad \omega_{1c}^{v}=0,\quad \omega_{2c}^{v}=0,\quad \omega_{10c}^{v}=0. \tag{11.2.26}$$

After several transformations, the following *non-linear differential equation* of third element T_2^C is obtained

$$\begin{aligned}
&A_{31}\ddot{\theta}_1+A_{32}\ddot{\theta}_2+A_{33}\ddot{\theta}_3+A_{34}\dot{\theta}_1\dot{\theta}_2+A_{35}\dot{\theta}_2\dot{\theta}_3+A_{36}\dot{\theta}_3\dot{\theta}_1\\
&\quad+A_{37}\dot{\theta}_1^2+A_{38}\dot{\theta}_2^2+A_{39}\dot{\theta}_3^2=M_1+M_2-Mgd\sin\theta_3,
\end{aligned} \tag{11.2.27}$$

where

$$\begin{aligned}
&A_{31}=A_{13},\quad A_{32}=A_{31},\quad A_{33}=J_2,\quad A_{34}=-A_{15}\\
&A_{35}=0,\quad A_{36}=0,\quad A_{37}=0.5A_{15},\quad A_{38}=A_{37},\quad A_{39}=0.
\end{aligned} \tag{11.2.28}$$

11.2.2.2 Lagrange Equations

A solution to the dynamics problem of the mobile robots can be developed based on the Lagrange equations for a mechanical system with non-holonomic constraints. The generalized coordinates of the robot are represented by six parameters $q_1 = x_{10}$, $q_2 = y_{10}$, $q_3 = \psi$, $q_4 = \theta_1$, $q_5 = \theta_2$, $q_6 = \theta_3$.

Lagrange equations with their multipliers $\lambda_1, \lambda_2, \lambda_3$, will be expressed by six differential relations with six generalized forces $\Phi_1 = 0$, $\Phi_2 = 0$, $\Phi_3 = 0$, $\Phi_4 = M_1$, $\Phi_5 = M_2$, $\Phi_6 = M_1 + M_2 - Mgd \sin\theta_3$, as follows:

$$\begin{aligned} &\frac{d}{dt}\left\{\frac{\partial E}{\partial \dot{q}_k}\right\} - \frac{\partial E}{\partial q_k} = \Phi_k + \sum_{s=1}^{3} \lambda_s c_{sk} \quad (k = 1, 2, \ldots, 5) \\ &\frac{d}{dt}\left\{\frac{\partial E}{\partial \dot{\theta}_3}\right\} - \frac{\partial E}{\partial \theta_3} = \Phi_6 \end{aligned} \tag{11.2.29}$$

Three kinematical conditions of constraint given by the relations (11.2.9)

$$\sum_{k=1}^{5} c_{sk}\dot{q}_k = 0 \quad (s = 1, 2, 3) \tag{11.2.30}$$

can be written in a matrix form $C\dot{\vec{q}} = 0$, where

$$C = \begin{bmatrix} \cos\psi & \sin\psi & 0 & -r/2 & -r/2 \\ -\sin\psi & \cos\psi & 0 & 0 & 0 \\ 0 & 0 & l & r & -r \end{bmatrix}, \quad \dot{\vec{q}} = \begin{bmatrix} \dot{x}_{10} & \dot{y}_{10} & \dot{\psi} & \dot{\theta}_1 & \dot{\theta}_2 \end{bmatrix}^T. \tag{11.2.31}$$

The components of the general expression of kinetic energy $E = \sum_{j=1}^{3} E_j$ are expressed as analytical functions of the generalized coordinates and their first derivatives with respect to time:

$$\begin{aligned} E_1 &= \frac{1}{8} mr^2 (6\dot{\theta}_1^2 + \dot{\psi}^2) \\ E_2 &= \frac{1}{8} mr^2 (6\dot{\theta}_2^2 + \dot{\psi}^2) \\ E_3 &= \frac{1}{2} m_3 (\dot{x}_{10}^2 + \dot{y}_{10}^2) + \frac{1}{2} (J_1 \cos^2\theta_3 + J_2 \sin^2\theta_3) + m_3 \dot{x}_{10} \dot{\theta}_3 d \cos\psi \cos\theta_3 \\ &\quad + m_3 \dot{y}_{10} \dot{\theta}_3 d \sin\psi \cos\theta_3 - m_3 \dot{x}_{10} \dot{\psi} d \sin\psi \sin\theta_3 + m_3 \dot{y}_{10} \dot{\psi} d \cos\psi \sin\theta_3. \end{aligned} \tag{11.2.32}$$

After elimination of the three multipliers, finally we obtain with certainty the same expressions (11.2.22), (11.2.24) for the torques M_1, M_2 required by the two driving wheels and the same differential equation (11.2.27).

If the above complete equations (11.2.22) and (11.2.24) do not take into account the rotation of the intermediate body of mass M, when $d = 0$, the expressions of active torques M_1 and M_2 can, for a symmetrical mobile robot, be written in the simplified form

$$\begin{aligned} M_1 &= \frac{1}{4}Mr^2(\ddot{\theta}_1 + \ddot{\theta}_2) + \frac{1}{2}mr^2[3\ddot{\theta}_1 + \rho^2(\ddot{\theta}_1 - \ddot{\theta}_2)] + J_1\rho^2(\ddot{\theta}_1 - \ddot{\theta}_2) \\ M_2 &= \frac{1}{4}Mr^2(\ddot{\theta}_1 + \ddot{\theta}_2) + \frac{1}{2}mr^2[3\ddot{\theta}_2 + \rho^2(\ddot{\theta}_2 - \ddot{\theta}_1)] + J_1\rho^2(\ddot{\theta}_2 - \ddot{\theta}_1). \end{aligned} \tag{11.2.33}$$

11.2.3 Simulations

Three important manoeuvres will be implemented:

1. *Rectilinear motion* with the angles $\psi = 0, \theta_1 = \theta_2$ is accomplished by equal torques applied to the wheels

$$M_1 = M_2 = \frac{1}{2}(3m + M)r^2\ddot{\theta}_1 + \frac{1}{2}Mrd(\ddot{\theta}_3 \cos\theta_3 - \dot{\theta}_3^2 \sin\theta_3) \tag{11.2.34}$$

and delivers the additional *differential equation*

$$J_2\ddot{\theta}_3 + Mgd\,\sin\theta_3 + Mrd\ddot{\theta}_1 \cos\theta_3 = 2M_1. \tag{11.2.35}$$

2. *Pure rotation* about the vertical axis passing through a fixed centre C_2 is developed with $\psi = -2\rho\theta_1$, $\theta_2 = -\theta_1$, $\theta_3 = 0$ and two equal torques, opposite in sign, applied to the wheels

$$M_1 = |M_2| = \left[mr^2\left(\frac{3}{2} + \rho^2\right) + 2\rho^2 J_1\right]\ddot{\theta}_1. \tag{11.2.36}$$

3. *Circular motion* when the centre A_2 describes during the first interval a horizontal circle of radius $l = A_2B_2$, around the vertical axis $B_2x_2^B$ with a permanent angle $\theta_2 = 0$.

As an application, we will analyze the motion of a rolling robot which has the following characteristics $m = 0.5\,\text{kg}$, $M = 5\,\text{kg}$, $J_1 = 0.2\,\text{kg m}^2$, $J_2 = 0.4\,\text{kg m}^2$, $r = 0.3\,\text{m}$, $d = 0.2\,\text{m}$, $\rho = 0.5$. The response of the dynamical system subjected to two different inputs is to study. These inputs will be represented by torques pulses (11.2.18), actuating the wheels for a duration of $\Delta t = 1\,\text{s}$ applied at $t_0 = 0$. Each simulation takes 60 s, but most of the entirety of the output plots will be reported in the time window that runs from 0 s to $\Delta t + 3\,\text{s}$, so as for example, to better show the transient non-actuated response as well.

The above three manoeuvres have been simulated: (1) rectilinear motion, while maintaining constant the orientation angle $\psi = 0$; (2) pure rotation about the vertical axis passing through C_2, with $\theta_3 = 0$, i.e. varying the orientation angle ψ only; (3) circular motion of centre A_2 around the axis $B_2x_2^B$, with $\theta_2 = 0$. In every simulation, the system is considered as initially at rest.

1. *Rectilinear motion*

In this simulation run, two equal torque pulses (11.2.18) of amplitude $M_1^\circ = M_2^\circ = 0.1$ Nm are applied to the two wheels. The output plots are reported in the following graphs: $\theta_1 = \theta_2$ (Figs. 11.12 and 11.13) and θ_3 (Figs. 11.14 and 11.15). The angles θ_1, θ_2 and their first derivatives are equal, since the load condition is symmetric.

From Fig. 11.15, where θ_3 is represented by a periodic signal, we can infer that the oscillation of the intermediate body in the steady state is between −0.0129 and 0.0129 rad.

The rectilinear trajectory is followed with high accuracy. Nevertheless, the wheels will never experience the same torque in reality, a consideration which calls for a suitable control algorithm for accomplishment of the locomotion task.

2. *Pure rotation*

Two equal torque pulses $M_1 = -M_2 = M_1^\circ \sin \pi t$ of amplitude $m_1^\circ = 0.1$ Nm, opposite in sign, are applied to the wheels, during $\Delta t = 1$ s. The output plots are displayed in two graphs: θ_1 (Figs. 11.16 and 11.17).

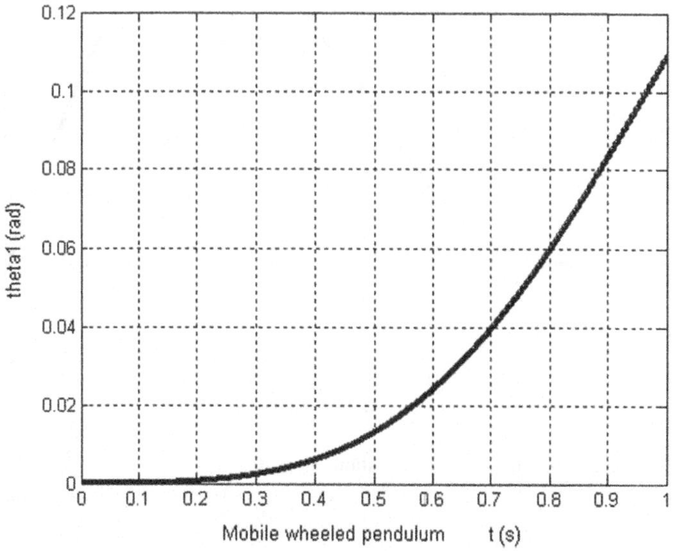

Fig. 11.12 Actuated rectilinear motion: rotation angle θ_1

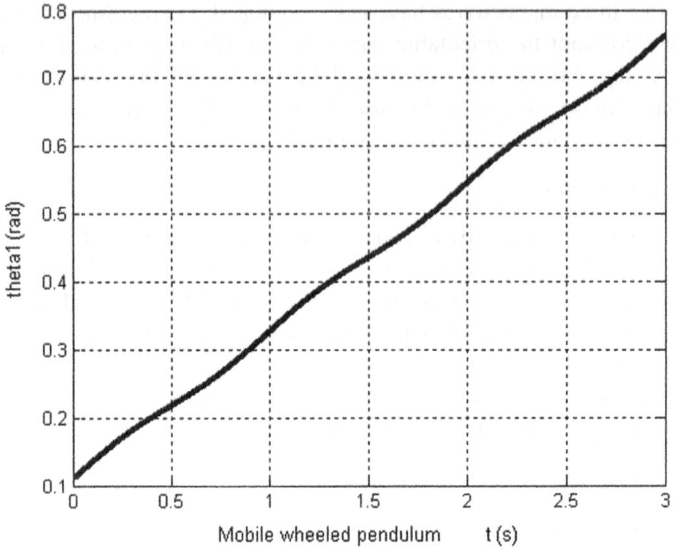

Fig. 11.13 Rectilinear motion: rotation angle θ_1

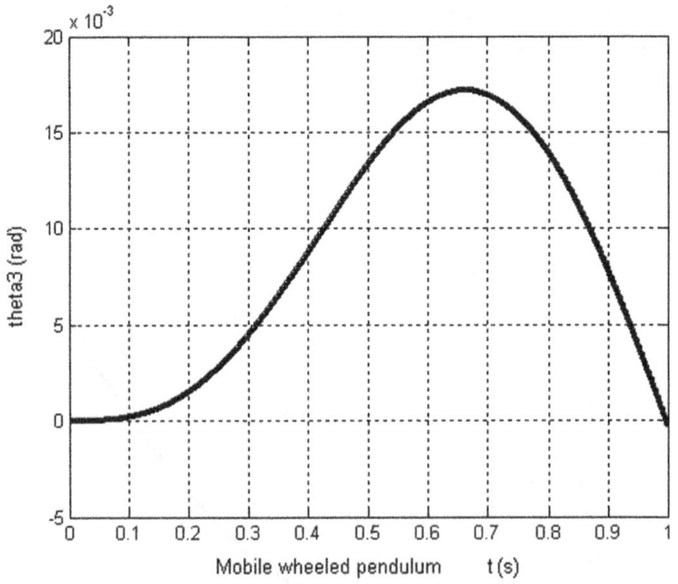

Fig. 11.14 Actuated rectilinear motion: rotation angle θ_3

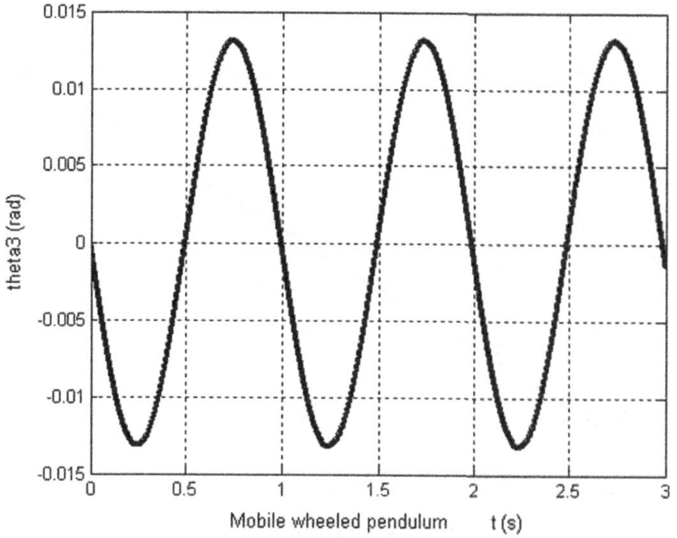

Fig. 11.15 Rectilinear motion: rotation angle θ_3

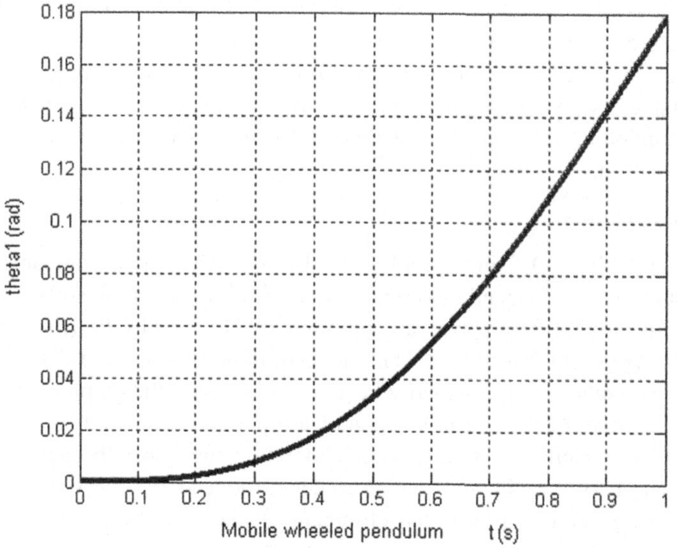

Fig. 11.16 Actuated pure rotation: rotation angle θ_1

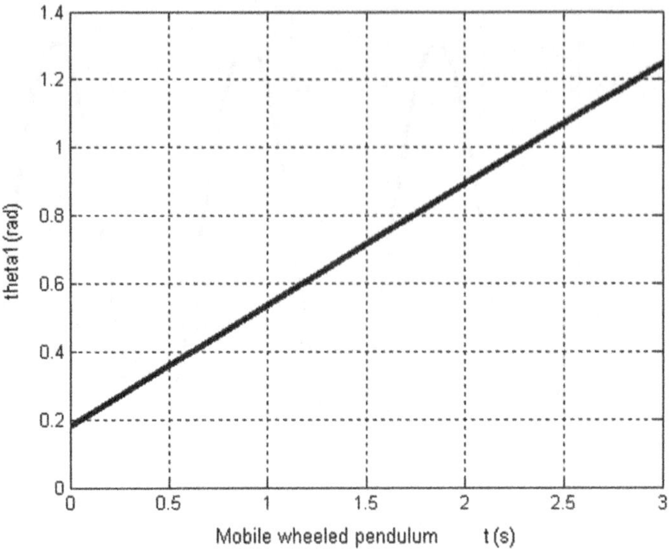

Fig. 11.17 Pure rotation: rotation angle θ_1

The trajectory of point C_2 reduces to a point coincident with the origin O of the inertial reference frame. Moreover, the angular velocity $\dot{\theta}_1$ will be constant in the steady state. For the assigned initial conditions and the type of input, the angle θ_3 and its first and second derivatives will remain theoretically equal to zero during the whole simulation, while θ_1,θ_2 and their derivatives are equal in amplitude and opposite in sign.

3. *Circular motion*

The torque $M_1(t) = 0.1 \sin \pi t$ applied to the first wheel is accompanied, in this third simulation, by a second unknown torque $M_2(t)$ applied to the other wheel. The output plots of this manoeuvre are reported by the graphs θ_1 (Figs. 11.18 and 11.19), θ_3 (Figs. 11.20 and 11.21), the evolution of the necessary torque M_2 (Fig. 11.22) applied to the second wheel and, of course, the graph of the angle θ_2 (Fig. 11.23) described in the last non-actuated period of simulation.

Dynamics modelling, taking into consideration the mass, the tensor of inertia and the action of weight and inertia force introduced by every element of the two-wheeled pendulum robot, can eliminate all the forces of internal joints and determine rapidly the time-history evolution of torques and powers required by the driving wheels, in an inverse dynamics problem.

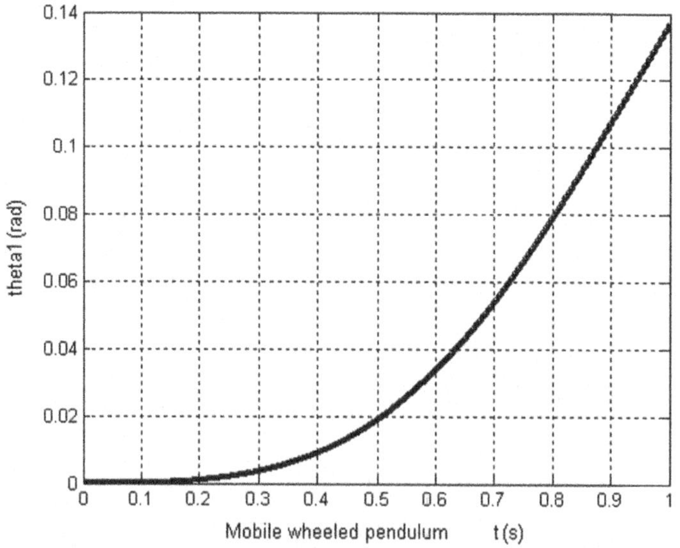

Fig. 11.18 Actuated circular motion: rotation angle θ_1

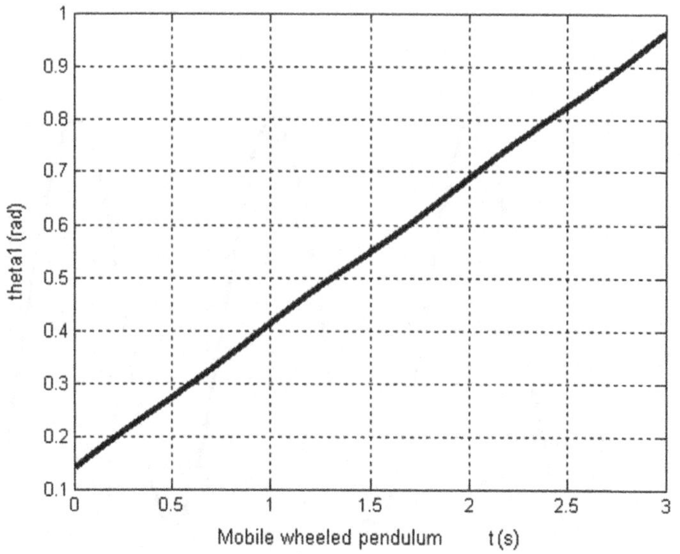

Fig. 11.19 Circular motion: rotation angle θ_1

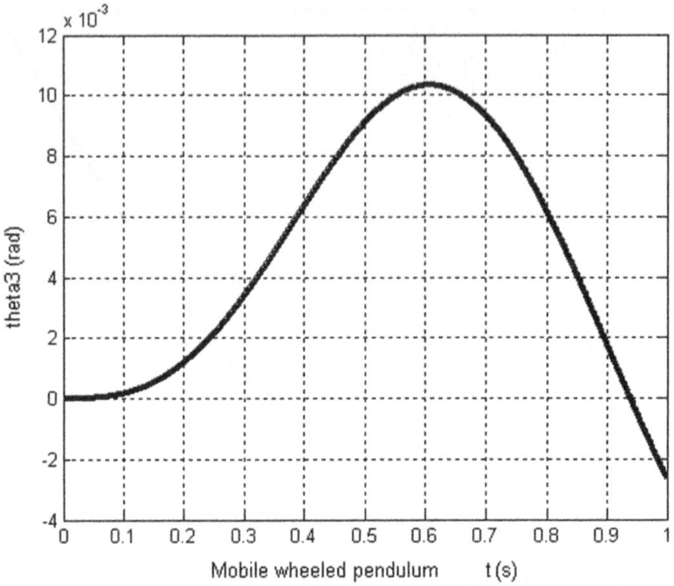

Fig. 11.20 Actuated circular motion: rotation angle θ_3

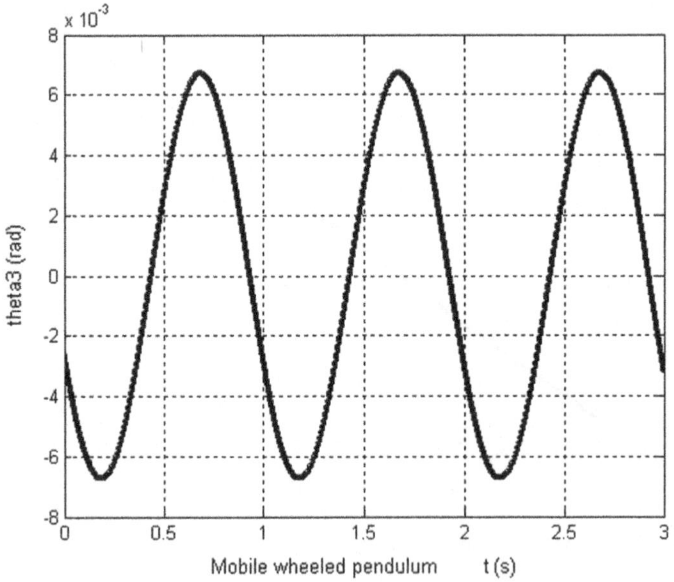

Fig. 11.21 Circular motion: rotation angle θ_3

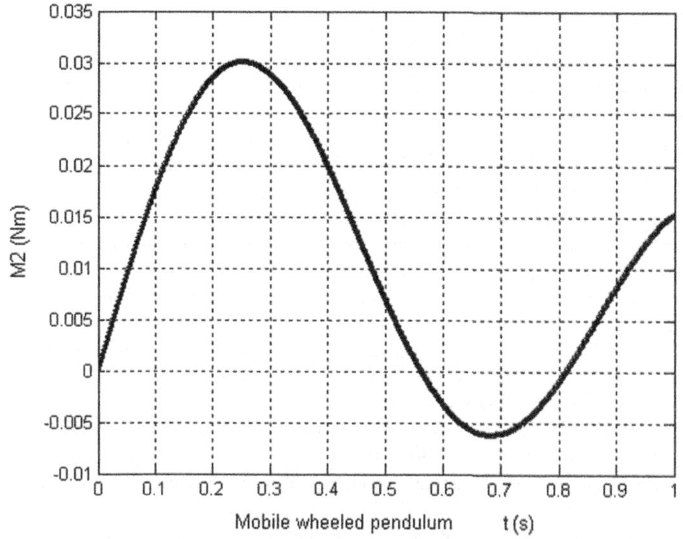

Fig. 11.22 Circular motion: torque of second wheel M_2

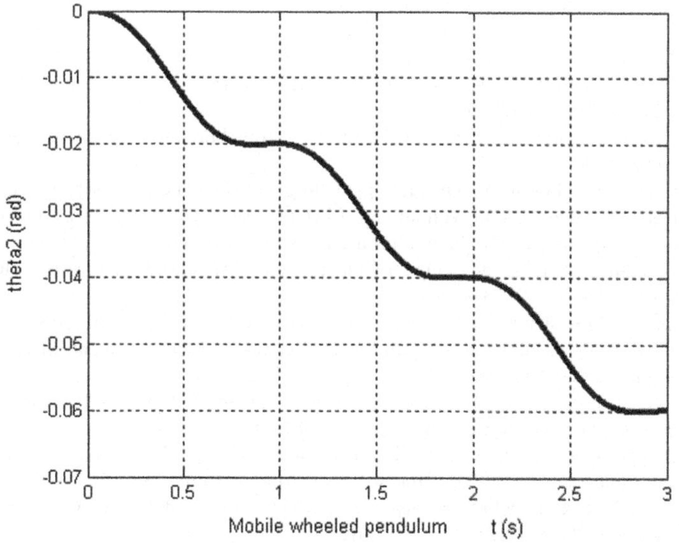

Fig. 11.23 Circular motion: rotation angle θ_2

References

1. Angeles, J.: Fundamentals of Robotic Mechanical Systems: Theory, Methods and Algorithms. Springer, New York (2002)
2. Papadopoulos, E., Poulakakis, E.: On motion planning of non-holonomic mobile robots. In: Proceedings of International Symposium on Robotics, Montréal, pp. 77–82 (2000)
3. Baron, L., Angeles, J.: The direct kinematics of parallel manipulators under joint-sensor redundancy. IEEE Trans. Robot. Autom. **16**(1), 12–19 (2000)
4. Velinski, S.A., Gardner, J.F.: Kinematics of mobile manipulator and implications for design. J. Robot. Syst. **17**(6) (2000)
5. Lerbet, J.: Mécanique des systèmes de solides rigides comportant des boucles fermées. Thesis, Paris (1987)
6. Abou-Samah, M., Krovi, V.: Cooperative frameworks for multiple mobile robots. In: Proceedings of CCToMM Symposium on Mechanisms, Machines and Mechatronics, Canadian Space Agency, Montréal (2001)
7. Agulló, J., Cardona, S., Vivancos, J.: Kinematics of vehicle with directional sliding wheels. Mech. Mach. Theory **22**(4), 295–301 (1987)
8. Muir, F.P., Neuman, C.P.: Kinematic modeling of mobile robots. J. Robot. Syst. **4**(2), 281–304 (1987)
9. Staicu, S.: Dynamics analysis of the Star parallel manipulator. Robot. Auton. Syst. **57**(11), 1057–1064 (2009)
10. Giergiel, J., Zylski, W.: Description of motion of a mobile robot by Maggie's equations. J. Theor. Appl. Mech., Warsaw **43**(3), 511–521 (2005)
11. Hendzel, Z.: An adaptive critic network for motion control of a wheeled mobile robot. Nonlinear Dyn. **50**(4), 849–855 (2007)
12. Pathak, K., Franch, J., Agrawal, S.K.: Velocity and position control of a wheeled inverted pendulum by partial feedback linearization. IEEE Trans. Rob. **21**(3), 505–513 (2005)
13. Gracia, L., Tornero, J.: Kinematic modelling and singularity of wheeled mobile robots. Adv. Robot. **21**(7), 793–816 (2007)
14. Chakraborty, N., Ghosal, A.: Kinematics of wheeled mobile robots on uneven terrain. Mech. Mach. Theory **39**(12), 1273–1287 (2004)
15. Chakraborty, N., Ghosal, A.: Dynamic modelling and simulation of a wheeled mobile robot for traversing uneven terrain without slip. J. Mech. Des. **127**(5), 901–909 (2005)
16. Salerno, A., Angeles, J.: On the nonlinear controllability of a quasi-holonomic mobile robot. In: Proceedings of IEEE International Conference on Robotics & Automation, ICRA'2003, Taipei, vol. 3, pp. 3379–3384 (2003)
17. Salerno, A., Ostrovskaya, S., Angeles, J.: The dynamics of a novel rolling robot: analysis and simulation. In: Proceedings of the 11th World Congress in Mechanism and Machine Science, Tianjin, pp. 1956–1960 (2004)
18. Staicu, S., Liu, X.-J., Wang, J.: Inverse dynamics of the HALF parallel manipulator with revolute actuators. Nonlinear Dyn. **50**(1–2), 1–12 (2007)
19. Cheng, H.H., Lee, J.J., Penkar, R.: Kinematic analysis of a hybrid serial-and-parallel-driven redundant industrial manipulator. Int. J. Robot. Autom. **10**(4), 159–166 (1995)
20. Staicu, S.: Matrix modeling of inverse dynamics of spatial and planar parallel robots. Multibody Syst. Dyn. **27**(2), 239–265 (2012)
21. Colbaugh, R., Trabatti, M., Glass, K.: Redundant non-holonomic mechanical system: characterization and control. Robotica. **17**(2), 203–217 (1999)

Chapter 12
Kinematics and Dynamics of a Hybrid Parallel Manipulator

Serial and parallel kinematical architectures, each of which has their own benefits and drawbacks, are the two basic kinds of construction used in robotics. Serial robots consist of a sequence of arms, whereas parallel robots with several prismatic or revolute joints take directly effect on the moving tool carrier. A general manipulating task in industry can be divided into two sub-tasks concerning position and orientation mechanism respectively. The *position mechanism* controls the general position, whereas the *orientation mechanism* manipulates the orientation of the end effectors. It is natural to merge the serial and parallel chains together to obtain a hybrid industrial robot, so as to retain the merits of both serial and parallel configurations—large workspace, high dexterity, high stiffness and high force-to-weight ratio—while their disadvantages are minimized. A simple way to do this is to utilize a parallel mechanism to position the end-effectors and connect serial joints to adjust the orientation, and vice versa.

A hybrid manipulator is a combination of two or more closed-chain and open-chain mechanisms or a sequence of parallel robots [1–3]. A serial-parallel manipulator has several modules with parallel structure that are connected serially [4–6]. With great potential in the field of industry, this kind of hybrid architectures possesses the advantages of both serial and parallel manipulators in terms of rigidity, accuracy and workspace considerations. Hybrid structures combining serial and parallel manipulators, which are a step forward in development, have presented the opportunity of achieving a compromise between the high rigidity of fully parallel robots and the extended workspace of serial manipulators. Some connections of parallel and serial mechanisms have already found various commercial applications such as high-speed milling, welding and component assembly in the aeronautical and automotive industries.

Recently, there has been an increasing level of interest in the hybrid manipulators, although comparatively few papers on these manipulators are available because, in contrast to open chain serial structures, the kinematics and dynamics analysis of hybrid manipulators present an inherent complexity due to the presence of kinematic constraints and singularities.

S. Staicu, *Dynamics of Parallel Robots*, Parallel Robots: Theory and Applications,
https://doi.org/10.1007/978-3-319-99522-9_12

Industry has demonstrated that hybrid constructions have been much more successful commercially than all other parallel kinematic machines combined. Many studies have focused on the analysis and modelling of several hybrid robots, by investigating the kinematics and dynamics of a type of hybrid robots constructed by serially- connected non-redundant parallel modules. Shahinpoor [7] presented a kinematic analysis and solved numerically a system of nonlinear equations for the inverse problem of a hybrid robotic system which consists of two serially connected parallel manipulators. Cheng et al. [8] used a numerical Newton-Raphson method to obtain the solution to the direct kinematics problem of a 10-DOF hybrid redundant manipulator, containing a closed-loop slider-crank mechanism and a parallel-driven mechanism. Considering the lower module as a positional mechanism and the upper module as an orientation device, Zhang and Song [6] analyzed the geometry and the position of a hybrid manipulator composed of two serially connected parallel robots, each mechanism having three degrees of freedom. Based on screw theory and the principle of virtual work, Gallardo et al. [9] addressed a complete kinematics analysis of a modular spatial hyper-redundant manipulator constructed with a variable number of serially-connected mechanical modules.

After a complete kinematics analysis, taking into consideration the frictionless revolute and prismatic joints, a recursive method based on explicit equations of parallel robots dynamics is applied to the *inverse dynamics* of a two-module spatial hybrid manipulator, to prove that the number of equations and computational operations can be reduced significantly by using a set of matrices for kinematics and dynamics modelling.

12.1 Structural Description of the Hybrid Parallel Manipulator

Considering the lower module as a positional mechanism and the upper module as an orientation device, we analyze first the geometry and the position of a hybrid manipulator composed of two distinct serially connected parallel robots, each mechanism having three degrees of freedom. Practically, the degrees-of-freedom value $F = 6$ of the hybrid mechanism is equal to the degrees of freedom associated with all the moving links $v = 24$ minus the total number of independent constraint relations $l = 18$ imposed by the joints.

The structure of the *lower parallel module* consists of a fixed circular base $A_1B_1C_1$ of radius R, a circular mobile platform $A_4B_4C_4$ of radius r, mass m_p and tensor of inertia $\hat{J}_p$, and three extensible legs with identical structure. Each leg connects the fixed base to the moving platform by two universal (U) joints interconnected through a prismatic ($\underline{P}$) joint made up of a cylinder and a piston. Hydraulic systems or electrical prismatic actuators with excellent dynamic behaviour and positioning accuracy can be used to vary the lengths of the prismatic joints and to control the location of the platform. The first platform achieves translation only if, along each leg, the first axis of the lower revolute joint is parallel to that of the last one of the upper joint and if, also, the two intermediate joint axes are parallel to one another. Thus, the first module has three active prismatic joints and six passive universal joints.

The architecture of the *upper parallel module* consists of a new identical mobile platform $D_4E_4F_4$ of radius r, mass m_p and inertia tensor J_p, connected to the lower moving base through a central spherical joint H and three identical extensible $U\underline{P}S$ legs, each comprising a prismatic actuator. The lower end of each leg is connected to the first moving platform by a universal joint situated on a circle of radius R (Fig. 12.1). We note that this spatial module is, sometimes, not considered a spherical mechanism, because the three legs and the moving platform do not have a common stationary point.

Considering that the two moving platforms initially start from a *central configuration* with respect to a fixed coordinate system $Ox_Oy_Oz_O$, we assign successively the systems of coordinates $Qx_Qy_Qz_Q(Q = O, G)$ and two mobile frames $Gx_Gy_Gz_G$ and $Hx_Hy_Hz_H$ to the moving platforms, with their centres $K = (G, H)$ initially located at an identical relative elevation $OG = GH = h$ above the base.

The active arbitrary leg $S = (A, B, C, D, E, F)$ of the hybrid parallel robot consists of a little cross of a universal joint $S_1x_1^Sy_1^Sz_1^S$, of negligible mass, which rotates with an angle φ_{10}^S and connects to a moving cylinder $S_2x_2^Sy_2^Sz_2^S$ of length l_2, mass m_2 and inertia tensor $\hat{J}_2$, having a relative rotation with the angle φ_{21}^S. An actuated

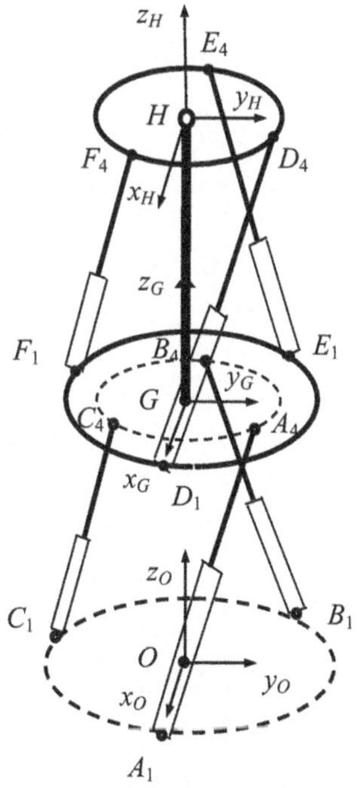

Fig. 12.1 Symmetric spatial hybrid parallel robot

piston of length l_3, mass m_3 and inertia tensor $\hat{J}_3$, linked to $S_3x_3^Sy_3^Sz_3^S$, has a relative displacement λ_{32}^S. Finally, a universal joint and a ball-joint S_4 are attached, respectively, at the edges of two circular moving platforms. The six pistons of the two modules initially start from the same position $S_2S_3 = l_1 = h/\sin\beta - l_3$ and the angles of orientation of the universal joints are given by

$$\begin{gathered} \alpha_A = \alpha_D = 0, \alpha_B = \alpha_E = -\alpha_C = -\alpha_F = \frac{2\pi}{3}, \nu = \frac{\pi}{6} \\ (R - r\cos\nu)\tan\delta = r\sin\nu, r\sin\nu\tan\beta = h\sin\delta, \end{gathered} \quad (12.1.1)$$

where the angle ν is defined as a chosen *twist* angle of a parallel robot and δ, β are two constant angles of rotation around the axes z_1^S and z_2^S, respectively (Fig. 12.2).

Since the hybrid manipulator is an assembly of links and joints, this can be symbolized in a more abstract form known as *equivalent graph* representation, using the associated graph to represent the topology of the mechanism (Fig. 12.3). In the kinematic graph representation, with *six independent loops*, the links are denoted by vertices, the prismatic joints by thick edges, the revolute and spherical joints by thin edges and, finally, the fixed link 0 by two small concentric circles.

Fig. 12.2 Kinematic scheme of the arbitrary leg S of the hybrid robot

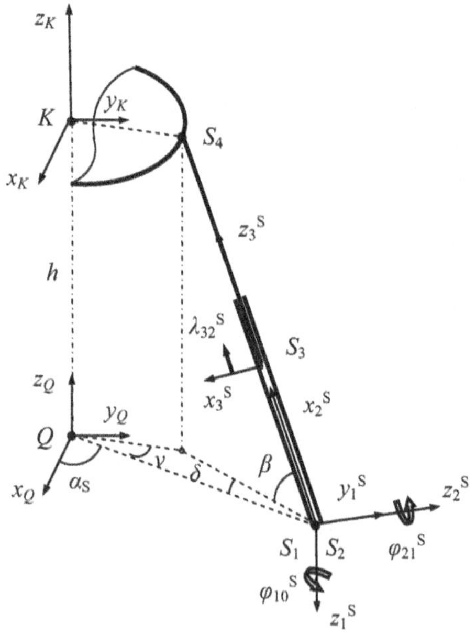

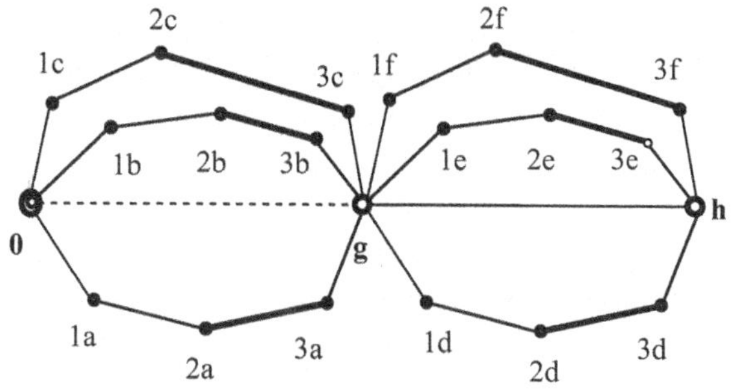

Fig. 12.3 Associated graph of the mechanism

12.2 Kinematics Analysis

Forward kinematics problem involves the mapping from a known set of input joint variables to a pose of the moving platforms that results from those given inputs, while the inverse kinematics problem involves a known position and orientation of the moving platforms to a set of input joint variables that will achieve that pose.

Starting from the two origins O, G and pursuing independent legs $OA_1A_2A_3A_4$, $OB_1B_2B_3B_4$, $OC_1C_2C_3C_4$, $GD_1D_2D_3D_4$, $GE_1E_2E_3E_4$ and $GF_1F_2F_3F_4$, we obtain the matrices of transformation

$$\begin{aligned}\sigma_{10} = \sigma_{10}^{\varphi} a_{\delta}\theta_3 a_{\alpha}^{S}, \quad \sigma_{21} = \sigma_{21}^{\varphi} a_{\beta}\theta_1^{T}, \quad \sigma_{32} = \theta_2, \quad \sigma_{20} = \sigma_{21}\sigma_{10}, \; \sigma_{30} = \sigma_{32}\sigma_{20} \\ S = (i,j) \;\; (i = A,B,C) \;\; (j = D,E,F) \;\; \sigma = (p,q) \;\; (p = a,b,c) \;\; (q = d,e,f),\end{aligned} \tag{12.2.1}$$

where we denote the matrices

$$\begin{aligned} a_{\alpha}^{S} &= rot(z, \alpha_S), \quad a_{\delta} = rot(z, \delta), \quad a_{\beta} = rot(z, \beta), \quad a_{v} = rot(z, v) \\ \sigma_{\tau,\tau-1}^{\varphi} &= rot(z, \varphi_{\tau,\tau-1}^{S}), \quad (\tau = 1, 2) \\ \theta_1 &= rot(x, \pi/2), \quad \theta_2 = rot(y, \pi/2), \quad \theta_3 = rot(y, \pi). \end{aligned} \tag{12.2.2}$$

Since all concurrent rotations of angles $\alpha_1, \alpha_2, \alpha_3$ around the centre H of the upper moving platform take place successively, the rotation matrix $h_{30} = h_{32}h_{21}h_{10}$ of this platform is obtained by multiplying three relative basic matrices

$$h_{10} = rot(x, \alpha_1), \quad h_{21} = rot(y, \alpha_2), \quad h_{32} = rot(z, \alpha_3). \tag{12.2.3}$$

We consider that, over the length of three seconds, the coordinates of G and the angles α_l can describe relative motions of two platforms through following analytical functions

$$\frac{x_0^G}{x_0^{G*}} = \frac{y_0^G}{y_0^{G*}} = \frac{h - z_0^G}{z_0^{G*}} = \frac{\alpha_l}{\alpha_l^*} = 1 - \cos\frac{\pi}{3}t \quad (l = 1, 2, 3). \tag{12.2.4}$$

where the values $2x_0^{G*}, 2y_0^{G*}, 2z_0^{G*}, 2\alpha_l^*$, denote the final positions of the moving platforms.

A set of 18 independent variables $\varphi^i_{\tau,\tau-1}, \lambda^i_{32}, \varphi^j_{\tau,\tau-1}, \lambda^j_{32}$, characterizing the kinematics of two modules, will be determined through 18 scalar equations, which are easily generated by some vector-loop constraint conditions

$$\vec{r}^{\,i}_{10} + \sum_{k=1}^{3} p^T_{k0} \vec{r}^{\,i}_{k+1,k} = \vec{r}^{\,G}_0 + \vec{r}^{\,i_4}_G, \quad \vec{r}^{\,j}_{10} + \sum_{k=1}^{3} q^T_{k0} \vec{r}^{\,j}_{k+1,k} = \vec{r}^{\,H}_G + h^T_{30} \vec{r}^{\,j_4}_H, \tag{12.2.5}$$

where

$$\begin{aligned}
&\vec{r}^{\,G}_0 = \begin{bmatrix} x_0^G & y_0^G & z_0^G \end{bmatrix}^T, \quad \vec{r}^{\,H}_G = h\vec{u}_3 \\
&\vec{r}^{\,S}_{10} = R a^{ST}_\alpha \vec{u}_1, \quad \vec{r}^{\,S}_{21} = \vec{0}, \quad \vec{r}^{\,S}_{32} = (l_1 + \lambda^S_{32})\vec{u}_1, \quad \vec{r}^{\,S}_{43} = l_3 \vec{u}_3, \quad \vec{r}^{\,S_4}_K = r a^{ST}_\alpha a^T_v \vec{u}_1 \\
&K = (G, H) \;\; S = (i, j) \\
&\vec{u}_1 = [1 \;\; 0 \;\; 0]^T, \quad \vec{u}_2 = [0 \;\; 1 \;\; 0]^T, \quad \vec{u}_3 = [0 \;\; 0 \;\; 1]^T.
\end{aligned} \tag{12.2.6}$$

From the vector equation (12.6) we obtain an inverse geometric solution for the spatial hybrid robot

$$\begin{aligned}
&-(l_1 + l_3 + \lambda^i_{32}) \cos(\varphi^i_{10} + \delta - \alpha_i) \cos(\varphi^i_{21} + \beta) = x_0^G + r\cos(\alpha_i + v) - R\cos\alpha_i \\
&(l_1 + l_3 + \lambda^i_{32}) \sin(\varphi^i_{10} + \delta - \alpha_i) \cos(\varphi^i_{21} + \beta) = y_0^G + r\sin(\alpha_i + v) - R\sin\alpha_i \\
&(l_1 + l_3 + \lambda^i_{32}) \sin(\varphi^i_{21} + \beta) = z_0^G \\
&-(l_1 + l_3 + \lambda^j_{32}) \cos(\varphi^j_{10} + \delta - \alpha_j) \cos(\varphi^j_{21} + \beta) = r\vec{u}^T_1 h^T_{30} a^{jT}_\alpha a^T_v \vec{u}_1 - R\cos\alpha_j \\
&(l_1 + l_3 + \lambda^j_{32}) \sin(\varphi^j_{10} + \delta - \alpha_j) \cos(\varphi^j_{21} + \beta) = r\vec{u}^T_2 h^T_{30} a^{jT}_\alpha a^T_v \vec{u}_1 - R\sin\alpha_j \\
&(l_1 + l_3 + \lambda^j_{32}) \sin(\varphi^j_{21} + \beta) = h + r\vec{u}^T_3 h^T_{30} a^{jT}_\alpha a^T_v \vec{u}_1.
\end{aligned} \tag{12.2.7}$$

The motion of the compounding elements of the hybrid robot are characterized by the following relative velocities of joints and relative angular velocities:

$$\vec{v}^{\,S}_{\tau,\tau-1} = \vec{0}, \quad \vec{v}^{\,S}_{32} = \dot{\lambda}^S_{32} \vec{u}_3, \quad \vec{\omega}^S_{\tau,\tau-1} = \dot{\varphi}^S_{\tau,\tau-1} \vec{u}_3, \quad \vec{\omega}^S_{32} = \vec{0} \; (\tau = 1, 2). \tag{12.2.8}$$

To describe the absolute kinematic state of each link of leg S, we compute the linear velocity $\vec{v}^{\,S}_{k0}$, the angular velocity $\vec{\omega}^S_{k0}$ and the *associate skew-symmetric matrix* $\tilde{\omega}^S_{k0}$ in terms of the vectors of the preceding body, using a recursive approach:

$$\begin{aligned}\vec{v}_{k0}^{S} &= \sigma_{k,k-1}\vec{v}_{k-1,0}^{S} + \sigma_{k,k-1}\tilde{\omega}_{k-1,0}^{S}\vec{r}_{k,k-1}^{S} + v_{k,k-1}^{S}\vec{u}_{3} \\ \vec{\omega}_{k0}^{S} &= \sigma_{k,k-1}\vec{\omega}_{k-1,0}^{S} + \omega_{k,k-1}^{S}\vec{u}_{3}, \quad \tilde{\omega}_{k0}^{S} = \sigma_{k,k-1}\tilde{\omega}_{k-1,0}^{S}\sigma_{k,k-1}^{T} + \omega_{k,k-1}^{S}\tilde{u}_{3}.\end{aligned} \tag{12.2.9}$$

Starting from the derivatives of geometrical constraints (12.6) and using the skew-symmetric matrix

$$\tilde{\omega}_{30}^{H} = h_{30}\dot{h}_{30}^{T} = \dot{\alpha}_{1}h_{32}h_{21}\tilde{u}_{1}h_{21}^{T}h_{32}^{T} + \dot{\alpha}_{2}h_{32}\tilde{u}_{2}h_{32}^{T} + \dot{\alpha}_{3}\tilde{u}_{3}, \tag{12.2.10}$$

which is associated with the angular velocity $\vec{\omega}_{30}^{H} = \dot{\alpha}_{1}h_{32}h_{21}\vec{u}_{1} + \dot{\alpha}_{2}h_{32}\vec{u}_{2} + \dot{\alpha}_{3}\vec{u}_{3}$ of the *upper* moving platform, we obtain the m*atrix conditions of connectivity* on the relative velocities [10]:

$$\begin{aligned}\vec{V}_{i} &= \begin{bmatrix}\omega_{10}^{i} & \omega_{21}^{i} & v_{32}^{i}\end{bmatrix}^{T} = [N_{i}]^{-1}\vec{P}_{i} \\ \vec{V}_{j} &= \begin{bmatrix}\omega_{10}^{j} & \omega_{21}^{j} & v_{32}^{j}\end{bmatrix}^{T} = [N_{j}]^{-1}\vec{P}_{j},\end{aligned} \tag{12.2.11}$$

where the following terms determine the contents of the 3 × 3 invertible square matrices $[N_{i}], [N_{j}]$ and the column matrices $\vec{P}_{i}, \vec{P}_{j}$:

$$\begin{aligned}&n_{m1}^{i} = u_{m}^{T}p_{10}^{T}\tilde{u}_{3}p_{21}^{T}\vec{r}_{42}^{i}, \quad n_{m2}^{i} = \vec{u}_{m}^{T}p_{20}^{T}\tilde{u}_{3}\vec{r}_{42}^{i}, \quad n_{m3}^{i} = \vec{u}_{m}^{T}p_{20}^{T}\vec{u}_{1}, \quad \vec{r}_{42}^{i} = \vec{r}_{32}^{i} + p_{32}^{T}\vec{r}_{43}^{i} \\ &n_{m1}^{j} = u_{m}^{T}q_{10}^{T}\tilde{u}_{3}q_{21}^{T}\vec{r}_{42}^{j}, \quad n_{m2}^{j} = \vec{u}_{m}^{T}q_{20}^{T}\tilde{u}_{3}\vec{r}_{42}^{j}, \quad n_{m3}^{j} = \vec{u}_{m}^{T}q_{20}^{T}\vec{u}_{1}, \quad \vec{r}_{42}^{j} = \vec{r}_{32}^{j} + q_{32}^{T}\vec{r}_{43}^{j} \\ &p_{m}^{i} = \vec{u}_{m}^{T}\dot{\vec{r}}_{0}^{G}, \quad p_{m}^{j} = \vec{u}_{m}^{T}h_{30}^{T}\tilde{\omega}_{30}^{H}\vec{r}_{H}^{j_4} \quad (m = 1, 2, 3) \\ &\qquad (i = A, B, C) \ (j = D, E, F) \ (p = a, b, c) \ (q = d, e, f).\end{aligned} \tag{12.2.12}$$

From these equations, we can obtain the *complete* Jacobian matrix of the hybrid manipulator. This matrix is a fundamental prerequisite for the analysis of the robot workspace and the particular configurations of singularities where the spatial hybrid robot will become uncontrollable.

Based on different conditions of connectivity, expressions of relative accelerations $\vec{\Gamma}_{i} = \begin{bmatrix}\varepsilon_{10}^{i} & \varepsilon_{21}^{i} & \gamma_{32}^{i}\end{bmatrix}^{T}$, $\vec{\Gamma}_{j} = \begin{bmatrix}\varepsilon_{10}^{j} & \varepsilon_{21}^{j} & \gamma_{32}^{j}\end{bmatrix}^{T}$ are obtained from the column matrix

$$\vec{\Gamma}_{i} = [N_{i}]^{-1}\vec{S}_{i}, \quad \vec{\Gamma}_{j} = [N_{j}]^{-1}\vec{S}_{j}, \tag{12.2.13}$$

where the following terms determine the contents of column matrices $\vec{S}_{i} = \dot{\vec{P}}_{i} - [\dot{N}_{i}]\vec{V}_{i}$, $\vec{S}_{j} = \dot{\vec{P}}_{j} - [\dot{N}_{j}]\vec{V}_{j}$:

$$
\begin{aligned}
s_m^i &= \vec{u}_m^T \ddot{\vec{r}}_0^G - \omega_{10}^i \omega_{10}^i \vec{u}_m^T p_{10}^T \tilde{u}_3 \tilde{u}_3 p_{21}^T \vec{r}_{42}^i - \omega_{21}^i \omega_{21}^i \vec{u}_m^T p_{20}^T \tilde{u}_3 \tilde{u}_3 \vec{r}_{42}^i \\
&\quad - 2\omega_{10}^i \omega_{21}^i \vec{u}_m^T p_{10}^T \tilde{u}_3 p_{21}^T \tilde{u}_3 \vec{r}_{42}^i - 2\omega_{10}^i v_{32}^i \vec{u}_m^T p_{10}^T \tilde{u}_3 p_{21}^T \vec{u}_1 - 2\omega_{21}^i v_{32}^i \vec{u}_m^T p_{20}^T \tilde{u}_3 \vec{u}_1 \\
s_m^j &= \vec{u}_m^T q_{30}^T \{\tilde{\omega}_{30}^H \tilde{\omega}_{30}^H + \tilde{\varepsilon}_{30}^H\} \vec{r}_H^{j_4} - \omega_{10}^j \omega_{10}^j \vec{u}_m^T q_{10}^T \tilde{u}_3 \tilde{u}_3 q_{21}^T \vec{r}_{42}^j - \omega_{21}^j \omega_{21}^j \vec{u}_m^T q_{20}^T \tilde{u}_3 \tilde{u}_3 \vec{r}_{42}^j \\
&\quad - 2\omega_{10}^j \omega_{21}^j \vec{u}_m^T q_{10}^T \tilde{u}_3 q_{21}^T \tilde{u}_3 \vec{r}_{42}^j - 2\omega_{10}^j v_{32}^j \vec{u}_m^T q_{10}^T \tilde{u}_3 q_{21}^T \vec{u}_1 - 2\omega_{21}^j v_{32}^j \vec{u}_m^T q_{20}^T \tilde{u}_3 \vec{u}_1 \\
\tilde{\omega}_{30}^H \tilde{\omega}_{30}^H + \tilde{\varepsilon}_{30}^H &= \ddot{\alpha}_1 h_{32} h_{21} \tilde{u}_1 h_{21}^T h_{32}^T + \ddot{\alpha}_2 h_{32} \tilde{u}_2 h_{32}^T + \ddot{\alpha}_3 \tilde{u}_3 \\
&\quad + \dot{\alpha}_1^2 h_{32} h_{21} \tilde{u}_1 \tilde{u}_1 h_{21}^T h_{32}^T + \dot{\alpha}_2^2 h_{32} \tilde{u}_2 \tilde{u}_2 h_{32}^T + \dot{\alpha}_3^2 \tilde{u}_3 \tilde{u}_3 \\
&\quad + 2\dot{\alpha}_1 \dot{\alpha}_2 h_{32} h_{21} \tilde{u}_1 h_{21}^T \tilde{u}_2 h_{32}^T + 2\dot{\alpha}_2 \dot{\alpha}_3 h_{32} \tilde{u}_2 h_{32}^T \tilde{u}_3 + 2\dot{\alpha}_1 \dot{\alpha}_3 h_{32} h_{21} \tilde{u}_1 h_{21}^T h_{32}^T \tilde{u}_3.
\end{aligned}
\tag{12.2.14}
$$

The following relations give the linear accelerations $\vec{\gamma}_{k0}^S$, the angular accelerations $\vec{\varepsilon}_{k0}^S$ and the *significant matrices* $\tilde{\omega}_{k0}^S \tilde{\omega}_{k0}^S + \tilde{\varepsilon}_{k0}^S$ of each link from the leg S

$$
\begin{aligned}
\vec{\gamma}_{k0}^S &= \sigma_{k,k-1} \vec{\gamma}_{k-1,0}^S + \sigma_{k,k-1} \{\tilde{\omega}_{k-1,0}^S \tilde{\omega}_{k-1.0}^S + \tilde{\varepsilon}_{k-1,0}^S\} \vec{r}_{k,k-1}^S \\
&\quad + 2 v_{k,k-1}^S \sigma_{k,k-1} \tilde{\omega}_{k-1,0}^S \sigma_{k,k-1}^T \vec{u}_3 + \gamma_{k,k-1}^S \vec{u}_3 \\
\vec{\varepsilon}_{k0}^S &= \sigma_{k,k-1} \vec{\varepsilon}_{k-1,0}^S + \varepsilon_{k,k-1}^S \vec{u}_3 + \omega_{k,k-1}^S \sigma_{k,k-1} \tilde{\omega}_{k-1,0}^S \sigma_{k,k-1}^T \vec{u}_3 \\
\tilde{\omega}_{k0}^S \tilde{\omega}_{k0}^S + \tilde{\varepsilon}_{k0}^S &= \sigma_{k,k-1} \{\tilde{\omega}_{k-1,0}^S \omega_{k-1,0}^S + \tilde{\varepsilon}_{k-1,0}^S\} \sigma_{k,k-1}^T \\
&\quad + \omega_{k,k-1}^S \omega_{k,k-1}^S \tilde{u}_3 \tilde{u}_3 + \varepsilon_{k,k-1}^S \tilde{u}_3 + 2\omega_{k,k-1}^S \sigma_{k,k-1} \tilde{\omega}_{k-1,0}^S \sigma_{k,k-1}^T \tilde{u}_3.
\end{aligned}
\tag{12.2.15}
$$

As an application let us consider a serial-parallel hybrid mechanism which has the following architectural and mechanical characteristics

$$
\begin{aligned}
&x_0^{G*} = 0.05\,\text{m},\quad y_0^{G*} = 0.05\,\text{m},\quad z_0^{G*} = 0.15\,\text{m} \\
&\alpha_1^* = \pi/12,\quad \alpha_2^* = \pi/18,\quad \alpha_3^* = \pi/36 \\
&R = 0.6\,\text{m},\quad r = 0.2\,\text{m},\quad l_2 = 1\,\text{m},\quad l_3 = 0.6\,\text{m},\quad h = 0.8\,\text{m} \\
&m_2 = 1\,\text{kg},\quad m_3 = 0.75\,\text{kg},\quad m_p = 5\,\text{kg},\quad \Delta t = 3\,\text{s},\quad g = 9.81\,\text{ms}^{-2} \\
&\hat{J}_2 = \begin{bmatrix} 0.01 & & \\ & 0.325 & \\ & & 0.325 \end{bmatrix},\quad \hat{J}_3 = \begin{bmatrix} 0.325 & & \\ & 0.325 & \\ & & 0.01 \end{bmatrix},\quad \hat{J}_p = \begin{bmatrix} 00.6 & & \\ & 0.6 & \\ & & 1.2 \end{bmatrix}.
\end{aligned}
$$

Assuming that the robot starts at rest from a central configuration, a program using MATLAB software was developed to solve first a complete inverse kinematics of the hybrid parallel mechanism.

For the first example, the lower platform translates along the *vertical direction* z_0, while the upper platform rotates about the same axis. The time-histories for the input displacements λ_{32}^i (Fig. 12.4) and λ_{32}^j (Fig. 12.5) are computed for a period of $\Delta t = 3$ s in terms of given the above analytical equation (12.5).

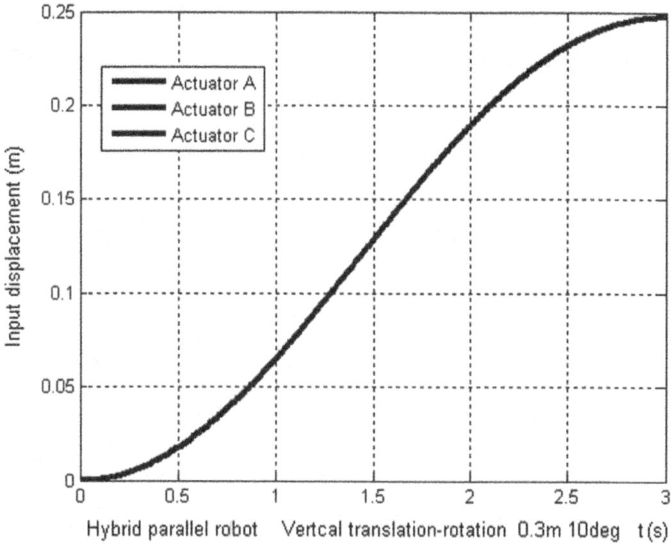

Fig. 12.4 Displacements λ_{32}^{i} of sliders (vertical *T-R*)

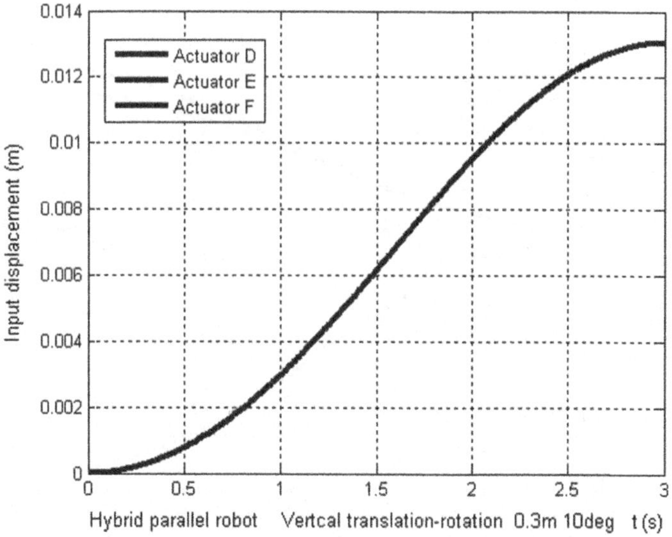

Fig. 12.5 Displacements λ_{32}^{j} of sliders (vertical *T-R*)

In the second case, when the lower platform moves along x_0, pursuing a *rectilinear horizontal trajectory*, and the upper platform rotates about x_H *horizontal axis* with a variable angular acceleration $\ddot{\alpha}_1$, the graphs are plotted and sketched in Figs. 12.6 and 12.7.

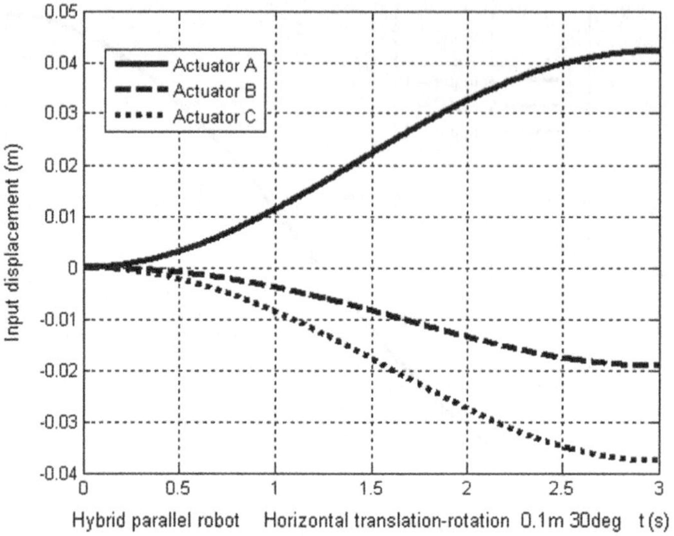

Fig. 12.6 Displacements λ_{32}^{i} of sliders (horizontal *T-R*)

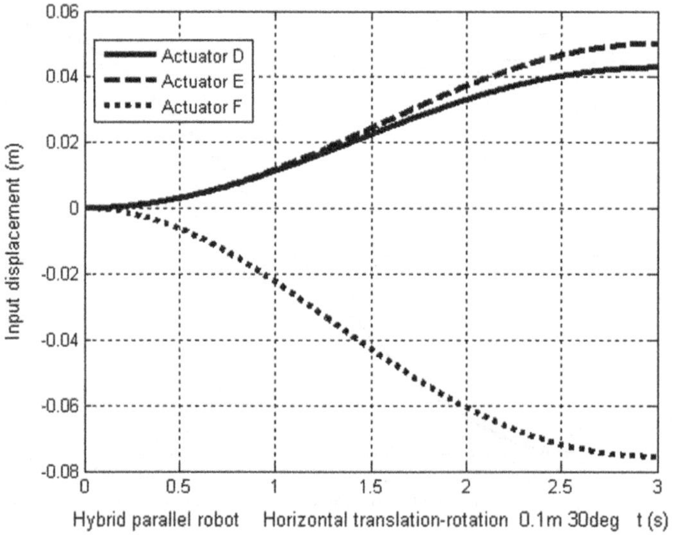

Fig. 12.7 Displacements λ_{32}^{j} of sliders (horizontal *T-R*)

12.3 Inverse Dynamics Model

The dynamics model for hybrid manipulators is crucial especially if we are aiming for high-speed and high-acceleration applications. Knowing the position and kinematics state of each link, as well as the external forces acting on the hybrid parallel manipulator, we apply some explicit recursive matrix equations to the inverse dynamic problem in order to obtain the input forces and powers required by all prismatic actuators.

Six independent pneumatic or hydraulic systems generate the internal active forces $\vec{f}^{\,i}_{32} = f^{i}_{32}\vec{u}_3, \vec{f}^{\,j}_{32} = f^{j}_{32}\vec{u}_3$ and control the motion of six moving pistons of the legs. The serial-parallel hybrid robot can artificially be transformed into a set of *seven open trees* by imaginarily *cutting* each joint for two moving platforms and taking its effect into account by introducing the corresponding constraining conditions (Fig. 12.8).

The inertia force and the moment of inertia forces on an arbitrary rigid body T_k^S $(S = A, B, C, D, E, F)$ of the tree S

$$\begin{aligned} \vec{f}_k^{\,in\,S} &= -m_k^S[\vec{\gamma}_{k0}^S + (\tilde{\omega}_{k0}^S\tilde{\omega}_{k0}^S + \tilde{\varepsilon}_{k0}^S)\vec{r}_{k0}^S], \\ \vec{m}_k^{\,in\,S} &= -(m_k^S\tilde{r}_k^{CS}\vec{\gamma}_{k0}^S + \hat{J}_k^S\vec{\varepsilon}_{k0}^S + \tilde{\omega}_{k0}^S\hat{J}_k^S\vec{\omega}_{k0}^S), \end{aligned} \tag{12.3.1}$$

are determined with respect to the centre of joint S_k. On the other hand, the wrench of two vectors $\vec{f}_k^{\,*S} = m_k^S g\sigma_{k0}\vec{u}_3$, $\vec{m}_k^{\,*S} = m_k^S g\tilde{r}_k^{CS}\sigma_{k0}\vec{u}_3$ $(\sigma = a, b, c, d, e, f)$ give the influence of the weight $m_k^S\vec{g}$ acting on the same element T_k^S of the manipulator.

Pursuing the leg S, two *significant recursive relations* generate the vectors

$$\vec{F}_k^S = \vec{F}_{k0}^S + \sigma_{k+1,k}^T\vec{F}_{k+1}^S, \quad \vec{M}_k^S = \vec{M}_{k0}^S + \sigma_{k+1,k}^T\vec{M}_{k+1}^S + \tilde{r}_{k+1,k}^S\sigma_{k+1,k}^T\vec{F}_{k+1}^S, \tag{12.3.2}$$

with the notations $\vec{F}_{k0}^S = -\vec{f}_k^{\,in\,S} - \vec{f}_k^{\,*S}$, $\vec{M}_{k0}^S = -\vec{m}_k^{\,in\,S} - \vec{m}_k^{\,*S}$.

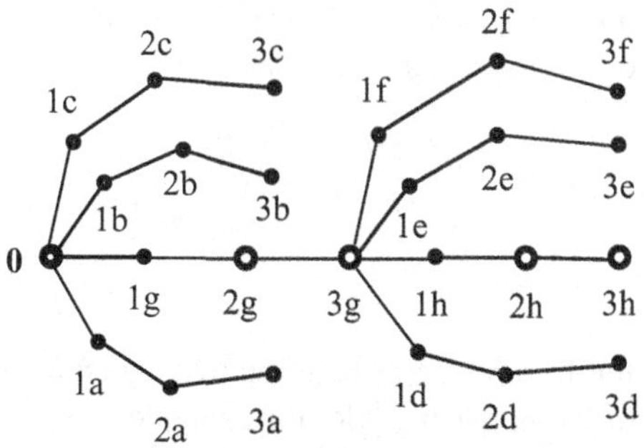

Fig. 12.8 Seven open trees of the hybrid parallel robot

Starting from (12.18), we develop a set of six recursive relations for each leg S:

$$\vec{F}_3^S = \vec{F}_{30}^S, \quad \vec{F}_2^S = \vec{F}_{20}^S + \sigma_{32}^T \vec{F}_3^S, \quad \vec{F}_1^S = \vec{F}_{10}^S + \tau_{21}^T \vec{F}_2^S$$
$$\vec{M}_3^S = \vec{M}_{30}^S, \quad \vec{M}_2^S = \vec{M}_{20}^S + \sigma_{32}^T \vec{M}_3^S + \tilde{r}_{32}^S \sigma_{32}^T \vec{F}_3^S, \quad \vec{M}_1^S = \vec{M}_{10}^S + \sigma_{21}^T \vec{M}_2^S + \tilde{r}_{21}^S \sigma_{21}^T \vec{F}_2^S$$
$$(S = A, B, C, D, E, F)(\sigma = a, b, c, d, e, f). \tag{12.3.3}$$

Specifically, for the two moving platforms, we have

$$\begin{aligned} &\vec{F}_3^H = m_p(gh_{30}\vec{u}_3 + \vec{\gamma}_{30}^H), \quad \vec{F}_2^H = h_{32}^T \vec{F}_3^H, \quad \vec{F}_1^H = h_{21}^T \vec{F}_2^H \\ &\vec{M}_3^H = \hat{J}_p \vec{\varepsilon}_{30}^H, \quad \vec{M}_2^H = h_{32}^T \vec{M}_3^H, \quad \vec{M}_1^H = h_{21}^T \vec{M}_2^H. \\ &\vec{F}_1^G = \vec{F}_2^G = \vec{F}_3^G = m_p(g\vec{u}_3 + \vec{\gamma}_{30}^G) + h_{10}^T \vec{F}_1^H + d_{10}^T \vec{F}_1^D + e_{10}^T F_1^E + f_{10}^T \vec{F}_1^F. \end{aligned} \tag{12.3.4}$$

The *virtual velocities* are expressed as functions of the pose of the hybrid robot by the matrix kinematical equations (12.12) and (12.13). Considering some *independent virtual motions* of the spatial mechanism, virtual velocities should be compatible with the virtual motions imposed by all kinematical constraints at any instant in time. Let us assume that the robot has successively six independent virtual motions, determined by the velocities:

$$\begin{aligned} &v_{32a}^{Av} = 1, \quad v_{32a}^{Sv} = 0\ (S \neq A); \quad v_{32b}^{Bv} = 1, \quad v_{32b}^{Sv} = 0\ (S \neq B); \quad v_{32c}^{Cv} = 1, \quad v_{32c}^{Sv} = 0\ (S \neq C) \\ &v_{32d}^{Dv} = 1, \quad v_{32d}^{Sv} = 0\ (S \neq D); \quad v_{32e}^{Ev} = 1, \quad v_{32e}^{Sv} = 0\ (S \neq E); \quad v_{32f}^{Fv} = 1, \quad v_{32f}^{Sv} = 0\ (S \neq F). \end{aligned} \tag{12.3.5}$$

Using the *equations of parallel robots dynamics* [11, 12], based on the principle of virtual work, the following compact matrix relations result

$$\begin{aligned} f_{32}^A = \vec{u}_3^T \{ &\vec{F}_3^A + \omega_{10a}^{Av} \vec{M}_1^A + \omega_{21a}^{Av} \vec{M}_2^A + \omega_{10a}^{Bv} \vec{M}_1^B + \omega_{21a}^{Bv} \vec{M}_2^B \\ &+ \omega_{10a}^{Cv} \vec{M}_1^C + \omega_{21a}^{Cv} \vec{M}_2^C \} + v_{10a}^{Gv} \vec{u}_1^T \vec{F}_1^G + v_{21a}^{Gv} \vec{u}_2^T \vec{F}_2^G + v_{32a}^{Gv} \vec{u}_3^T \vec{F}_3^G, \end{aligned} \tag{12.3.6}$$

for the *input force* required by the first active prismatic joint A_3 along the axis $A_3z_3^A$ of the lower module, and

$$\begin{aligned} f_{32}^D = \vec{u}_3^T \{ &\vec{F}_3^D + \omega_{10d}^{Dv} \vec{M}_1^D + \omega_{21d}^{Dv} \vec{M}_2^D + \omega_{10d}^{Ev} \vec{M}_1^E + \omega_{21d}^{Ev} \vec{M}_2^E \\ &+ \omega_{10d}^{Fv} \vec{M}_1^F + \omega_{21d}^{Fv} \vec{M}_2^F \}. + \omega_{10d}^{Hv} \vec{u}_1^T \vec{M}_1^H + \omega_{21d}^{Hv} \vec{u}_2^T \vec{M}_2^H + \omega_{32d}^{Gv} \vec{u}_3^T \vec{M}_3^H \end{aligned} \tag{12.3.7}$$

for the *input force* required by the first active prismatic joint D_3 along the axis $D_3z_3^D$ of the upper module, for example.

The relations (12.18)–(12.23) represent the *inverse dynamics model* of the hybrid parallel robot.

Considering the weights of all links as passive external forces acting on the spatial hybrid manipulator, the time-history evolution for the input forces f_{32}^{S} and input powers $P_{32}^{S} = v_{32}^{S} f_{32}^{S}$ (Figs. 12.9, 12.10, 12.11 and 12.12) of six prismatic actuators are computed into the *first evolution* of two platforms, for the same period $\Delta t = 3$ s. As can be seen, it is proved to be true that all active forces and powers of the lower module are permanently equal to one another but are different from the forces and the powers of the upper module.

In the case where the platforms simultaneously make a *horizontal translation-rotation* about x_0 axis, the graphs are plotted and illustrated in Figs. 12.13, 12.14, 12.15 and 12.16.

The simulation through the program certifies that the current matrix recursive formulation can easily be transformed in a model which is successfully expected to be deployed for the automatic robotized control of hybrid parallel robots, and that one of its major advantages is the effectiveness and accuracy of numerical computation.

The concept and the procedure developed above can directly be extended to the analysis of a complex robotic system composed of a multitude of serially-arranged similar parallel modules, but they are also applicable to the study of hybrid robots composed of different structures of parallel modules.

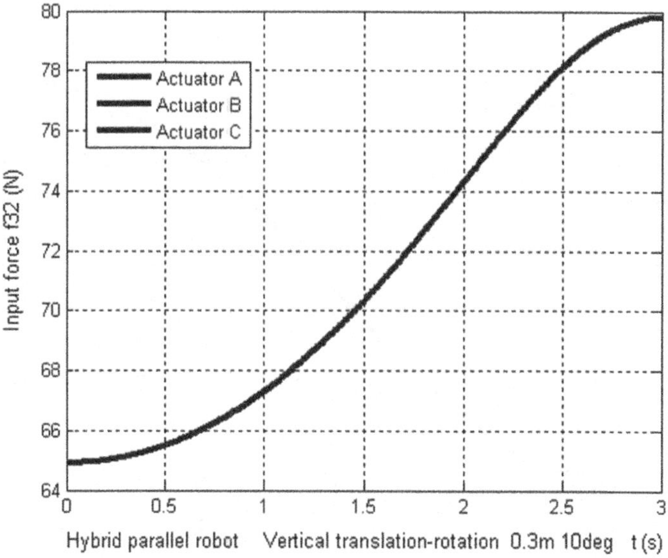

Fig. 12.9 Input forces f_{32}^{i} of actuators (vertical *T-R*)

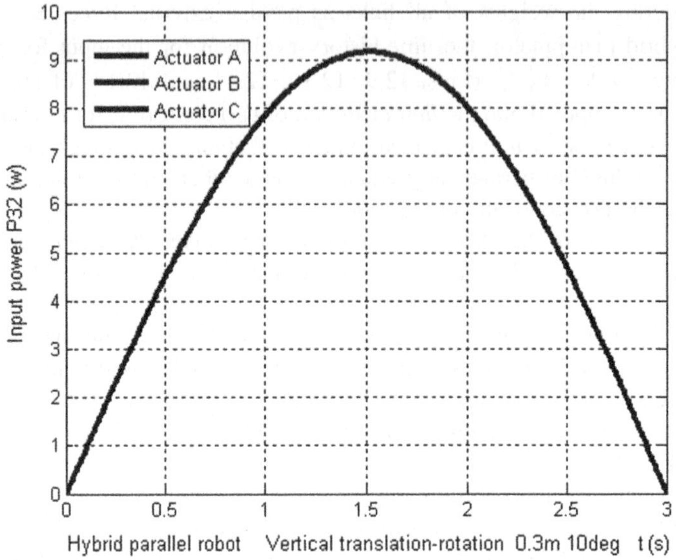

Fig. 12.10 Input powers P^i_{32} of actuators (vertical *T-R*)

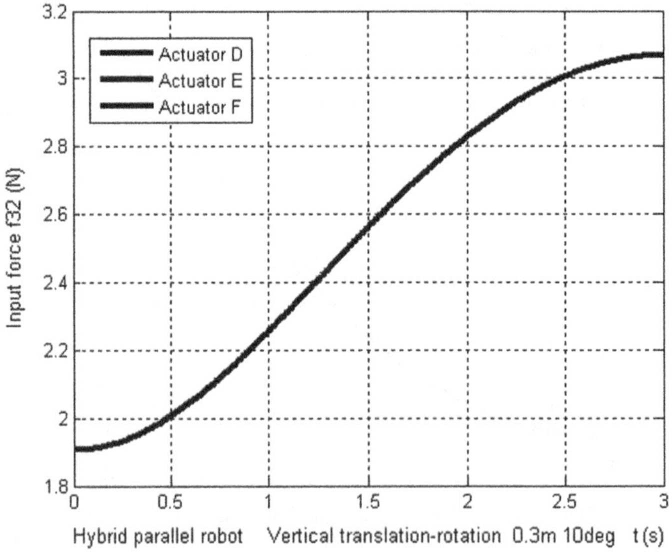

Fig. 12.11 Input forces f^j_{32} of actuators (vertical *T-R*)

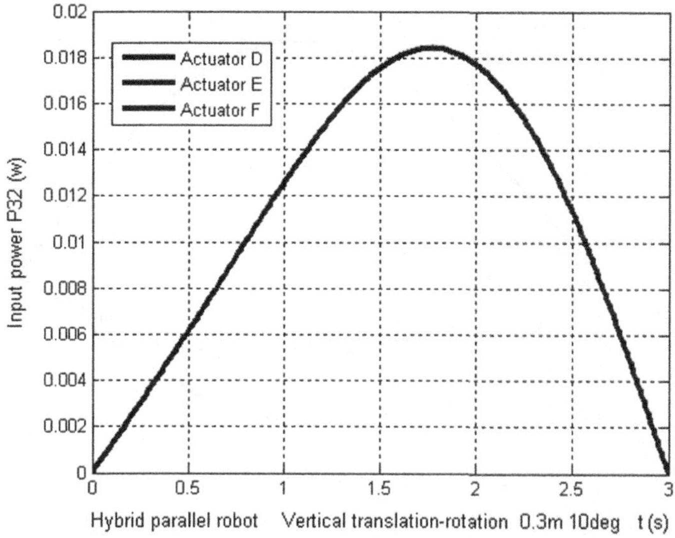

Fig. 12.12 Input powers P_{32}^j of actuators (vertical *T-R*)

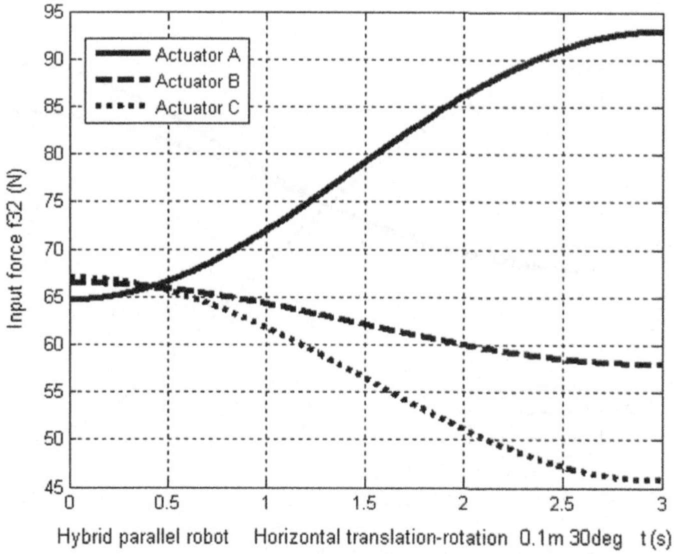

Fig. 12.13 Input forces f_{32}^i of actuators (horizontal *T-R*)

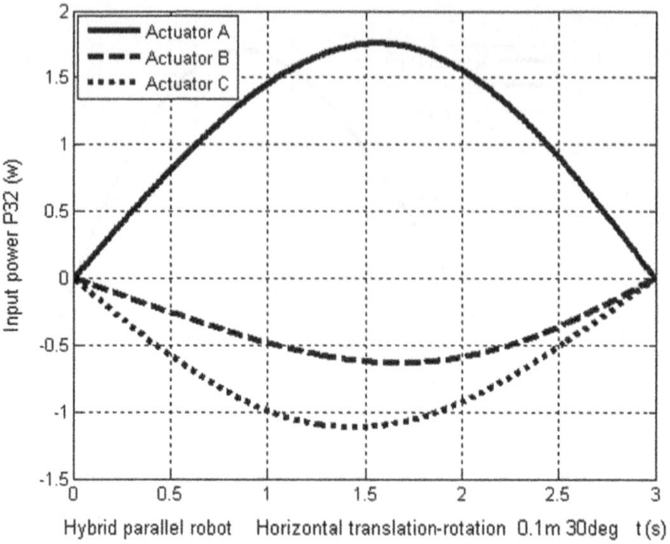

Fig. 12.14 Input powers P_{32}^{i} of actuators (horizontal *T-R*)

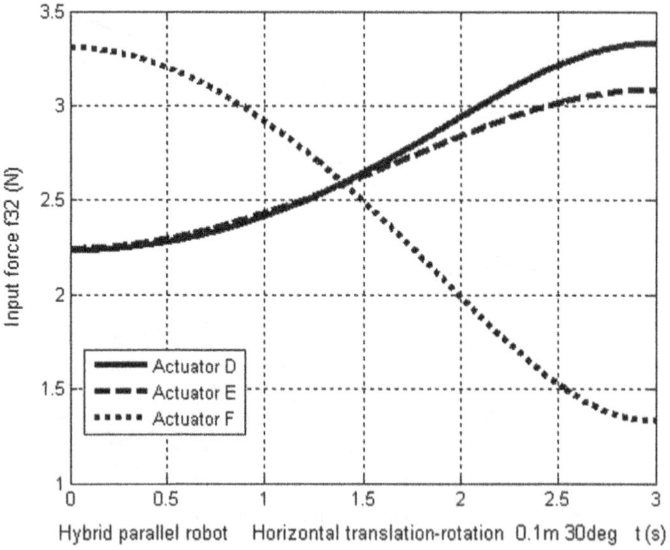

Fig. 12.15 Input forces f_{32}^{j} of actuators (horizontal *T-R*)

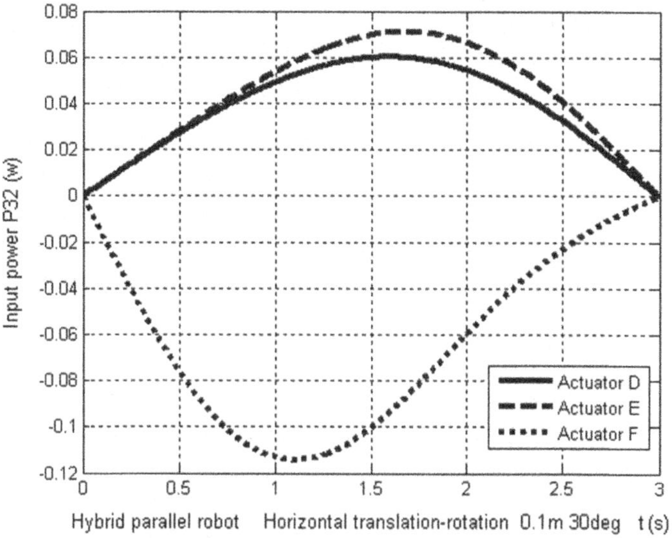

Fig. 12.16 Input powers P_{32}^{j} of actuators (horizontal *T*-*R*)

References

1. Liu, N., Wu, J.: Kinematics and application of a hybrid industrial robot Delta-RST. Sens. Transducers **169**(4), 186–192 (2014)
2. Tanev, T.K.: Kinematics of a hybrid (parallel-serial) robot manipulator. Mech. Mach. Theory **35**(9), 1183–1196 (2000)
3. Tsai, L.-W., Joshi, S.: Kinematics analysis of 3-DOF position mechanisms for use in hybrid kinematic machines. J. Mech. Des. **124**(2), 245–253 (2002)
4. Campos, A., Budde, C., Hesselbach, J.: A type synthesis method for hybrid robot structures. Mech. Mach. Theory **43**(8), 984–995 (2008)
5. Ibrahim, O., Khalil, W.: Inverse and direct dynamic models of hybrid robots. Mech. Mach. Theory **45**(4), 627–640 (2010)
6. Zhang, C.-D., Song, S.-M.: Geometry and position analysis of a novel class of hybrid manipulators. In: Proceedings of ASME Design Technical Conference, Minneapolis, DE, vol. 72, pp. 1–9 (1994)
7. Shahinpoor, M.: Kinematics of a parallel-serial (hybrid) manipulator. J. Robot. Syst. **9**(1), 17–36 (1992)
8. Cheng, H.H., Lee, J.J., Penkar, R.: Kinematic analysis of a hybrid serial-and-parallel-driven redundant industrial manipulator. Int. J. Robot. Autom. **10**(4), 159–166 (1995)
9. Gallardo, J., Lesso, R., Rico, J.M., Alici, G.: The kinematics of modular spatial hyper-redundant manipulators formed from RPS type limbs. Robot. Auton. Syst. **59**(1), 12–21 (2011)
10. Li, Y., Staicu, S.: Inverse dynamics of a 3-$\underline{P}$RC parallel kinematic machine. Nonlinear Dyn. **67**(2), 1031–1041 (2012)
11. Staicu, S., Zhang, D.: A novel dynamic modelling approach for parallel mechanisms analysis. Robot. Comput.-Integr. Manuf. **24**(1), 167–172 (2008)
12. Staicu, S.: Dynamics modelling of a Stewart-based hybrid parallel robot. Adv. Robot. **29**(14), 929–938 (2015)

Uncited References

13. Staicu, S.: Dynamics modelling of a Stewart-based hybrid parallel robot. Adv. Robot. **29**(14), 929–938 (2015)
14. Wittenberg, J.: Dynamics of System of Rigid Bodies. B. G. Teubner, Stuttgart (1977)

The manufacturer's authorised representative in the EU is Springer Nature Customer Service Centre GmbH, Europaplatz 3, 69115 Heidelberg, Germany. If you have any concerns regarding our products, please contact ProductSafety@springernature.com

Printed and bound by CPI Group (UK) Ltd, Croydon, CR0 4YY
30/09/2026
02225601-0016